SURVEY OF PROGRESS IN CHEMISTRY

VOLUME 8

CONTRIBUTORS TO VOLUME 8

EVAN H. APPELMAN
ROBERT L. BURWELL, JR.
STANLEY DAGLEY
H. A. JOHANSEN
J. KONECNY
FELIX SCHREINER
ARTHUR F. SCOTT
MICHAEL SZWARC

Survey of Progress in Chemistry

EDITED BY

ARTHUR F. SCOTT

Department of Chemistry
Reed College
Portland, Oregon

Volume 8

1977

ACADEMIC PRESS
New York San Francisco London
A Subsidiary of Harcourt Brace Jovanovich, Publishers

ACADEMIC PRESS, INC.
111 Fifth Avenue, New York, New York 10003

United Kingdom Edition published by
ACADEMIC PRESS, INC. (LONDON) LTD.
24/28 Oval Road, London NW1

LIBRARY OF CONGRESS CATALOG CARD NUMBER: 63–22332

ISBN 0–12–610508–1

PRINTED IN THE UNITED STATES OF AMERICA

Contents

Microbial Degradation of Organic Compounds in the Biosphere

STANLEY DAGLEY

The Thermochemical Generation of Hydrogen from Water

FELIX SCHREINER AND EVAN H. APPELMAN

Theoretical and Practical Aspects of Immobilized Enzymes

J. KONECNY

A Book Review of an Early (1781) Survey of Progress in Chemistry

ARTHUR F. SCOTT

Contributors

EVAN H. APPELMAN, Chemistry Division, Argonne National Laboratory, Argonne, Illinois

ROBERT L. BURWELL, JR., Ipatieff Catalytic Laboratory, Department of Chemistry, Northwestern University, Evanston, Illinois

STANLEY DAGLEY, Department of Biochemistry, College of Biological Sciences, University of Minnesota, St. Paul, Minnesota

H. A. JOHANSEN,* Westinghouse Research Laboratories, Pittsburgh, Pennsylvania

J. KONECNY, Pharmaceutical Division, CIBA-GEIGY Ltd., Basel, Switzerland

FELIX SCHREINER, Chemistry Division, Argonne National Laboratory, Argonne, Illinois

ARTHUR F. SCOTT, Reed College, Portland, Oregon

MICHAEL SZWARC, Department of Chemistry, State University of New York, College of Environmental Science and Forestry, Syracuse, New York

*Present address: Rt. 1, Box 89, McMinnville, Oregon 97128.

Preface to Volume 1

It is nowadays almost a commonplace that a chemist finds it practically impossible to keep abreast of important developments in chemistry generally, not to mention developments in his own field of special interest. College teachers as a group are perhaps more conscious of this information problem than are some other segments of the chemistry profession. This is really not surprising. Because of their responsibility for the education of chemists-to-be, college teachers feel obliged not only to keep their course materials in step with new developments but also to have some understanding of advances in other areas of chemistry. Yet, as those of us who have been teaching during the past forty years know so well, there has been no significant change in the time available to the teacher for study and assimilation of the steadily increasing flow of new ideas and information. In moments of despair, when a teacher tries to picture the situation which could lie ahead, he is likely to conclude that chemistry instruction might soon lag hopelessly behind the real chemical world—a state of affairs which would be without precedent in academic history. Ways of forestalling such an outcome appear to be rather limited at this point in time. One way would be to improve the means of communication between the research frontier and the teacher; another would be to increase the "study" time available to the teacher; the best of all, however, would be the combination of improved transmission of information to the teacher and the deliberate increase in his "study" time.

The *Survey of Progress in Chemistry* which is being launched with this volume is an attempt to improve the transmission of new material to the college chemistry teacher. Its aim can be expressed neatly in terms of a simile borrowed from J. D. Bernal. We can view the traditional dissemination of chemical information through scientific journals as being on a par with the most primitive level of wind-pollination in certain plants. The more pollen grains released—above a certain number—the more miss the stigmata waiting to catch them. What is needed in the case of chemistry teachers is a more advanced mechanism corresponding to the selective stage of insect-pollination where, with far less pollen, more gets to the right flowers. It is our hope that the *Survey* will be a definite step toward the more advanced and efficient mechanism, and will serve to bring to the teacher authoritative discussions of recent developments in chemistry that he, as a teacher, should be acquainted with but does not have time to glean from the published literature by himself.

Although the primary "target" of the *Survey* is the chemistry teacher, it is hoped, of course, that the *Survey* will meet the needs of many other groups of chemists who would like to know what is happening in chemistry outside their own fields of activity. For obvious reasons, the series should be helpful also to the beginning graduate student and the ambitious undergraduate who are seeking an introduction to some topic not yet covered in the typical textbook.

In putting together this first volume of the *Survey* the editor naturally faced some interesting policy questions with respect to topics, contributors, and the "level of presentation." In considering topics there seemed to be no good reason to adhere to the traditional subdivisions of chemistry and devote one chapter to physical chemistry, another to organic chemistry, and so on. Indeed, on the frontier of chemical research today there are many points where attack is by a combination of two or more of these division; witness, for example, such new fields as "organic photochemistry" and "inorganic biochemistry." Although it is our intention to include, in a single volume, chapters dealing with several areas of chemical activity, the selection of topics for any one volume is dependent to some extent on who can spare the time to write for that particular volume; for it is axiomatic these days that a chemist who is at the forefront of his field of research is an exceedingly busy individual. In preparing his chapter each contributor has kept the college teacher in mind and has employed whatever mode of presentation he deemed best. This is another way of stating that the articles in the *Survey* are not intended to be the usual type of review article which serves another, quite different purpose. To enable the authors to "see" the same group of readers the editor advised them to assume the typical reader to be a college teacher who had completed his graduate studies about ten years ago. It is believed that this description of the potential teacher-reader takes in the major fraction of the present group of chemistry teachers and likewise the major fraction of active chemists.

In conclusion, it is noted with both pleasure and gratitude that many active research chemists have given generous support to the *Survey* project: by encouragement, by suggestions of topics and authors, and by consenting to prepare chapters for this or future volumes of the *Survey*. The editor is especially indebted to the contributors to this first volume because of their willingness to join in the experiment and also for their full cooperation in carrying it out.

September, 1963 ARTHUR F. SCOTT

SURVEY OF PROGRESS IN CHEMISTRY

VOLUME 8

Heterogeneous Catalysis

ROBERT L. BURWELL, JR.

Ipatieff Catalytic Laboratory,
Department of Chemistry,
Northwestern University,
Evanston, Illinois

I. Introduction

Many chemical reactions in the laboratory are catalytic, a large fraction of industrial processes involve catalysis, and one may suspect that catalysis will dominate the future of preparative chemistry. It seems almost certain that we know a much larger fraction of those reactions which proceed spontaneously than of those which require catalysts since reactions in the latter category must be both more numerous and more difficult to discover.

The following is from "Definitions, Terminology, and Symbols in Colloid and Surface Chemistry—Part II. Heterogeneous Catalysis," International Union of Pure and Applied Chemistry (1976):

> Catalysis is the phenomenon in which a relatively small amount of a foreign material, called a catalyst, augments the rate of a chemical reaction without itself being consumed. A catalyst provides for sets of elementary processes (often called elementary steps) which link reactants and products and which do not occur in the absence of the catalyst. For example, suppose the reaction
>
> $$A = C$$
>
> to proceed at some rate which might be measurable but might be essentially zero. The addition of X might now provide a new pathway involving the intermediate B,
>
> $$A + X \rightarrow B$$
> $$B \rightarrow C + X$$
>
> If reaction by this pathway proceeds at a rate significant with respect to the uncatalyzed rate such that the total rate is increased, than X is a catalyst. In this sense, a *catalytic reaction* is a closed sequence of elementary steps similar to the propagation steps of a gas-phase chain reaction.

The catalyst enters into reaction but it is regenerated at the end of each reaction cycle. Thus, if the catalyst can be protected against poisoning and deactivation, one unit of catalyst results in the conversion of many units of reactants.

Catalysis is usually divided into three categories: *homogeneous*, *heterogeneous*, and *enzyme*. In homogeneous catalysis, all the reactants and the catalyst are molecularly dispersed in one phase. In heterogeneous catalysis, the catalyst constitutes a separate phase. In the usual case, the catalyst is a solid and the reactants and products are in one or more gaseous or liquid phases. The catalytic reaction occurs at the surface of the solid, and ideally, its rate is proportional to the area of the catalyst. The significance of the restrictive term, "ideally," will appear later. Enzyme catalysis may share some of the characteristics of both preceding types. Enzymes are macromolecules, usually small enough to be molecularly dispersed with the reactants and products in one phase, but large enough so that one may speak of active sites at the surface of the enzyme molecule.

For many years the three branches of catalysis developed independently, but recently interactions among them have been developing (Basolo and Burwell, 1973) and it is likely that such interactions will increase.

As an example of catalysis, let us consider the metathesis of olefins illustrated by the following reaction:

$$^{*}CH_2{=}CH{-}CH_3 + {}^{*}CH_2{=}CH{-}CH_3 \rightleftharpoons {}^{*}CH_2{=}{}^{*}CH_2 + H_3CHC{=}CHCH_3$$

The methylene groups are labelled (*) to indicate that the reaction is to result in actual cleavage of C=C bonds followed by cross reformation of new C=C bonds. The change in energy is almost zero and the reaction is driven only by an increase in entropy. A decade ago few would have imagined such a reaction to be possible at lower temperatures.

However, catalysts for the reaction have been discovered, some of which cause rapid metathesis at room temperature. The reaction was first reported by Banks and Bailey as a heterogeneously catalyzed reaction. Heterogeneous catalysts known to be effective are compounds of molybdenum, tungsten, and rhenium deposited on the surface of alumina or silica, e.g., MoO_3/Al_2O_3 (where / indicates a phase boundary) or $Mo(CO)_6/Al_2O_3$. Both materials need a suitable pretreatment (or *activation*) before they become active as catalysts. The first homogeneous catalyst for the metathesis of olefins was reported only three years after the report of Banks and Bailey. A number of homogeneous catalysts have been reported [see Haines and Leigh (1975) and Calderon *et al.* (1976) for reviews of the metathesis reaction]: WCl_6 + LiBu, $EtAlCl_2$ or BuMgCl; WCl_4 + $AlCl_3$; $Mo(NO_2)(PPh_3)_2Cl_2$ + $EtAlCl_2$; and $W(CO)_3(PPh_3)_2Cl_2$ + $EtAlCl_2$. The most active homogeneous catalysts are compounds which contain exactly those elements listed above for heterogeneous catalysts. One may suspect that both types of catalysts involve similar processes.

II. General Survey of Heterogeneous Catalysis

A. Reactors

Among the important types of reactors both in the laboratory and in industry are *plug flow*, *batch*, and *gradientless*. Examples of these reactors are illustrated in Fig. 1.

In the plug flow reactor illustrated, gases pass through a bed of catalyst. The conversion (or fraction reacted) depends upon the quantity and activity of the catalyst, the flow rate, the partial pressures of reactants and products, and the temperature. In the recirculation reactor (one of the class of gradientless reactors), a pump recirculates effluent from the catalyst back

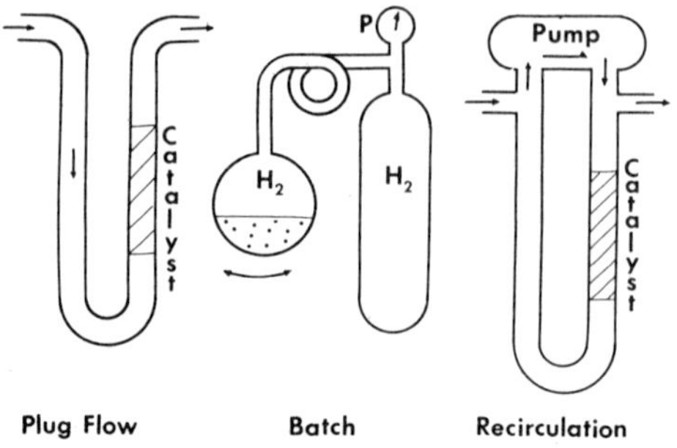

FIG. 1. Some of the types of reactors used in heterogeneous catalysis.

through the bed. The recirculation rate must be made larger—(100-fold or so) than the rate of addition and removal. If this provision is met, the concentrations of products and reactants change little between the top and bottom of the bed in contrast to the situation in the plug flow reactor. In both reactors, the concentrations at any particular point in the catalyst bed are constant with time (provided the catalyst activity is constant). The batch reactor is exemplified by one which could be used for hydrogenation of a liquid olefin at 25°. The reactor contains liquid olefin, a finely divided heterogeneous catalyst, and provision for the supply of hydrogen. Agitation, provided here by a rocking motion, is needed to keep the catalyst in suspension and to make a large liquid–gas interface to facilitate solution of hydrogen in the liquid. The concentrations at the surface of the catalyst change with time—unlike the situation in a flow reactor.

B. Heterogeneous Catalysts

1. *Supported Catalysts*

Catalysts come in a variety of forms, e.g., wires, pellets, and powder. In principle, the catalytic activity should be proportional to the area of the catalytic ingredient. Thus, one might think it best to use the finest powder possible. For example, if one takes a cube of platinum 1 cm on edge, the surface area is 6 cm^2. If this cube is divided into cubes 100 Å (10^{-6} cm) on edge the total area would be 6×10^6 cm^2, or 600 m^2. The catalytic activity would have increased 10^6 times. However, the material would be unusable in flow reactors. Even at large pressure drops, flow through a bed of such fine particles would be very slow. The 100 Å cubes would work well in the batch reactor but it would be difficult to recover such fine particles after reaction. There are other problems. In the bed of the flow reactor, platinum particles would adhere to each other and sinter with loss of surface area. If the platinum became covered with carbonaceous material during use, one could not regenerate the catalyst by burning off the carbon at 500°C. In such an operation, the area would be reduced to a few square meters. What is the solution to this problem of preparing a catalyst of high area, resistant to sintering, and of a large enough granule size to use in a fixed bed? There is no perfect solution, but quite often one deposits the tiny particles of platinum upon high area, porous silica or alumina. The silica and alumina are called *supports*.

Silica gel is a porous material which results from the partial consolidation of little spheroids of silica about 100 Å in diameter. Silica gels have, then, a "texture" which resembles that of a cemented, loose gravel bed. One now puts some kind of platinum compound onto the internal surface of the silica gel, reduces it in hydrogen, and obtains silica gel with tiny platinum particles distributed throughout the pore volume. For example, one could ion exchange $Pt(NH_3)_4^{2+}$ onto the surface by treatment of silica gel with a slightly basic solution of $Pt(NH_3)_4Cl_2$. Silica gel is covered with surface silanol groups (symbolized σ-OH), about 5/nm² (100 Å²), which are weakly acidic, $pK_a \sim 7$–8. The following reaction

$$2\sigma\text{-OH} + Pt(NH_3)_4Cl_2 + 2NH_4OH \rightarrow 2\sigma\text{-O}^- \ Pt(NH_3)_4^{2+} + 2NH_4Cl + 2H_2O$$

distributes $Pt(NH_3)_4^{2+}$ widely throughout the gel. By proper treatment and reduction, one can obtain silica gel containing platinum particles throughout the gel with a platinum percentage exposed of over 50% (J. R. Anderson, 1975). *Percentage exposed* or *dispersion* is defined as the percent of total platinum atoms which are on the surface. Electron microscopy supports the structural statements made above. A schematic view of a 1% platinum on silica catalyst is shown in Fig. 2.

If one uses granules of silica gel, one to a few millimeters in diameter,

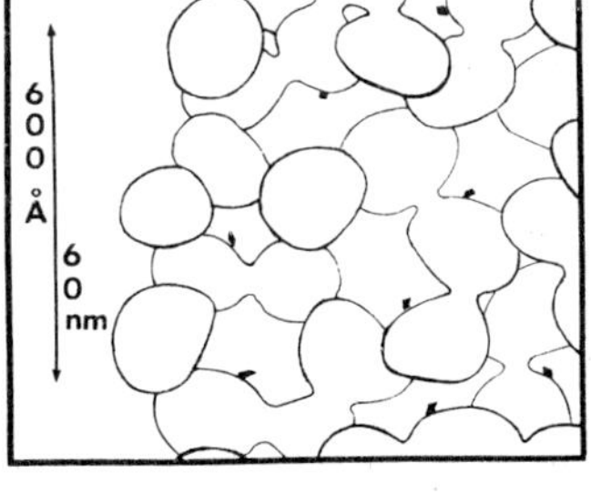

FIG. 2. Schematic of a platinum/silica gel catalyst. The larger spheroids are about 10–15 nm in diameter, composed of silica, and fused to form the silica gel agglomerate. The small black particles are little crystallites of platinum. They would usually be in the range of 1.5–4 nm in diameter (10 Å = 1 nm).

one can pack a bed with the resulting catalyst and get good rates of flow through the bed. A large fraction of the precious metal is available for catalytic action and the catalyst does not sinter at a serious rate below 450°.

Aluminas can be prepared which have textures much like those of silica gels. Such aluminas are microcrystalline materials in the γ- or η-structures. They are widely used as catalyst supports. Pt/Al_2O_3 is probably more resistant to sintering than Pt/SiO_2. Silica gel itself loses surface area above 500°, whereas aluminas are more refractory and suffer little loss in area even at 800°. A common Pt/Al_2O_3 catalyst of high percentage exposed is made by impregnating pellets of γ-alumina with H_2PtCl_6(aq) followed by drying, calcining (heating in air), and reduction by hydrogen.

Other catalytically active ingredients can be deposited on alumina, for example, Pd, Rh, Cr_2O_3, and MoO_3. If alumina is impregnated with $Mo(CO)_6$ and heated in an inert atmosphere to 260°, a very active catalyst for the metathesis of olefins results. In this last example, the percentage exposed of the molybdenum surface complex is 100%.

The preparation of the best catalysts has until recently been almost exclusively a proprietary art, but recently methods of preparing many effective catalysts have appeared in the literature. A partial but substantial understanding of the factors involved in the preparation of a number of classes of catalysts has been developed (J. R. Anderson, 1975).

As a final example of supported catalysts we consider the catalysts used in combustion of carbon monoxide and unburned hydrocarbons in automobile exhaust. Additional air will be added, if necessary, to the exhaust and the resulting gas will be passed over the catalyst. Two particular problems may be noted. The catalyst bed must permit very high flow rates at very low pressure drops, otherwise engine performance and fuel economy suffer extensively, and the catalyst must endure operating temperatures in excess of 1000°C. Catalysts in current use are of two forms, platinum group metals deposited on alumina pellets which are resistant to high temperatures and similar metals deposited on monolithic supports made of a porcelain-like material. The monolithic support is in the form of a cylinder about 15 cm in diameter and 8 cm long and it is pierced perpendicular to its circular ends by many channels a few millimeters in diameter. The monolithic support is coated with a thin film of porous alumina which is then impregnated with the precious metals.

2. *Concentration and Temperature Gradients*

The supported catalysts of the usual type have the important advantages already described. They also have a serious disadvantage: restricted rates of transport of heat and of reactants and products in and out of the catalyst

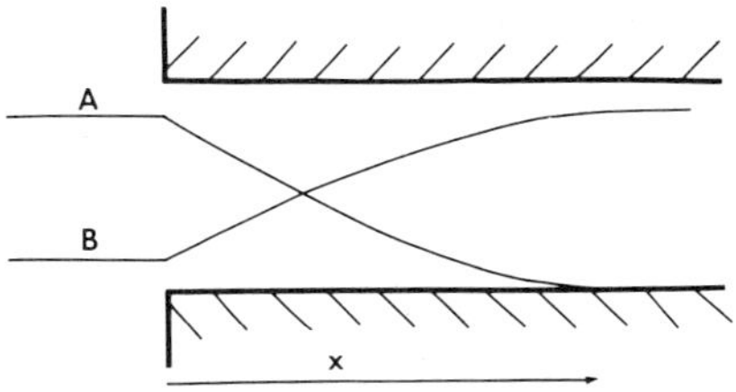

FIG. 3. Variation in the concentrations or partial pressures of A and B with depth of penetration x into a pore during the reaction A → B on a porous catalyst. Concentrations or pressures are represented by the vertical distance above the bottom of the pore.

granules or pellets. The problem of mass transport puts an upper limit upon the effective rate of reaction which can be obtained with a given catalyst. In order for reactants to diffuse into a catalyst granule, there must be a concentration gradient of reactants such that their concentrations decrease in going from the outside of a granule to its center. Consider a very simple reaction A → B. If $(d[A]/dx)/x$ (where x is the distance from the surface toward the center of a granule) is large, [A] will be zero within a central zone of the catalyst and some of the catalyst will not effect reaction. This situation is shown in Fig. 3. On the other hand if $(d[A]/dx)/x$ is small enough, all zones of the catalyst granule are nearly equally active. One can test for the presence of serious concentration gradients by crushing the catalyst into smaller particles. If the rate resulting from a given weight of catalyst increases, the concentration gradients were serious within the original granules. Where concentration gradients are serious, temperature gradients are apt also to be serious since by their nature, porous silica and alumina are poor conductors of heat. If the reaction being studied is substantially exothermic, the temperature of the center of a catalyst pellet can be many degrees higher than that of its outer surface.

In scientific studies, one should provide conditions such that $(d[A]/dx)/x$ is very small. Industrial reactors are often operated under conditions such that the ratio is substantial since such conditions may be the most economic. An important part of the subject of *reactor engineering* in chemical engineering is devoted to this matter. In the rest of this article, however, we shall assume that the studies involve a low value of the ratio.

3. *Unsupported Catalysts*

Despite these problems of mass and heat transport, the advantages of supported catalysts are so substantial that they are widely used both in the laboratory and in industry. However, other forms of catalysts are used. Examples are platinum powders obtained by reducing PtO_2 (the Adams catalyst); nickel powders obtained by dissolving the aluminum from an aluminum–nickel alloy by a solution of sodium hydroxide (Raney nickel); low-area metallic iron plus *promoters* in the synthesis of ammonia;

$$N_2 + 3H_2 = 2NH_3$$

and gauze of a platinum–rhodium alloy used in the oxidation of ammonia to nitric oxide at high temperatures (the following reaction then converts nitric oxide to nitric acid: $2NO + \frac{3}{2}O_2 + H_2O \rightarrow 2HNO_2$) (Thomas, 1970).

C. Promoters

Sometimes the addition of small amounts of various compounds increases the activity, retards deactivation, or increases the selectivity of catalysts. Such added materials are called *promoters*. For example, the addition to the ammonia synthesis catalyst of small amounts of what can be represented analytically as $K_2O + Al_2O_3$ (or similar combinations) results in the first two effects described above.

D. Selectivity

A key aspect of a heterogeneous catalyst is *selectivity*. Hardly any reaction of $N_2 + H_2$ is thermodynamically favored, except that to ammonia. With

TABLE I

Reactions between Carbon Monoxide and Hydrogen

Catalyst	Conditions	Products
Ni[a]	100–200°C, 1–10 atm	$CH_4 + H_2O$
$ZnO \cdot Cr_2O_3$[b]	400°C, 500 atm	$CH_3OH + H_2O$
Co/ThO_2	190°C, 1–20 atm	CH_4, C_2H_6, and higher alkanes + H_2O
Fe + alkali	315°C, 15 atm	As with Co/ThO_2 plus olefins, alcohols, and acids
Ru	200°C, 200 atm	High-molecular-weight alkanes + H_2O
ThO_2	400°C, 200 atm	Branched-chain alkanes + H_2O

[a]This is the *methanation* reaction of current interest in the manufacture of *synthetic natural gas* (SNG). It converts carbon monoxide, which has a relatively low heat of combustion, to methane, which has a high one.

[b]More modern versions of this catalyst operate at conditions of less severity and accordingly at less expense.

most other combinations of reactants, more than one and often myriads of reactions are thermodynamically possible. Thus, the reaction of $CO + H_2$ to form hundreds of thousands of organic compounds of C, H, and O is thermodynamically possible. Although heterogeneous catalysts rarely exhibit the high level of selectivity exhibited by many enzymes, they are usually useful only if they exhibit a substantial level of selectivity. We exemplify in Table I the kind of selectivity which is available in catalysts which catalyze the hydrogenation of carbon monoxide. No catalyst is known which gives a yield of 100% in the reaction

$$3CO + 6H_2 = n\text{-}C_3H_7OH + 2H_2O$$

but catalysts are available which give high yields of methanol and others, which give substantial yields of mixed alcohols.

Stereoselectivity is also important in heterogeneous catalysis. Thus, palladium catalysts lead to the *selective* hydrogenation of $RC \equiv CR$ to olefins (as opposed to alkanes) and to the *stereoselective* formation of *cis*-olefins.

E. Poisoning and Deactivation

The presence of any foreign compound which strongly adsorbs on the surface of the catalyst can reduce the rate of a catalytic reaction. Such substances are called *poisons*, and the phenomenon appears also in homogeneous and enzyme catalysis. If the poison is adsorbed so strongly on the effective surface of the catalyst that it cannot be removed by eliminating the poison from the fluid (i.e., gas or liquid) phase, the poison is called a *permanent* poison. If eliminating the poison from the fluid phase causes its desorption from the catalyst, it is called a *temporary* poison. If the poison is adsorbed still more weakly (about as strongly as the reactants), the poison may be called a *competitive inhibitor*. Poisoning is a pervasive problem in heterogeneous catalysis (and, indeed, in all types of catalysis), and even very small amounts of strongly held poisons can cause serious reductions in rates [for further details, see the article by C. Kemball, in Basolo and Burwell (1973)].

The surface of a metal is *soft* in the *soft* and *hard* classification of acids and bases of Pearson (1973). Thus, effective poisons for metallic catalysts are soft: carbon monoxide, sulfur compounds like mercaptans or hydrogen sulfide, hydrogen cyanide, or phosphines. On the contrary in oxide catalysis where a catalytic site is a coordinatively unsaturated ion at the surface (for example, Al_2O_3 or Cr_2O_3) the most effective poisons are apt to be hard bases like water or amines. In addition a compound which changes the most favorable oxidation number of a catalytic site can be a poison. Water poisons

TABLE II

EXAMPLES OF HETEROGENEOUS CATALYTIC REACTIONS

Catalyst	Reaction	Temperature[a]
	Hydrogenation and Related Reactions	
Pt, Pd, Rh, Ni metals as powders or supported on SiO_2, Al_2O_3, C	$H_2 + C_2H_4 = C_2H_6$	−100°
	$H_2 + C_2H_2 = C_2H_4$	−100°
	$H_2 + D_2 = 2HD$	−190°
	$C_3H_8 + D_2 = C_3H_7D, C_3H_6D_2, \ldots, + HD$	50°
	Cyclopropane + H_2 = propane	40°
	$C_2H_6 + H_2 = 2CH_4$	200°
Pd supported	H_2 + benzyl acetate = toluene + acetic acid	25°
Cr_2O_3 activated to generate Cr^{3+}(cus) (coordinatively unsaturated surface)	$H_2 + C_2H_4 = C_2H_6$	−100°
	$D_2 + C_3H_8 = C_3H_7D + HD$	200°
	$H_2 + D_2 = 2HD$	−180°
Metallic Fe	$3H_2 + N_2 = 2NH_3$	400°
Pt, Cr_2O_3	Heptane = toluene + $4H_2$	450°
Pd, Cr_2O_3	1-Butene = 2-butene accompanying hydrogenation	0°
Pt, Ni, Cu	Acetone + H_2 =2-propanol	25°, 25°, 100°
Various catalysts	CO + nH_2 = . . . (See Table I)	
Basic oxides	Acetone + *sec*-butyl alcohol = 2-propanol + butanone	150°
	Formation of Linear Polyethylene	
Cr^{2+}/SiO_2	Polymerization of C_2H_4 to linear polyethylene	50°
	Olefin Metathesis	
Mo compounds on a support	2-Propylene = ethylene + butylene	25°
	Dehydration	
Al_2O_3	$C_2H_3OH = C_2H + H_2O$	300°

TABLE II (*continued*)

Catalyst	Reaction	Temperature[a]
	Oxidation	
Metals, oxides, etc.	$H_2O_2 = H_2 + O_2$	25°, and up
Pt, many transition metal oxides	$2H_2 + O_2 = 2H_2O$	0°, 100° +
	$2CO + O_2 = 2CO_2$	50°, 200° +
	$CH_4 + 2O_2 = CO_2 + 2H_2O$	200°, 350° +
Pt, V_2O_5/SiO_2[b]	$2SO_2 + O_2 = 2SO_3$	400°
Metallic Ag	$C_2H_4 + \frac{1}{2}O_2 =$ ethylene oxide	250°
Bismuth molybdate	$CH_3CH{=}CH_2 + O_2 = CH_2{=}CH{-}CHO$	400°
	$CH_3CH{=}CH_2 + NH_3 + O_2 = CH_2{=}CHCN$	450°
V_2O_5/SiO_2	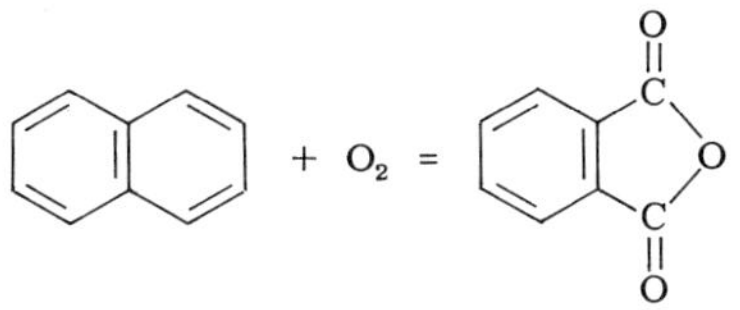	350°
Fe_3O_4	$H_2O + CO = H_2 + CO_2$	450°
$CuCl_2$/pumice	$2HCl + O_2 = Cl_2 + H_2O$	375°
	Acidic Catalysts	
Silica–aluminas, amorphous and crystalline (zeolites), exhibit strongly acidic surface sites as shown by bonding of bases and by indicators. They catalyze a variety of reactions which are usually considered to involve intermediates of a carbonium ion-like nature. Examples are catalytic cracking of long chain alkanes (450°), skeletal isomerization of alkanes (100° +), alkylation of aromatic hydrocarbons by olefins (25° +), and polymerization of butylenes to branched-chain low-molecular-weight materials.		

[a]The listed temperature (in °C) is the lowest at which a reasonable rate is obtained in a flow reactor. It is obviously just a crude indicator of catalytic reactivity.

[b]The actual operating catalyst consists of a molten solution of vanadium pentoxide, potassium sulfate, and sulfur trioxide spread on the surface of silica (Thomas, 1970, Chapter 18).

the iron catalyst of the ammonia synthesis reaction by converting its surface from Fe(0) to Fe(II).

Deactivation of catalysts may also be observed in the absence of the *extrinsic* poisoning discussed above. In reactions at higher temperatures, this can result from sintering with resulting loss in area of the catalytic ingredient. In reactions involving hydrocarbons, it may result from deposition on the surface of the catalyst of slowly reacting hydrocarbon residues loosely called *coke*. In the second case, it is often possible to regenerate the catalyst by burning off the coke in air at temperatures of 500°C or so. In the process catalytic cracking used in the refining of petroleum, the lifetime of the catalyst is only a few minutes and regeneration proceeds continuously.

F. Heterogeneous Catalytic Reactions

So far several examples of heterogeneous catalytic reactions have been given. At this point it may be useful to examine Table II which gives examples designed to illustrate the present scope of descriptive heterogeneous catalysis.

G. Basic Mechanism

So far we have been concerned with the mechanical as well as some fundamental descriptive aspects of heterogeneous catalysis. We must now inquire into the nature of catalytic action. A basic postulate of heterogeneous catalysis is that at least one reactant must be chemisorbed. Exceptions to this principle appear to be rare, but one example is provided by the *ortho*- to *para*-hydrogen conversion (Hall, 1975). In this, *para*-hydrogen reacts to form an equilibrium mixture of *ortho*- and *para*-hydrogen (3:1 at 0°C and above) upon merely being physisorbed upon surfaces which contain paramagnetic surface atoms. There is a large number of catalysts which are effective below 0°C, for example, Group VIII metals, organic free radicals, zinc oxide, activated alumina, activated chromia, and Fe^{3+}/Al_2O_3. However, the reaction, which occurs merely upon the presence of *para*-hydrogen in the strong inhomogeneous magnetic field of surface, paramagnetic atoms, is in a sense not a chemical reaction since it involves merely inverting the spin of the proton. This leads to interconversion between *para*-hydrogen in which the spins of the two protons are opposed and *ortho*-hydrogen in which they are parallel.

In the usual case, five processes can be recognized in an overall heterogeneous catalytic reaction:

(a) Diffusion of reactants from the bulk fluid phase to the catalytic site.
(b) Chemisorption of one or all of the reactants.

(c) Reaction between adsorbed species (*surface* reaction) or between a molecule in the fluid phase and an adsorbed species.

(d) Desorption of products.

(e) Diffusion of products to the bulk fluid phase.

III. Adsorption

A. General

Since adsorption is a key step in heterogeneous catalysis, a general survey of this phenomenon must be given. Adsorption is divided in two types, *physisorption* and *chemisorption.* The former usually is considered only at the gas–solid interface and it results from those physical forces of attraction which lead to the condensation of vapors into liquids. Strengths of physisorption usually increase with boiling point and commonly ΔH_{ads} is roughly twice ΔH of condensation to the liquid. Both ΔH's are negative quantities. Thus at a given temperature and partial pressure, nitrogen and carbon monoxide, having nearly the same boiling point, will be physisorbed to nearly equal extents, whereas the higher boiling carbon dioxide will be adsorbed more extensively. Physisorption is *nonspecific* in the sense that it depends primarily upon boiling point and not upon the chemical nature of the *adsorptive* (i.e., the compound in the gas phase; upon adsorption it becomes the *adsorbate*).

B. Physisorption

Physisorption is of interest to heterogeneous catalysis for two reasons. First, it may constitute a precursor state which is an intermediate between the gas phase and a chemisorbed state. Second, it provides the most important method for determining the surface area and the texture of a catalyst. By *texture* one means the detailed geometric description of the pore space of a catalyst, whereas *structure* refers to the detailed geometric description of the solid portion of a catalyst.

Figure 4 exhibits the adsorption isotherm of nitrogen at its boiling point upon nonporous samples of a number of oxides such as silica and alumina (Gregg and Sing, 1967). On the *y*-axis we plot θ, the average number of layers of nitrogen molecules adsorbed; on the *x*-axis we plot P/P_0, the ratio of the pressure to the saturation pressure of nitrogen at the temperature of the isotherm. A large number of experiments indicate that a monolayer of nitrogen corresponds to a close-packed layer in which each nitrogen atom occupies an area of 0.162 nm^2 (16.2 $Å^2$). As one increases P_{N_2}, the amount of adsorbed nitrogen increases. It reaches a monolayer ($\theta = 1$) at $P/P_0 = 0.09$,

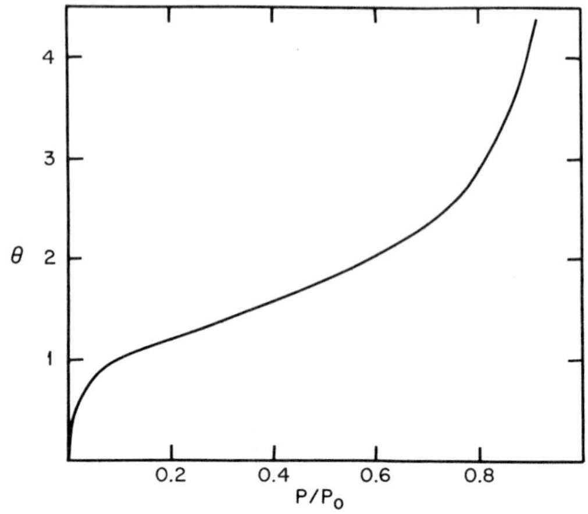

FIG. 4. The average number of monolayers θ on the surface of a nonporous solid as a function of the relative pressure of the adsorptive, P/P_0.

$\theta = 2$ at $P/P_0 = 0.57$, and $\theta = 3$ at 0.80. Thus, measurement of the volume (STP) of nitrogen adsorbed at $P/P_0 = 0.09$ gives the surface area. In addition, the Brunauer–Emmett–Teller (BET) equation

$$\theta = \frac{x}{x_m} = \left(\frac{P/P_0}{1 - P/P_0}\right)\left(\frac{c}{1 + (c-1)\,P/P_0}\right) \tag{1}$$

(where c is a constant characteristic of the gas, adsorbent, and temperature; x is the amount adsorbed; and x_m is the amount adsorbed at a monolayer) is usually obeyed for P/P_0 between about 0.06 and 0.3. One can use the BET equation to obtain x_m and then the surface area from data in this pressure range.

Pores are classified according to size: *micropores*, less than 2 nm; *mesopores*, 2–20 nm; *macropores*, greater than 20 nm. If the catalyst contains only small micropores of a diameter between that of nitrogen and twice its diameter, the heat of adsorption is augmented by the increased interaction of nitrogen with itself or with two pore walls as shown in Fig. 5. This leads to an isotherm resembling that in Fig. 6.

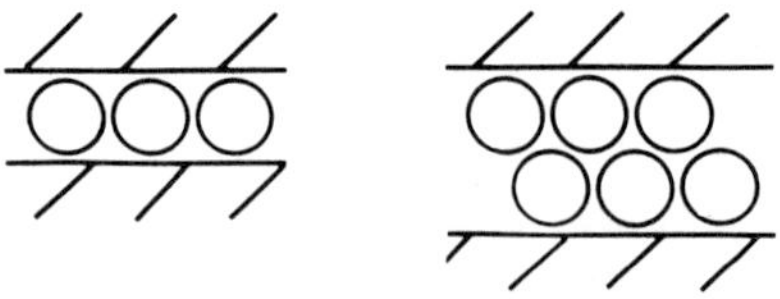

FIG. 5. Adsorption in micropores.

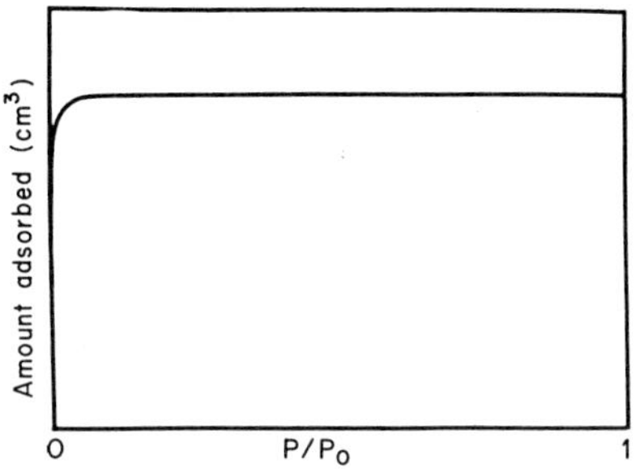

FIG. 6. Possible isotherm for strong adsorption in micropores: the amount adsorbed vs the relative pressure of the adsorptive.

The BET equation fails and the exact meaning of surface area becomes hazy. The measured isotherm may cease to be a true isotherm, that is, reduction in pressure may not lead to desorption if $-\Delta H_{ads}/RT$ is large enough.

If the pores are in the mesopore range, they will fill at a pressure lower than $P/P_0 = 1$. The curvature of a meniscus in a pore results in a lower vapor pressure than that in equilibrium with a planar surface, as shown in Fig. 7. According to the Kelvin equation, the vapor pressure P of the liquid in the capillary of radius r when the liquid fully wets the wall of the capillary is

$$\ln P/P' = \frac{2\bar{V}\gamma}{rRT}$$

where P' is the vapor pressure from a flat surface, $\bar{V}$ is the molar volume of the liquid, and γ is its surface tension.

The isotherm of a mesoporous solid might look like that in Fig. 8. The curve rises above that of Fig. 4 when condensation occurs in the pores. The exact shape of the isotherm provides information as to the distribution

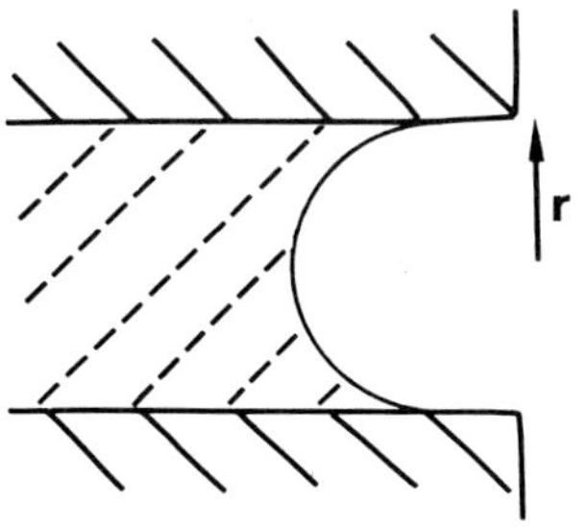

FIG. 7. Meniscus of a liquid which wets the walls of a capillary of radius r.

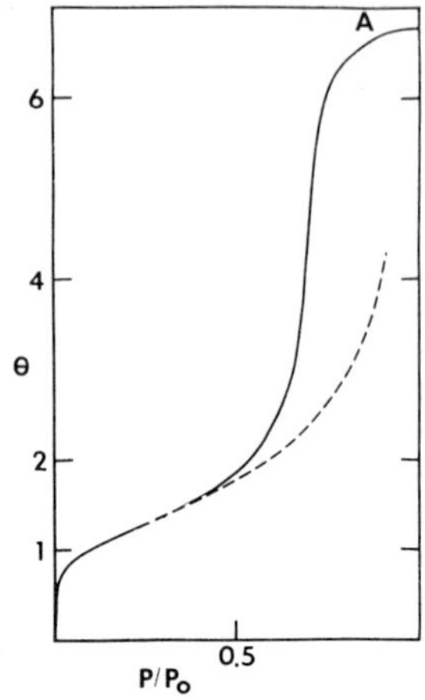

FIG. 8. Isotherm for physisorption in a mesoporous adsorbent is given by the solid line. The dashed line represents the isotherm of Fig. 4 for an adsorbent devoid of micropores or mesopores.

of pore diameters, information which is needed in evaluating concentration gradients in catalyst pores during catalytic reactions. After all the pores are filled, further adsorption at the internal surface becomes impossible and thus one can calculate the total pore volume from the value at point A in Fig. 8 and the volume occupied by one molecule of nitrogen in liquid nitrogen at $-196°C$.

More details about physisorption are provided by Gregg and Sing (1967).

C. Chemisorption

1. *Types*

A chemisorbed species and the atom or atoms of the surface to which it is bound constitutes a surface chemical compound, the *adsorption complex*, and one may expect to find that the chemical nature of the binding in adsorption complexes is at least as varied as in molecular compounds in general. Table III provides examples of various chemisorbed species according to current interpretation. Some of the entries are reasonably well founded, others are somewhat speculative.

A represents the product of the reaction of ammonia with a *coordinatively unsaturated surface* (cus) ion of Al^{3+} at the surface of alumina or of the amorphous solid solution of silica and alumina. The bonding is that between a Lewis acid and a Lewis base. The Al^{3+} (cus) would be called the *adsorption*

TABLE III

VARIOUS CHEMISORBED SPECIES

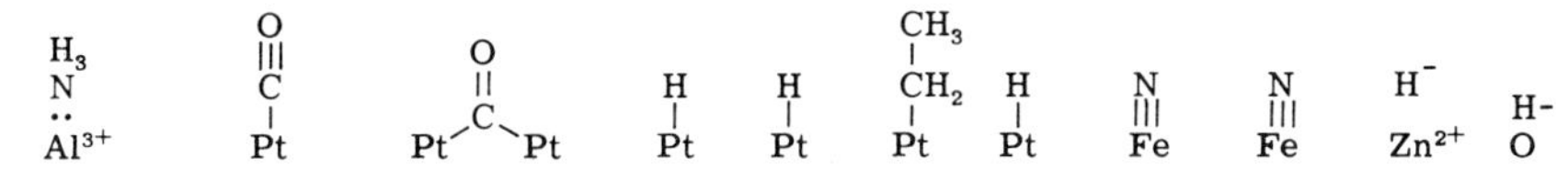

A B C D E F G

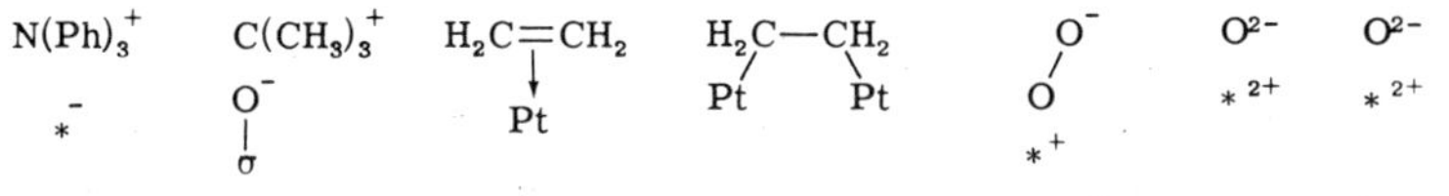

H[a] I[c] J K L[a,b] M[a,b]

[a]The * in **H**, **L**, and **M** represent some unspecified electron donor or acceptor sites.
[b]The location of the positive charges in **L** and **M** is unspecified here. It might correspond to a change in the oxidation number of transition metal ions, it might correspond to removal of electrons from the metallic electrons of a metal, etc.
[c]The carbonium ion in **I** results from transfer of a proton to isobutylene from an acidic σ-OH on a silica–alumina.

site. **B** and **C** are *linear* and *bridged* adsorbed carbon monoxide, respectively, are named in the same way as in molecular coordination complexes, and are presumably of similar chemical natures. **D** represents hydrogen and **E**, ethane *dissociatively adsorbed* on platinum. **D** is a surface hydride and **E** involves also a surface alkyl. The dissociative adsorption is homolytic since each hydrogen atom formally takes one of the two electrons in the H:H bond (and similarly in $CH_3-\overset{H}{\underset{H}{C}}{:}H$). **F** represents nitrogen dissociatively adsorbed on metallic iron. The binding is probably much like that in solid iron nitride. In **G** hydrogen has undergone *heterolytic dissociative adsorption* on a pair site of Zn^{2+} (cus) and O^{2-} (cus). It is heterolytic because one atom of hydrogen formally takes both electrons from the H:H bond. **H** represents a charge-transfer complex (*charge-transfer adsorption*). The solid can be silica–alumina. **I** represents adsorption of isobutylene on silica–alumina in which a proton from a strongly acidic surface OH (σ-OH) has been transferred to isobutylene to form the *t*-butyl carbonium ion.

Both **J** and **K** represent ethylene adsorbed on the surface of platinum metal. **J** is a π-complex, whereas **K** can be considered to be ethane disubsti-

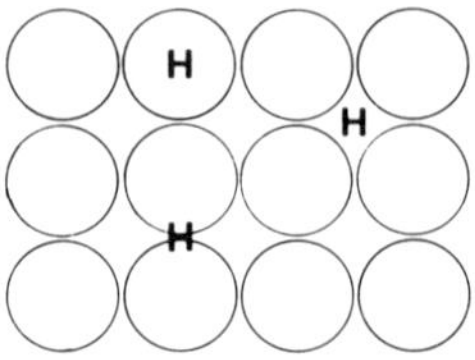

FIG. 9. Possible locations of adsorbed hydrogen atoms on the (100) surface of platinum or other fcc metal.

tuted by platinum atoms. The matter of which best represents adsorbed ethylene has been under debate. **L** is a charge-transfer complex formed by the adsorption of oxygen on, for example, the oxide of an altervalent metal. If the oxide is Cr_2O_3, the complex might be represented as Cr^{4+}, O_2^-. Finally, **M** is dissociatively adsorbed oxygen, a surface oxide.

In **D**, **E**, and **F**, for example, we rarely know whether the adsorbed species is attached to one, two, or several atoms of metal. Each of the three positions for hydrogen in Fig. 9 is conceivable.

In examples **B–F**, all of the surface of the metal, if it is clean, will be capable of adsorption. In **D**, at 25°C, each surface atom of platinum can adsorb irreversibly (i.e., the adsorption is not reversed by evacuation at 25°C) one atom of hydrogen. This is the usual basis for the measurement of the percentage exposed of supported platinum (or palladium, etc.) catalysts (J. R. Anderson, 1975). The surface of platinum behaves similarly in **M**. The concept of a monolayer is simple here; it corresponds to one atom of adsorbate (H or O) per surface atom of platinum. But if this is simple, it may be inadequate because hydrogen can adsorb reversibly beyond an H/Pt ratio of unity. In **J** the adsorbate is larger and even if all surface atoms of platinum are capable of catalytic activity, $CH_2{=}CH_2$/Pt must be less than unity and $C_5H_{11}CH{=}CH_2$/Pt, much less than unity. The situation in **A** is more complicated. The *areal* (per unit area) concentration of Al^{3+} (cus) depends on the temperature of pretreatment of the alumina. Chromia prepared from a hydrous oxide exhibits a similar variation in Cr^{3+} (cus) with the temperature of activation. Activated Cr^{3+} (cus) adsorbs irreversibly at −78°C oxygen as in **L** or carbon monoxide as $Cr^{3+}CO$. The quantities of the adsorbed carbon monoxide and oxygen are equal and they measure the areal concentration of Cr^{3+} (cus). At lower temperatures of activation, the areal concentrations correspond to but a small fraction of the total surface. In sum, sites for adsorption and catalysis may correspond to the entire surface of the catalytic phase or but to a small portion of it.

The determination of the exact structure of an adsorption complex is more difficult than that of molecular chemical compounds. One cannot prepare a pure crystal of adsorption complex and determine its structure by X-ray diffraction. To date, one has had to deduce the structure from chemical and spectroscopic data. However, there is another difficulty—the adsorption complex is apt to exist in more than one geometric form. Thus **D** and **E** must be different in detail upon different crystal faces of platinum and there may even be different forms of **D** and **E**, in the sense of Fig. 9, upon just one crystal face of platinum.

Chemisorption is a chemical reaction and, like chemical reactions in general, chemisorption may be slow or fast. Thus, the adsorption of hydrogen and oxygen upon the Group VIII metal particles of catalysts is very rapid when the metal is clean and a substantial fraction of the collisions by hydrogen and oxygen leads to adsorption. In such cases the activation energy of adsorption is very small (<1 kcal/mole). On the other hand, the adsorption of an alkane on a Group VIII metal (**E** in Table III) or of nitrogen on iron (**F** in Table III) is slow. Here the activation energy is substantial. Such adsorption is called *activated adsorption*. E_a for desorption must equal E_a for adsorption minus ΔH_{ads}. Since $\Delta H_{ads} < 0$ for adsorptions of practical interest, the E_a for desorption is at least as large as $-\Delta H_{ads}$. When $-\Delta H_{ads}$ is large , the rate of desorption becomes significant only at rather high temperatures.

2. *Adsorption Isotherms*

Determination of adsorption isotherms provides information about chemisorption. ΔH for the process of chemisorption must be a negative quantity if any substantial coverage is to occur, and we shall confine our attention to cases in which $\Delta H_{ads} < 0$ in what follows. If ΔH_{ads} is the same on all sites (i.e., ΔH_{ads} is independent of θ) and if $-\Delta H$ for chemisorption is much greater than $-\Delta H$ for physisorption on top of the chemisorbed layer, the following equation represents the process for simple adsorption–desorption

$$A(g) + * = A*$$

where A is the adsorptive, * represents the adsorption site, and A* is the adsorption complex. At equilibrium,

$$K = \frac{[A*]}{[*]P_A} = \frac{\theta}{(1-\theta)P} \tag{2}$$

Here K is the adsorption equilibrium constant, [A *] and [*] are the areal concentrations of adsorption complex and free adsorption sites, P_A or P is the pressure of A(g), and θ is the fraction of the adsorption sites which are covered. The equation may be rearranged to

$$\theta = KP/(1 + KP) \tag{3}$$

It is called the Langmuir adsorption isotherm and θ versus P is shown in

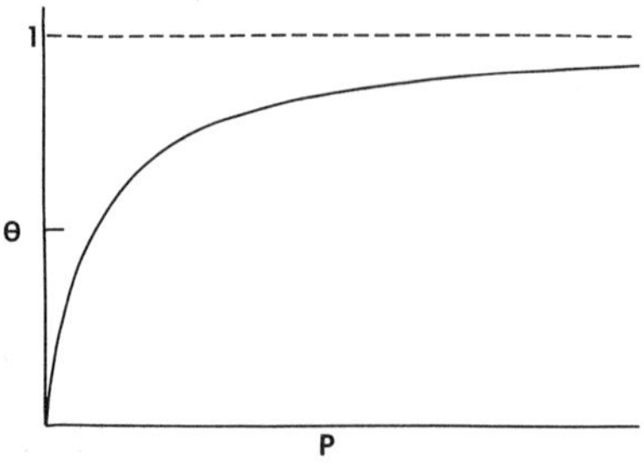

FIG. 10. The Langmuir adsorption isotherm: the fraction of a monolayer vs the pressure of the adsorptive.

Fig. 10. Initially, where $1 \gg KP$, $\theta = KP$ (the Henry's law region). At high values of P, θ approaches unity. If $-\Delta H$ for chemisorption is much larger than $-\Delta H$ for physisorption, then, if one operates under conditions such that the isotherm corresponds to an equilibrium process, as one must if the isotherm is to have significance, the temperature must be so high that physisorption is negligible. Therefore θ will not significantly exceed unity at higher pressures. If isotherms are determined at several temperatures, one can obtain ΔH_{ads} from

$$\left(\frac{\partial \ln P}{\partial T}\right)_\theta = \frac{\Delta H_{ads}}{RT^2} \quad \text{and} \quad \ln P_\theta = \frac{-\Delta H_{ads}}{RT} + \text{constant} \tag{4}$$

Thus, in Fig. 11, the values of P at any set value of θ provide the value of ΔH_{ads} at that value of θ by means of Eq. (4). If the Langmuir isotherm is valid, ΔH_{ads} will be independent of θ. In general, however, ΔH_{ads} decreases with θ. Figure 11 corresponds to this situation since here P'/P'' (where P' and P'' are equilibrium pressures for a set value of θ at two different temperatures) decreases with θ. In general, we must expect observed isotherms usually to be of a form more complicated than that of the Langmuir adsorption isotherm since K will be a function of θ.

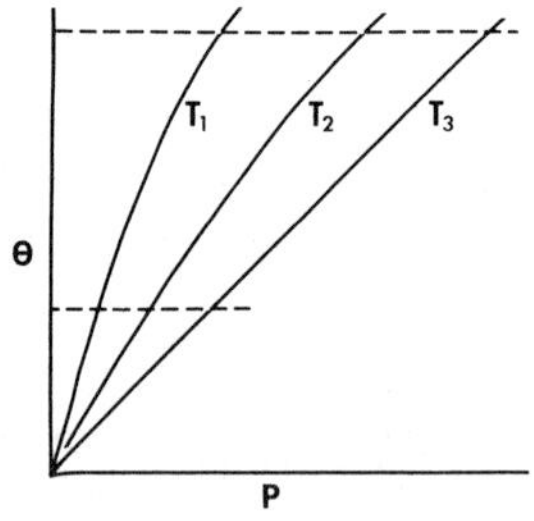

FIG. 11. Isotherms at three different temperatures: $T_3 > T_2 > T_1$.

TABLE IV

TECHNIQUES FOR STUDYING CHEMISORPTION

Surface chemical physics	Surface physical chemistry
	Techniques
Flash filament	Measurement of adsorption isotherms
LEED[a]	Temperature-programmed desorption
Molecular beams impinging on surfaces	Optical spectroscopy (IR, visible, and UV)
Field emission microscopy	Resonance spectroscopy (EPR[a], NMR[d])
Work function measurement	Mössbauer spectroscopy
Auger spectroscopy, XPS[b]	XPS
	Area of typical sample
$0.5\ cm^2$	$> 1\ m^2$
	Typical region of operating pressures
10^{-11}–10^{-4} torr[e]	10^{-6}–10^{4} torr

[a]LEED = low-energy electron diffraction.
[b]XPS = X-ray photoelectron spectroscopy.
[c]EPR = electron paramagnetic resonance.
[d]NMR = nuclear magnetic resonance.
[e]Torr = unit of pressure of 1 mm Hg.

3. *Methods of Studying Chemisorption*

Techniques for the study of chemisorption may be classified under *surface physical chemistry* and *surface chemical physics*. Examples of techniques in these two areas are in Table IV as well as some operating parameters typical of the two areas of investigation. Descriptions of many of these techniques appear in R. B. Anderson (1968); see also Gager and Hobson (1975) for Mössbauer spectroscopy.

The techniques of surface chemical physics (see Somorjai, 1972, and Drauglis and Jaffee, 1975) often involve the flow of electrons (LEED, field emission microscopy, work-function measurements, Auger and XPS spectroscopy) or of molecules through distances of several centimeters. This puts an upper limit of $10^{-5} - 10^{-4}$ torr on most techniques of surface chemical physics since at higher pressures molecules in the gas phase scatter the moving electrons or molecules. The experiments of surface physical chemistry can be extended to pressures of 1 atm, or even higher.

Most experiments in surface chemical physics involve surfaces with areas of less than 1 cm^2, and field emission microscopy involves a spheroidal metallic tip about 500–1000 Å in diameter. If molecules are adsorbed on nearly every collision, then, at a pressure of 10^{-6} torr, a surface is covered in 1 sec. Thus, partial pressures of reactive components (e.g., oxygen, carbon monoxide, and water vapor) must be kept below 10^{-10} torr for an experiment lasting only 1000 sec. This means that the vacuum during pretreatment of the surface must be in the ultrahigh-vacuum (UHV) range, 10^{-10} torr. The active areas in a surface physical chemical experiment are apt to be in the range of several square meters. This much larger area has an important effect. Suppose that we have an active area of 1 $m^2 = 10^{18}$ $nm^2 = 10^{20}$ $Å^2$ and that the surface is covered at 4 adsorbed molecules per square nanometer. Then, 1 m^2 of surface can adsorb 4×10^{18} molecules, but at a pressure of 7.6×10^{-6} torr $= 10^{-8}$ atm, a volume of 1 liter holds only 3×10^{14} molecules. Thus, at 7.6×10^{-6} torr in this volume, only small coverages could result on areas of 1 m^2, whereas serious coverages would result on areas of 0.5 cm^2. As a result, if a clean surface can be prepared, it can be kept clean much more easily in the experiments of surface physical chemistry than in those of surface chemical physics.

The techniques of surface physical chemistry can readily be applied to conventional catalysts including supported catalysts. Indeed, most of the techniques require relatively large areas so as to obtain adequate sensitivity. On the other hand, the techniques of surface chemical physics cannot be readily applied to most conventional catalysts and particularly not to supported catalysts. It is not easy for example for beams of electrons to reach the tiny crystallites of platinum in the interior of the Pt/SiO_2 catalysts of Fig. 2. Further, electrons which are given off or adsorbed in many of the techniques of surface chemical physics tend to charge the platinum particles electrically since the support is an insulator. This causes trouble and, although there are means of mitigating this problem, the problem remains serious.

In addition, it will be noted that the probes in surface physical chemistry are usually of relatively low energy (IR, EPR), whereas those in surface chemical physics are often of high energy (LEED, Auger, field emission) and are likely to alter the system under investigation, particularly if the system involves adsorbed organic species.

It is not surprising, then, that the techniques of surface physical chemistry have so far been more useful in application to heterogeneous catalysis. However, we may expect more application and influence of surface chemical physics in the future. The surfaces studied in the latter technique are so small that they can readily be single-crystal faces, whereas the surfaces used in the techniques of surface physical chemistry are almost always polycrystalline. Thus, surface chemical physics provides information about much more precisely defined surfaces. In particular, in LEED, electrons can be diffracted by crystals exposing a single, well characterized crystal plane. The wave properties of the electrons cause the diffraction and information about structure much like that in X-ray diffraction results. However, because of their low energy, ~100 eV, the electrons whose diffraction is examined have penetrated only a few atomic layers. Thus, it is possible to examine chemisorbed layers by LEED. In the past it has been very difficult to extract more than a partial surface structure from LEED data but it is likely that complete structures will be generally obtainable in the not too distant future. Experiments with LEED indicate that atoms or molecules are adsorbed on many metallic surfaces in regular structures and often in structures whose geometry changes as θ increases. One of the challenges of the future will be in applying this information to heterogeneous catalysis. There is an apparent conflict between the case of a crystal face on which θ changes discontinuously at a particular pressure and that of a polycrystalline sample which follows an adsorption isotherm which resembles at least to some degree the Langmuir adsorption isotherm.

The contributions of surface chemical physics may be expected first in the area of metallic catalysts. Insulating oxides are apt to benefit only later. It is doubtful if applications of surface chemical physics to such catalysts as Cr_2O_3/Al_2O_3 or $Mo(CO)_6/Al_2O_3$ will come during the lifetime of anyone but the youngest reader and certainly not of the writer. Substantial contributions to an understanding of many catalysts, and of the latter two catalysts in particular, have come recently from the areas of coordination chemistry and organometallic chemistry and the scale of such contributions will probably increase, at least in the near future. Heterogeneous catalysts and the reactions catalyzed by them are so varied that many areas will contribute to the advancement of our understanding: coordination chemistry, organometallic chemistry, surface chemical physics, surface physical chemistry, solid state physics, and organic chemistry, to name the more obvious. Chemisorption and heterogeneous catalysis have needed a theoretical (i.e., quantum mechanical) framework which would provide a general understanding of surface compounds in a fashion similar to that in coordination chemistry and organic chemistry. The problem of surface compounds is much more difficult than that of simpler molecular systems and, in the past, few persons have worked on theories of surface. Recently, however, a number of theoreticians, both physicists and chemists, have begun to work in the area of surfaces of metals and chemisorption (Gomer, 1975, and Drauglis and Jaffee, 1975).

IV. Mechanism in Heterogeneous Catalysis

A. General

It should be clear by now that heterogeneous catalysis is a branch of chemistry, that it can involve a very wide scope of descriptive chemistry, and that the basic difference between homogeneous catalysis and heterogeneous catalysis is that the intermediates of the former are species dissolved in solution whereas those of the latter are surface compounds. The nature of the intermediates of heterogeneous catalysis must be deduced from chemical and physical evidence, but in practice this has been true of most homogeneous catalytic reactions also. In the opinion of the author, the two major features which in general have made heterogeneous catalysis more "difficult" than homogeneous reactions are (a) the absence of a quantum mechanical framework to which allusion has just been made and (b) the probable multiplicity of surface sites. The coordinatively unsaturated species at the surface of a metal, an oxide, etc., almost inevitably will be found in several forms at the surface of polycrystalline materials, at least one form on every particular crystal face exposed, although, as in Fig. 9, more than one chemisorbed form may exist even on a single crystal face. Presumably the sites would be of the same general form but would differ in the exact geometric relationship of the atoms surrounding the site. One might expect that the sites in different crystal faces and at edges would differ in rate and in selectivities. We know little about this but the matter is one of substantial current interest.

In sum, a heterogeneous catalyst may be and probably often is a mixture of catalytic species of somewhat differing properties. Although, in fact, many homogeneous catalysts which are coordination complexes also appear to involve more than one catalytically active species, the problem is surely more serious with heterogeneous catalysts.

B. Kinetics of Heterogeneous Catalytic Reactions

A large literature exists which deals with rates of catalytic reactions. The accumulation of precise data in this area is more difficult than in the case of homogeneous reactions for several reasons.

(1) Successive batches of catalyst will vary in activity and successive samples of catalyst from the same batch exposed to what was hoped to be the same conditions of activation will vary in activity. With time, this problem is becoming less serious but the problem of exact reproducibility in a material as complicated as many catalysts is clearly a difficult one. It should be recognized that heterogeneous catalysts and particularly sup-

ported catalysts are "materials" rather than chemicals. That is, the catalyst involves highly particular distributions and structures of its ingredients and these are usually not the equilibrium ones. The Pt/SiO_2 catalyst of Fig. 2 is clearly just one of the myriad of nonequilibrium arrangements of the ingredients; the exact reproduction of such a nonequilibrium arrangement may clearly require great care.

(2) The rates of many catalytic reactions on particular catalysts decline with time (*catalyst deactivation*). This makes it difficult to ascertain the exact functionality of the rate in terms of the temperature and the partial pressures (or concentrations) of reactants and products. Poisons are also a continual problem.

(3) If temperature gradients and concentration gradients in the catalyst pores $[(dC_i/dx_i)/C_i]$ are not negligible, the observed rate is a complicated quantity resulting from exposure of various layers of the catalyst pellets or granules to differing temperatures and concentrations of reactants and products. Many data in the literature suffer from the last problem, but this is less common in current literature.

It is the feeling of this writer that these problems have combined with the usual presence of multiple sites to restrict the scientific information obtained from kinetic studies in heterogeneous catalysis and that kinetics has so far been of less utility in heterogeneous catalysis than in the study of homogeneous reactions. There are clearly some who would disagree with this view.

This is not to argue that kinetics in heterogeneous catalysis is useless in mechanistic studies. It is essential, but it is far from enough (Boudart and Burwell, 1974).

A few examples of kinetics for surfaces with uniform sites are derived below. The examples given are for batch reactors, for a thin slice perpendicular to the flow in a flow reactor, or for differential reactors in which the conversion in a flow reactor is so small that changes in P_is are negligible. Extension of these results to flow reactors requires one to integrate along the length of the catalyst bed.

Consider the very simple reaction of Section I,

$$A(g) \rightarrow B(g)$$

for example,

$$CH_2{=}C(CH_3){-}CH_2{-}CH_3 \rightarrow CH_3{-}C(CH_3){=}CH{-}CH_3$$

to proceed via the sequence,

(1) $A(g) + * \rightleftarrows A*$
(2) $A* \rightarrow B*$
(3) $B* \rightleftarrows B(g) + *$

Step (2) is the rate limiting process (r.l.p.), i.e., all preceding steps are fast and essentially at equilibrium. As a further simplification, it is assumed that $([B(g)]/[A(g)])_{eq} = K_{eq}$ is large so that the reverse reaction can be neglected, and that K_B (the Langmuir adsorption equilibrium constant) is small, so that $\theta_B \ll \theta_A$ and 1. This is what is meant by the unequal arrows in step (3). A model of this form, in which a *surface reaction* step like (2) is rate limiting, is called a *Langmuir-Hinshelwood mechanism.*

Given these assumptions, the overall rate is equal to the rate of step (2)

$$r = \text{rate} = \text{rate of reaction (2)} = ak_2\theta_A \tag{5}$$

where k_2, the rate constant for the surface reaction, can be expressed in moles per unit area, per unit mass, or per unit volume of catalyst and a is the area, mass, or volume of the catalyst. By Eq. (2),

$$r = ak_2 \frac{K_A P_A}{1 + K_A P_A} \tag{6}$$

The variation in rate with P_A is exactly of the form shown in Fig. 10 since r is just θ_A multiplied by ak_2. At low values of P_A, $r = ak_2K_AP_A$, that is, the reaction is first order in A. At high values of P_A, $r = ak_2$, that is, the reaction is zero order in A and the rate of the reaction is independent of the pressure of A.

Suppose that K_BP_B is not negligible with respect to unity. Equation (5) still applies but the value of θ_A becomes

$$\theta_A = \frac{K_A P_A}{1 + K_A P_A + K_B P_B} \tag{7}$$

and

$$r = ak_2 \frac{K_A P_A}{1 + K_A P_A + K_B P_B} \tag{8}$$

Here, the product B inhibits the reaction, not because the reverse reaction B → A proceeds at a significant rate but because B competes for surface.

If K_AP_A and K_BP_B are negligible vs unity but a foreign compound Q is rather strongly but reversibly adsorbed, then

$$r = ak_2 \frac{K_A P_A}{1 + K_Q P_Q} \tag{9}$$

Here Q inhibits, i.e., rather weakly poisons, the reaction. Expressions of the form of Eqs. (6), (8), and (9) are unusual in standard homogeneous reactions but they are common in enzyme kinetics because the basic form of the mechanisms can be the same as in heterogeneous catalysis.

However, there is no *a priori* requirement that adsorption-desorption steps constitute a preequilibrium and that the surface reaction be a r.l.p. Suppose that the following sequence occurs:

(1) $A(g) + * \rightarrow A*$
(2) $A* \rightleftarrows B*$
(3) $B* \rightleftarrows B(g) + *$

Assume as before that K_{eq} for the reaction A → B is large, but that K_B is relatively large, that the rate of desorption of A from A* is negligibly small, and that the equilibrium constant for reaction (2) makes **(A*)/(B*)** very small (i.e., once adsorbed, **A** is rapidly converted to **B***). Then step (1) can be called a r.l.p. since step (1) is essentially irreversible and the rate of reaction is the rate of adsorption of **A**, which is proportional to the free surface, $1 - \theta_A - \theta_B$.

$$r = ak_1(1 - \theta_A - \theta_B)P_A = ak_1(1 - \theta_B)P_A$$

$$= ak_1\left(1 - \frac{K_BP_B}{1 + K_BP_B}\right)P_A = ak_1\frac{P_A}{1 + K_BP_B} \tag{10}$$

If $1 \gg K_BP_B$,

$$r = ak_1P_A$$

This is identical in form to Eq. (6) when $K_AP_A \ll 1$, but the rate constants have a different meaning. It is this kind of situation which illustrates why one can usually not derive a complete mechanism from kinetic data alone.

The equations derived above could be extended to cases in which K_{eq} is not large by allowing for the reverse reaction.

Of course, however, reactions are rarely as simple as the examples above. We will consider just one example of greater complexity:

$$A(g) + B(g) \rightarrow C(g)$$

We assume that it goes by a Langmuir-Hinshelwood mechanism:

(1) $A(g) + * \rightleftarrows A*$
(2) $B(g) + * \rightleftarrows B*$
(3) $A* + B* \rightarrow C* + *$ (r.l.p.)
(4) $C* \longrightarrow C(g) + *$

where we assume that K_{eq} is large and that K_C is very small

$$r = k_3\theta_A\theta_B = k_3\frac{K_AP_AK_BP_B}{(1 + K_AP_A + K_BP_B)^2} \tag{11}$$

Further details on the kinetics of catalytic processes are available in Boudart and Burwell (1974) and Boudart (1968).

In all of the above examples we have assumed that the catalytic sites are identical and non-interacting so that we could use Langmuir adsorption isotherms. In fact, of course, catalytic sites are not apt to be so accommodating and more complicated isotherms apply. Further, one will not know *a priori* the form of the proper isotherms nor the values of the two or more constants needed to calculate values of the θ_i's. However, the Langmuir form often leads to a reasonably good approximation to the kinetics (Boudart, 1968), and this simplified treatment can often provide useful mechanistic conclusions. For example, the finding that a reaction involving A as a reactant is zero or negative order in A indicates that A is strongly adsorbed. The finding that a reaction involving hydrogen is half-order in hydrogen suggests that a surface reaction involving H∗ is a r.l.p. since the Langmuir isotherm for the dissociative adsorption of hydrogen $H_2(g) + 2* \rightarrow 2H*$, is

$$K_H = \theta_H^2/(1 - \theta_H)^2 P_{H_2}$$

$$\theta_H = \frac{K_H^{1/2} P_{H_2}^{1/2}}{1 + K_H^{1/2} P_{H_2}^{1/2}} \quad (12)$$

Effect of temperature upon rates. Given the reaction and assumptions leading to Eq. (6),

$$k_2 K_A = A \exp -E_a/RT \quad (13)$$

where E_a is the *apparent* activation energy of the catalytic reaction. The term activation energy is restricted to single elementary processes. In Eq. (13), the effect of temperature appears in both k_2 and K_A, i.e., E_a involves both the activation energy for step (2) and ΔH_{ads} of A. The quantity A is also composite. If P_A is large enough, $K_A P_A \gg 1$, then Eq. (6) reduces to $r = k_2$. Under this condition, one can measure the activation energy of step (2). In general, however, activation energies are complicated. Unless one knows the *exact* mechanism (a situation which is rare), interpretation of the E_a's can be seriously misleading. There have been and still are enthusiasts for determining apparent activation energies, but in the view of the writer, such work has not greatly illumined our understanding. Apparent activation energies can be very useful in designing and controlling catalytic reactors, but there they merely provide an interpolation formula.

C. Other Mechanistic Probes

The writer has argued that kinetics provides useful but drastically inadequate mechanistic information. What else can be useful? The following kinds of studies have been useful: Boudart and Burwell (1974) give a general review of this matter, and more detailed reviews are listed later.

(1) *Structure variation.* For example, a study of ethylene hydrogenation alone is not apt to lead to a general understanding of the hydrogenation of olefins. One will need to study the hydrogenation of other olefins, such as propylene, 2-butene, cyclopentene, and *nor*bornene. The particular structural features of the different olefins each provides important information.

(2) *Isotopic tracers.* So far deuterium and ^{13}C and ^{14}C have been the most used. For example, we can add to the exemplified case of ethylene hydrogenation an investigation using deuterium rather than hydrogen and considerable further information results (for reviews, see Emmett, 1972; Burwell, 1972a).

(3) *Stereochemistry.* For example, in studying hydrogenation, *cis*-2-butene can be used and the accompanying isomerization to *trans*-2-butene examined. Or, the hydrogenation of methylallene can be studied and stereoselectivities for formation of *cis*- and *trans*-2-butene found. Or, optically active compounds can be studied and the reactant examined for racemization at lower conversion. Stereochemical data put strong limitations upon mechanism. For reviews of this subject, see Siegel (1966) and Burwell (1972b).

(4) *Spectroscopic investigations* (IR, optical, EPR, X-ray) can be made during the actual catalytic reaction.

(5) *Chemisorption studies.* Where possible, chemisorption of reactants and products can be studied by the techniques listed in Table IV.

(6) *Physical characterization* of the catalyst: measurement of percentage exposed or of areal concentrations of sites, particle size by X-ray diffraction, characterization of active sites by EPR, XPS, optical spectroscopy, Mössbauer or chemical analysis. For reviews, see R. B. Anderson (1968) and J. R. Anderson (1975).

D. Determination of Mechanism

In principle, of course, all conceivable mechanisms which do not violate fundamental principles proceed at some finite rate. The hope of a mechanistic study is to find that one mechanism which overwhelmingly dominates. But, one can never be sure *a priori* that two quite different mechanisms may not contribute significantly. The derivation of a mechanism from experimental data does not resemble a derivation in geometry. In practice, one should test the prediction of all imaginable mechanisms against the data. Each type of data will eliminate some mechanisms and with enough data one may be able to confine the mechanism to one or a small group of possibilities. But there is no logical procedure for proving that a mechanism you have not considered would not be better.

In the remainder of this article we discuss a selection of catalytic reactions in some detail. One of the purposes of the next section is to exemplify the application of the ideas presented in Section IV.

V. Sample Catalytic Reactions

Sections V,A and B are chosen primarily to exhibit methods for the study of mechanism. The reactions of Sections V,C and D represent more complicated reactions chosen to provide some information about technologically important examples of catalysis. Section V,E deals with oxidation catalysis, an increasingly important area of heterogeneous catalysis.

A. Reactions between Hydrogen and Hydrocarbons on Group VIII Metals

Group VIII metals, either supported or as powders, are widely used for hydrogenation of acetylenes, dienes, and allenes to olefins with high to medium selectivity, of benzene to cyclohexane, and of olefins to alkanes (often with accompanying double-bond migration and cis–trans isomerization).

1. *Hydrogenation of Acetylenes*

As an example, let us consider the hydrogenation at about 20°C of 2-butyne on palladium/alumina, although other palladium catalysts will give similar results. Experimental observations are summarized in the following scheme (Germain, 1969, Chapter 3).

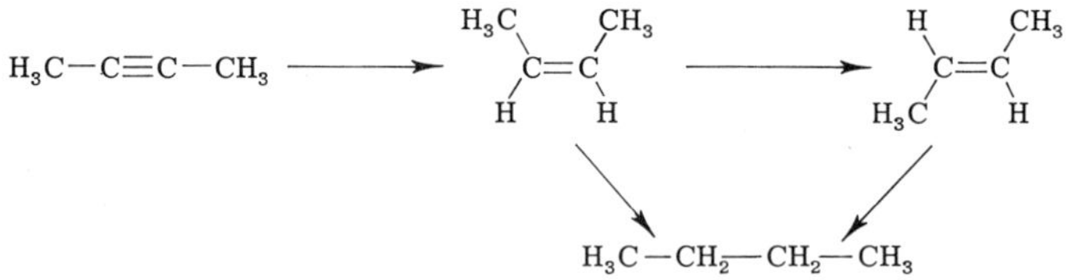

Cis-2-butene with only traces of accompanying *trans*-2-butene and butane is the initial product. The reaction is highly selective and stereoselective. The butyne is strongly adsorbed and inhibits further reaction of *cis*-2-butene by preventing its access to the surface of palladium. Once all butyne has been hydrogenated, *cis*-2-butene hydrogenates and isomerizes as shown. The hydrogenation reactions are fast with turnover numbers of about 50 sec^{-1} The *turnover number* is defined as the number of molecules reacting per surface atom of palladium per second. Other Group VIII metals are usually less selective in that some trans-ene and alkane are initial products but further reaction of the trans-enes is inhibited by an acetylene until all of the acetylene has disappeared.

If one adds a small amount of dimethyl sulfide to the reactant stream, the first step of the scheme is only slightly inhibited. The subsequent steps, however, are strongly inhibited. With the inhibited catalyst, one does not need to stop the reaction sharply in order to avoid loss of cis-alkenes. It has been suggested in such cases that the strongly adsorbing acetylene displaces sulfur from the surface of the metal but that the sulfur returns to the surface as soon as the acetylene has disappeared. Lindlar's catalyst, a supported palladium poisoned by quinoline and lead acetate, is a more practical version of a catalyst which hydrogenates acetylenes, but not most olefins.

The mechanism of the hydrogenation of 2-butyne on palladium must closely resemble the following model; all the steps are assumed to be irreversible.

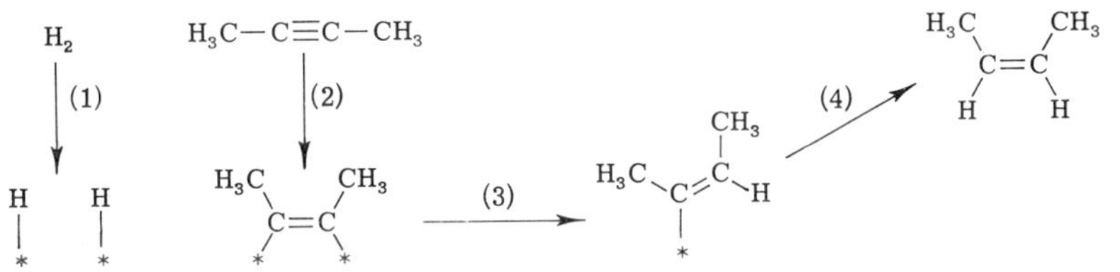

Adsorbed butyne formed in step (2) is written as a diadsorbed *cis*-2-butene, where *diadsorbed* signifies that two hydrogen atoms of the butene have been substituted by surface sites. There may well also be some contribution of π-bonding. Although the geometry of the hydrocarbon portion of the adsorbed intermediates shown above is reasonably sure, the exact nature of the bonding to palladium and whether the species are located over and bonded to just one atom of palladium or whether they are situated between two or more atoms of palladium is not known. This model is consistent with the following facts.

(a) *Cis*-2-butene is the initial product.

(b) When deuterium is substitued for hydrogen, the initial product is *cis*-$CH_3 - CD = CD - CH_3$. Thus, the methyl groups seem not to be involved in the reaction, or at least no C—H bond is cleaved.

(c) The reaction is of zero or negative order with respect to 2-butynes. The surface must be strongly covered by hydrocarbon intermediates.

(d) In the absence of butyne, the reaction $H_2 + D_2 \rightarrow 2HD$ is very fast by random associative desorption from $H* + D*$. The presence of butyne almost eliminates the reaction; again, hydrocarbon residues nearly cover the surface and when hydrogen adsorbs, it reacts rapidly and has little chance of desorbing.

(e) If 2-butyne is hydrogenated with an equimolar mixture of H_2 and D_2, *cis*-2-butene is formed as a random mixture of $CH_3-CH{=}CH-CH_3$, $CH_3-CD{=}CH-CH_3$, and $CH_3-CD{=}CD-CH_3$. Thus, hydrogen is dissociatively adsorbed and surface migration of adsorbed hydrogen atoms is fast (Burwell, 1969).

2. *Hydrogenation of Olefins*

Forty years ago Horiuti and Polanyi proposed a mechanism for the hydrogenation of olefins which when extended and amplified appears still to be essentially correct. It is shown here in its simplest form for the deuterogenation of ethylene (Germain, 1969, Chapter 3).

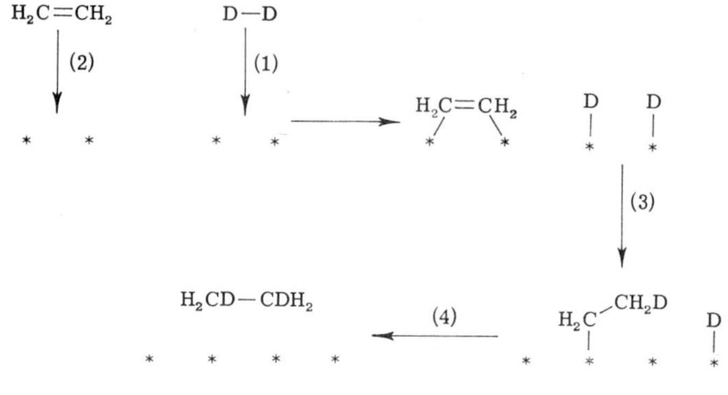

If all steps are irreversible as in the mechanism for the hydrogenation of 2-butyne, the only product would be CH_2DCH_2D. In fact, since ethylene, like 2-butyne, almost completely inhibits the reaction $H_2 + D_2 \rightarrow 2HD$, step (1) is nearly irreversible. However, with substituted ethylenes, the inhibition is incomplete, presumably because the larger olefins cannot pack as completely in order to block the surface. Compared to the overall rate of hydrogenation, desorption of most olefins is slow on platinum but rather rapid on palladium. Step (−4) is very slow since alkane added to the reacting system does not undergo isotopic exchange. What then about the degree of reversibility of step (3)? Isotopic analysis by mass spectroscopy of the ethane produced in the deuterogenation of ethylene shown that it is not even predominantly CH_2DCH_2D. Rather it contains all ethanes from ethane-d_0 to ethane-d_6. This must arise from the ratio of the rates of (−3)/(4) being rather large. Step (−3) operating on the monoadsorbed ethane at the right of the mechanism can give * − CH_2 − CHD − * + H*. Since HD desorption is very slow, the surface ratio H*/D* acquires a value of about unity. Thus, if step (−3) is fast enough, all hydrogen atoms in the product ethane will have been equilibrated with the surface (H,D) − * pool, and the observed isotopic smear results (Burwell, 1969).

If 1-butene is substituted for ethylene, the product of step (3) can be

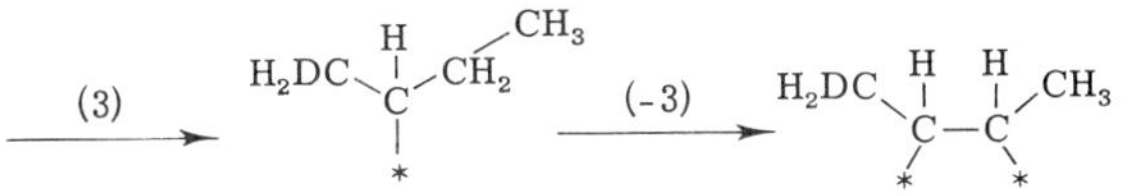

and it can revert by step (−3) to 2,3-diadsorbed butane as shown. On palladium this species must be assumed to desorb rather rapidly (i.e., the rate of steps (−2) and (4) are of the same order of magnitude). Thus, during the hydrogenation of 1-butene large amounts of the 2-butenes accumulate and finally disappear during the later stages of the hydrogenation. If the reactant is *cis*-2-butene, it adsorbs initially in the form at the right of the preceding scheme. It then reacts to form the 2-monoadsorbed butane at the left. This can revert to the form at the right, but since free rotation is possible about the C−C bonds, it can revert to adsorbed *trans*-2-butene and desorb. The occurrence of this set of reactions was described in Section V,A,1. On the contrary, with platinum, the rates of double-bond migration and of cis-trans isomerization are slow vs that of hydrogenation and there is only a small accumulation of isomeric olefins. We must assume, then, that step (−2) is relatively slow on platinum.

The matter is complicated by the demonstration that different sites on the surface of the metals promote somewhat different reactions (Burwell, 1969).

Evidence derived from studies of olefins of various structures indicates that the product of step (2), adsorbed olefin, must be in an eclipsed conformation. In the scheme, adsorbed olefin is shown as an α,β-diadsorbed alkane, but it might be a π-complexed olefin. The geometry of the hydrocarbon part is nearly the same in both forms. Any contribution from π-complexing would require an eclipsed conformation. As one example, consider the deuterogenation of *nor*bornene.

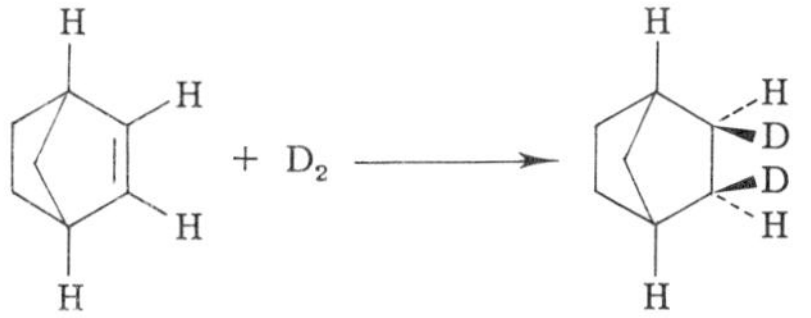

Here the reaction seems to follow the simplest (completely irreversible) form of the Horiuti-Polanyi mechanism presented at the beginning of this section. This results from step (−3) to a bridgehead hydrogen atom being impossible because it would involve a staggered conformation. Thus, step (−3) can only revert to the original species and this is without consequence. This reaction, because of the absence of confusing multiple exchange, demonstrates that in the hydrogenation of olefins in the gas phase or in nonpolar solvents, two hydrogen atoms are added cis to each other. The reaction illustrates another point. Both deuterium atoms are added on the least hindered side of the double bond. Therefore, the olefin adsorbs with the least hindered side of the plane of the double bond facing the catalyst. There are a large number of similar instances with other cyclic olefins and also with arenes.

3. *Isotopic Exchange Between Deuterium and Alkanes*

Step (−4) of the Horiuti–Polanyi mechanism is very slow during hydrogenation, but if one starts with deuterium and an alkane, the occurrence of the step can readily be demonstrated (Kemball, 1971; Burwell, 1969). That is, isotopic exchange occurs between alkane and deuterium. Temperatures of 40 to 100°C are convenient in flow reactors but lower temperatures can be used in batch reactors.

Consider the case in which *neo*pentane is the alkane. The product of step (−4) would be

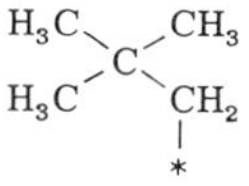

If step (−4) is followed by step (4), the only initial product would be monodeutero*neo*pentane. This is the major process on a number of Group VIII metals at lower temperatures. Here, of course, there is no possible 1,2-diadsorbed alkane. We must conclude that formation of 1,3-diadsorbed alkane from a 1-monoadsorbed alkane is relatively unfavorable.

With 3,3-dimethylpentane, step (−4) followed by step (−3) would lead to

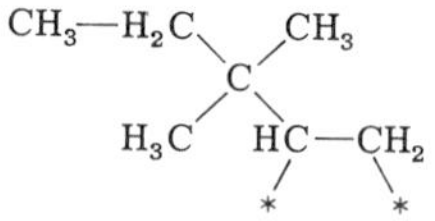

A sequence of steps (3) and (−3) followed by final desorption would give $(CH_3)_2(C_2H_5)C-CD_2-CD_3$ and this species is a major constituent of the initial product on several Group VIII metals. During isotopic exchange between deuterium and alkanes, the rate of desorption of HD is large. Thus, the surface is swept clean of hydrogen atoms and the surface ratio D*/H* is very large unlike the case during the reaction between deuterium and olefins. This makes the results of isotopic exchange with alkanes easier to interpret than those of reactions of olefins with deuterium.

Isotopic exchange between deuterium and cyclopentane provides further information. Initial adsorption of cyclopentane will generate

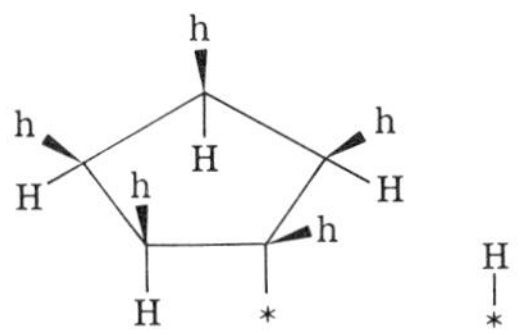

where we distinguish the hydrogen atoms on the lower side of the cyclopentane ring by H and those on the upper side by h. Alternation between steps (−3) and (3) will lead to a monoadsorbed cyclopentane in which all H atoms have been exchanged by D. Step (4) will then lead to $C_5H_5D_5$. This is a major initial product on platinum, rhodium, and palladium. There are, however, substantial initial yields of C_5D_{10}, particularly on palladium. These appear to result from some kind of a "roll-over" process in which diadsorbed cyclopentane changes its position of attachment from one side of the ring to the other. Desorption of cyclopentene followed by readsorption would accomplish this result, but the process appears to occur to a substantial degree without actual desorption of cyclopentene (Burwell, 1969).

On palladium and platinum catalysts, the exact nature of the initial distribution of the various isotopic species, cyclopentane-d_1 through cyclopentane-d_{10}, varies when different catalysts of the same metal but of differing particle sizes and conditions of pretreatment are used under otherwise identical conditions. In such a case the reaction is said to be *structure sensitive* on the particular catalyst. On the contrary, the turnover number for the hydrogenation of ethylene is nearly independent of the nature of platinum catalysts (Boudart, 1969; Boudart and Schlatter, 1972). The hydrogenation reaction is then nearly *structure insensitive*. Of course, judged merely by rate, a reaction might appear structure insensitive, but judged by selectivity factors, it might appear structure sensitive. Such a situation is apt to occur when selectivity results from the competing steps which follow a rate limiting process.

B. Reactions of Hydrocarbons on Oxides

1. *Hydroxylation and Dehydroxylation*

The matter of hydroxylation and dehydroxylation of the surface of oxides is intimately involved in the ability of such oxides as Cr_2O_3, Co_3O_4, and ZnO to catalyze reactions of hydrocarbons. Suppose that one cleaves a

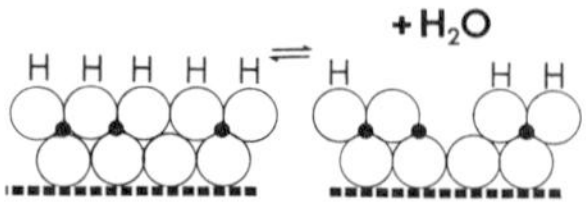

FIG. 12. Schematic representation of hydroxylation (reaction from right to left) and dehydroxylation (reaction from left to right). The large circles represent oxygen ions (as oxide or hydroxide) and the small, filled circles represent Al^{3+} or Cr^{3+}. The surface of the oxide is at the top.

crystal of an oxide along an electrically neutral surface. One obtains two electrically neutral fragments, each with a cleaved surface. The metal ions and the oxide ions on these cleaved surfaces will have coordination numbers less than those in the bulk. Now place water molecules on the cleaved surfaces at locations where oxide ions were before the crystal was cleaved, and transfer one of the hydrogen atoms as a proton to an adjacent surface O^{2-} ion. The surface will now be covered with OH^- and one will have replaced one surface O^{2-} by two OH^-. The crystal is still electrically neutral but the coordination number of the metal ion has been restored to its value in bulk, for Cr_2O_3, for example, to six. This addition of water to the surface is clearly favored. Thus, oxides of M^{2+} and M^{3+} prepared in water or exposed to water vapor are usually fully or at least extensively hydroxylated. Suppose one heats a sample of hydroxylated oxide in vacuum or in a stream of inert gas. The reverse process, dehydroxylation, $2\,\sigma{-}OH^- \rightarrow \sigma{-}O^{2-} + H_2O(g)$, will occur. This process exposes pairs of M^{n+} (cus) and O^{2-} (cus) ions. Hydroxylation and dehydroxylation are shown schematically in Fig. 12. These pair (cus) sites have a strong tendency to react not only with water but with other kinds of molecules, and they appear to be the catalytic sites for many reactions.

The areal concentration of Cr^{3+} (cus) on activated, hydrous chromia gel can be determined by "titration" with carbon monoxide or oxygen at $-78\,°C$. In this, one measures the extent of irreversible chemisorption of these compounds at $-78\,°C$. At low degrees of dehydroxylation, the adsorption complexes are probably $Cr^{3+}CO$ and $Cr^{4+}O_2^-$. After activation in a flowing inert gas to 250°C there are 0.2 Cr^{3+} (cus) per nm^2 (per 100 $Å^2$) and after activation to 400°C, 1.0 (Burwell *et al.*, 1969).

2. *Zinc Oxide*

Zinc oxide has catalytic activity for reactions of hydrocarbons rather similar to those of chromia. It is particularly suitable for infrared absorption studies of chemisorbed species and it has been possible clearly to characterize a number of catalytic intermediates (Kokes, 1973).

If zinc oxide is activated to about 450°C *in vacuo* or in oxygen and then exposed to hydrogen, two types of chemisorption of hydrogen can be observed at about room temperature. Type II chemisorption is rather slow and it does not participate in $H_2 + D_2 \rightarrow 2HD$ or $H_2 + C_2H_4 \rightarrow C_2H_6$. Type I participates in these reactions, it is much the more rapid and the weaker chemisorption, and it is reversible above $-40°C$. It is blocked completely by pre-exposure to water. Type I is characterized by two infrared absorption bands. The one at $1709\ cm^{-1}$ is due to $Zn^{2+}H^-$, the one at 3489 to OH^-. Of course, in $Zn^{2+}H^-$ there is considerable charge redistribution and the species is neither as ionic nor as basic as NaH. Thus, Type I chemisorption is heterolytic dissociative adsorption (species **G** of Table III).

As shown also by infrared absorption, propylene adsorbs by heterolytic dissociative adsorption to give

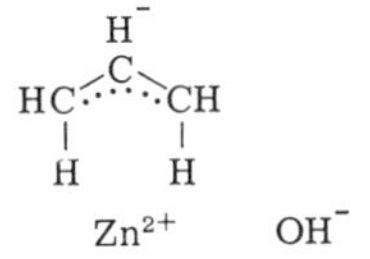

The adsorption of $D_3C-CH=CH_2$ or $D_2C=CH-CH_3$ leads to the appearance of the infrared spectrum of $D_2C\!\cdots\!C\!\cdots\!CH_2$, but this is accompanied by that of OD^- in the first case and of OH^- in the second.

Upon the adsorption of 1-butene, the following allyl species is formed

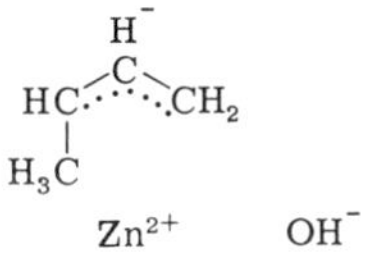

Associative desorption can give either 1-butene or 2-butene depending upon whether a proton is added to $C-1$ or to $C-3$. One indeed observes that zinc oxide is a good catalyst for the isomerization of 1- to 2-butene and that the cis/trans ratio in the 2-butene formed is about 10, a value characteristic of isomerization by the same mechanism in catalysis by dissolved organometallic complexes. In this mechanism of double–bond migration, one H is removed and then readded at a different position. In the Horiuti–Polanyi mechanism of Section V,A,2, an H atom is first added and then removed from a different position.

At $-196°C$, Type I chemisorption is too slow to be observed, but a molecular chemisorption of hydrogen can be observed by infrared spectroscopy. Since it is blocked by preadsorbed water, adsorption involves the same sites as Type I. The dipole moment of this adsorbed form must vary during the vibration but the observed frequency is not very far from the frequency calculated for the vibration in the gas phase, 4019 vs $4161\ cm^{-1}$. The adsorption corresponds to an unusually weak chemisorption since it can be removed by evacuation at $-196°C$. A similar chemisorbed form of hydrogen is observed on activated chromia, but this species cannot be removed at that temperature by evacuation. However, it is displaced as H_2 by a pulse of D_2 passing over the catalyst. Whether these molecular form of chemisorbed hydrogen have catalytic significance is as yet unclear.

Ethylene cannot adsorb allylicly on zinc oxide. It adsorbs as a π-complex and rather weakly since it is desorbed upon evacuation at 25°C. It reacts with $Zn^{2+}H^-$ to give $Zn^{2+}C_2H_5^-$ which then reacts with OH^- to give $C_2H_6(g)$.

On chromia, isotopic exchange between ethane and deuterium at about 200°C gives C_2H_5D as the initial product, not the isotopic smear observed on Group VIII metals. This process must involve preliminary heterolytic dissociative adsorption to give $Cr^{3+}(C_2H_5)^- + OH^-$. D_2 also appears to adsorb by heterolytic dissociative adsorption. Proton–deuteron interchange followed by associative desorption gives C_2H_5D. There can be no step equivalent to step (-3) of the Horiuti–Polanyi mechanism of Section V,A,2 (Burwell *et al.*, 1969).

C. Reactions on Acidic Catalysts

Silica gel is covered with silanol groups which are weakly acidic, pK_a of the order of 8. Hydroxylated γ-alumina is very weakly acidic. However, one can prepare high area, amorphous gels representable as SiO_2, Al_2O_3 which are homogeneous solid solutions if the mole ratio SiO_2/Al_2O_3 is about 4 or greater. As judged by Hammett indicators, such materials, after heating to about 400°C to remove adsorbed water, have surface sites as strongly acidic as concentrated sulfuric acid (Tanabe, 1970).

Silica gel can be considered to be formed by condensation of $Si(OH)_4$:

$$2Si(OH)_4 \rightarrow (HO)_3Si-O-Si(OH)_3 + H_2O.$$

Further condensation leads to $(HO)_3Si-O-Si(OH)_2-O-Si(OH)_3$ and finally to little particles of silica with diameters of about 10 nm (100 Å) which are covered with silanol groups. These primary particles agglomerate to form a gel which after drying is silica gel (see Fig. 1). In the primary particles, each atom of silicon is coordinated tetrahedrally with atoms of oxygen and each oxygen is bound to two atoms of silicon. In α-Al_2O_3, the aluminum ions are octahedrally coordinated to atoms of oxygen. In order to enter the silica structure, Al^{3+} must enter in tetrahedral coordination. One can consider, then, that $Al(OH)_3OH_2$ is the counterpart of $Si(OH)_4$ in the condensation process which leads to the gel. Thus, each Al^{3+} is accompanied by one H^+ so that the charge is the same as that of Si^{4+}, i.e., one can consider that one replaces Si^{4+} in silica gel by $Al^{3+} + H^{+}$. It is these H^{+} which are strongly acidic. The exact location of the protons is open to some doubt but the structure on the left will give the general idea (Tanabe, 1970).

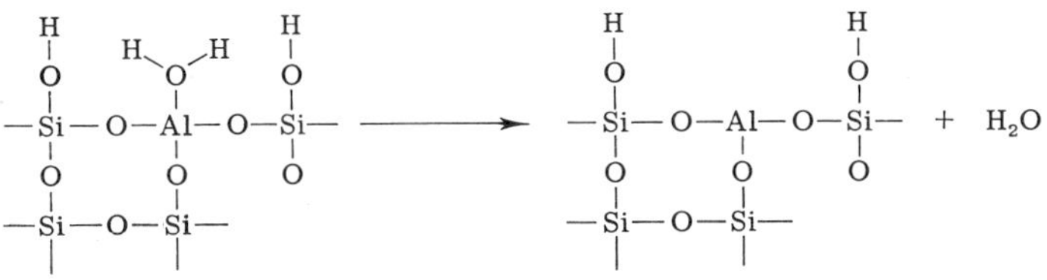

Heating to temperatures in the vicinity of 500 °C can lead to the loss of water with the formation of Al^{3+} (cus) at the surface. In the form on the left, silica–alumina is a proton acid; in the form on the right, it is a Lewis acid. There was considerable controversy for some years as to which form was responsible for the catalytic activity, but it now appears that the proton acid form is responsible for most of the interesting activity.

The proton form is a strong enough acid to convert propylene to an adsorbed species with the properties of a carbonium ion.

$$CH_2{=}CH-CH_3 + H^+ \rightarrow CH_3-\overset{+}{C}H-CH_3$$

Thus, silica–alumina can lead to the polymerization of olefins and to the alkylation of benzene by propylene to give isopropylbenzene. At higher temperatures, silica-alumina catalyzes the dealkylation of isopropylbenzene,

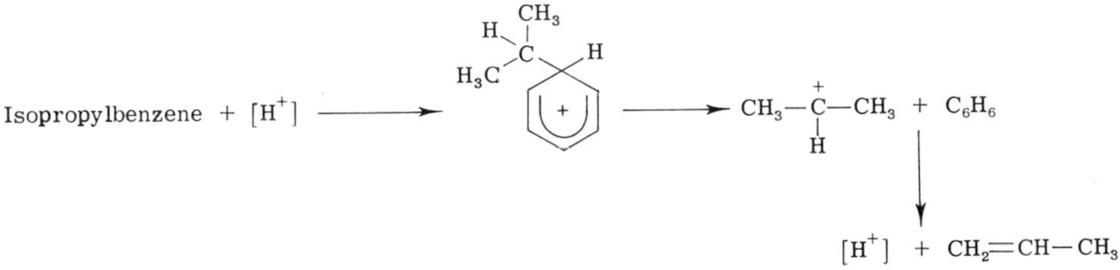

The major use of silica–alumina has been in catalytic cracking (Germain, 1969). To exemplify this process suppose that hexadecane is passed over silica–alumina at 450–470 °C. The alkane is initially converted to an adsorbed carbonium ion. Alkyl carbonium ions derived from straight-chain alkanes are known to isomerize to branched-chain carbonium ions one of which could be

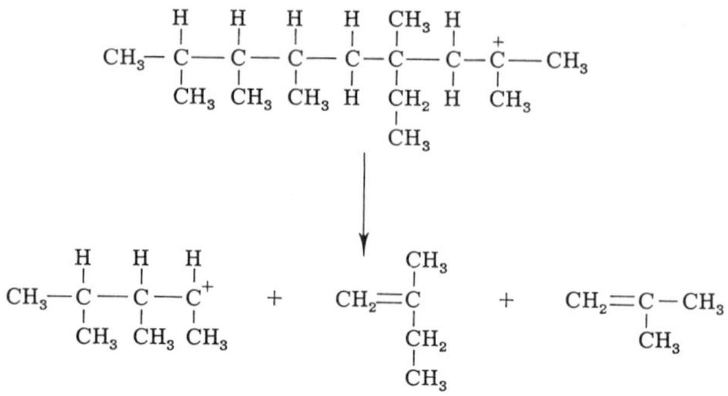

This carbonium ion will fragment as shown. The C_7 carbonium ion can be saturated by hydride ion transfer from an unreacted molecule of hexadecane on the surface,

to generate a new hexadecane carbonium ion which continues the chain. Hydrogen transfer reactions occur which saturate the olefins and deposit a hydrogen-poor hydrocarbon residue on the silica-alumina. This material, called *coke*, has a composition of about $(CH)_n$. The process is run in a reactor in which the catalyst is in a fluidized bed through which the hydrocarbon feed passes: Catalyst is continuously withdrawn, the coke burned off, and the catalyst returned to the reactor (Thomas, 1970).

One carbonium ion formed initially leads to the reaction of many molecules of alkane. The initial carbonium ions are probably formed, at least to a substantial extent, from small amounts of olefins originally in the feed or generated by some dehydrogenation process at the rather high temperature of the reaction.

Catalytic cracking, then, converts long, largely straight-chain, low octane hydrocarbons into smaller and branched-chain hydrocarbons. In practice much of the product is formed in the gasoline range and it is of a fairly good octane number. The process is a very valuable one in petroleum technology since in the United States and some other countries, the fraction of hydrocarbon in crude oil which is in the gasoline range is much below market requirements.

Amorphous silica-alumina has been replaced as the catalyst for catalytic cracking by synthetic zeolites, the so-called molecular sieves. These are crystalline aluminosilicates whose structure involves myriads of channels about 1 nm in diameter and whose volume occupies a substantial fraction of the total volume of the crystallites of the zeolite. All aluminum ions are in coordination number four. As the zeolite is prepared, the necessary charge balance is preserved by the presence in the channels of one Na^+ for each Al^{3+} in the structure. To make this material into a catalyst the sodium ions are partially exchanged with other ions, e.g., ions of rare earths plus ammonium. The tiny crystallites of zeolites are then imbedded in a matrix of silica-alumina or alumina. As compared with silica–alumina, the behavior is really only slightly changed; the sieve catalysts are more active and they give a more desirable product distribution. However, in a 100,000 barrel per day catalytic cracker, a few percent improvement in yield and in octane number can be worth millions of dollars per year.

The sodium ion in a molecular sieve can be ion exchanged by the ammonium ion and the sieve then heated. Ammonia is evolved and the residual proton becomes attached to an oxygen ion of the structure of the sieve. Such materials have very high surface acidity.

D. Bifunctional Catalysis

So far we have considered catalysts which possess basically just one kind of site (ignoring the relatively small variation in catalyst sites classified as nonuniformity). In this section we consider catalysts which possess two quite different types of sites. Suppose one deposits tiny crystallites of platinum on an acidic support of the type described in the preceding section. What properties would this catalyst possess?

In fact, this platinum *reforming catalyst* (see Thomas, 1970) possesses very valuable properties. The first example was developed by V. Haensel. Such catalysts consist of about 0.5% platinum by weight on silica–alumina or on acidic alumina. Since the surface of ordinary alumina is insufficiently acidic, the acidity is increased by treatment of the alumina or of the catalyst with Cl^- or F^-.

The naphtha (the hydrocarbon mixture which has the boiling point range of gasoline) obtained in the distillation of crude oil has a very low octane number, much too low to permit its use in current gasoline engines. During *catalytic reforming*, the following processes occur and all increase the octane number of the product. It will be recalled that aromatic hydrocarbons and highly branched-chain alkanes have good octane numbers whereas straight-chain alkanes have poor ones.

(1) Dehydrogenation of cyclohexanes to benzenes.
(2) Dehydrocyclization of heptane to toluene, hexane to benzene, etc.

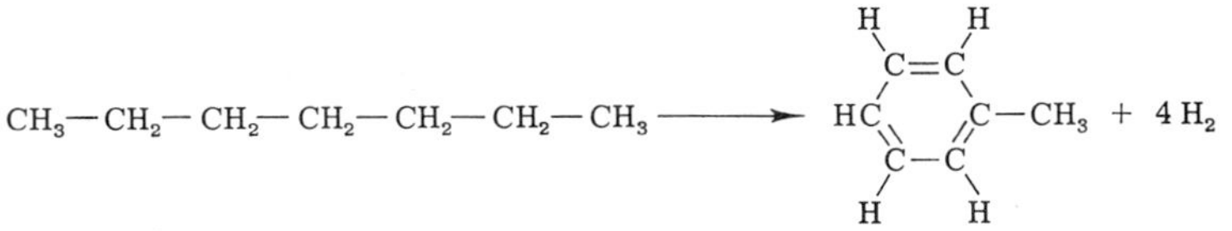

(3) Isomerization of straight-chain hydrocarbons to branched-chain hydrocarbons.

(4) Isomerization of cyclopentanes to cyclohexanes followed by their conversion to benzene by (1).

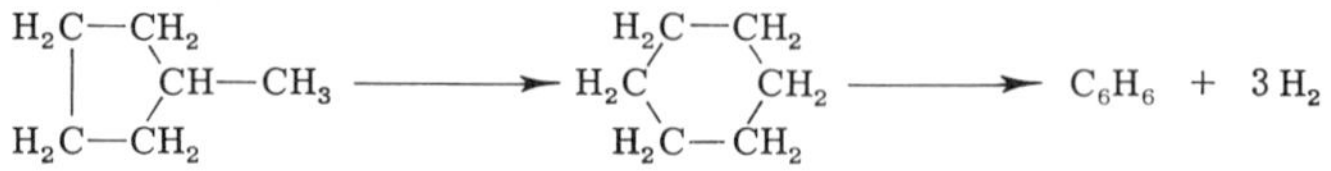

(5) Optionally, the conversion of alkanes to small hydrocarbons below the boiling range of gasoline.

In sum, the feed is naphtha of very low octane number and the product in good yield is gasoline of high octane number. The process is of such value that it can readily support the use of platinum. During its lifetime, the catalyst will convert more than 100,000 liters of naphtha per kilogram of catalyst, i.e., per 5 g or \$30 worth of platinum. The dead catalyst is returned to the catalyst manufacturer and the total loss of platinum is only 1%.

Operating temperatures are about 500°C. Choice of such high temperatures is not primarily dictated by considerations of rate. Reactions (1) and (2) are endothermic and temperatures nearly as high as the operating temperatures are required, so that the equilibrium constants have become favorable for these reactions. The isomerizations have quite small heats of reaction. Reaction (5) is strongly exothermic but the reaction is slow and severe conditions are needed if this reaction is to contribute much to the overall process.

The platinum serves another important purpose, it keeps the surface of the catalyst clean. The catalyst would be useless if it coked as fast as the acidic component alone. In the presence of hydrogen, the platinum catalyzes the hydrogenation of the coke or of coke precursors. The matter is not understood in detail. However, rather high pressures of hydrogen are needed to confine coking to an acceptable level. A typical operating pressure of hydrogen is about 20 atm.

Reaction (1) would, of course, go on platinum alone. The acidic component is not needed. Reaction (2) goes on platinum alone but the exact contribution if any of the acidic sites of dehydrocyclization has not been clearly unraveled.

Reaction (3) goes on acidic sites alone, but rather slowly. The platinum is essential to keep the catalyst clean but it probably contributes to the reaction by causing a small amount of dehydrogenation (the equilibrium constant for dehydrogenation of alkanes has become significant by 500°C). The olefin so produced reacts as follows:

(a) At Pt: alkane → olefin + H_2.

(b) At acidic sites: olefin → carbonium ion → branched chain carbonium ion → branched-chain olefin.

(c) At Pt: branched-chain olefin → branched-chain alkane.

This mechanism is supported by the observation of Weisz that the isomerization reaction proceeds readily upon a mixture of two materials, platinum upon an inert support and silica–alumina (Germain, 1969, Section IV,2,B).

Recently, catalysts containing rhenium as an additional ingredient have been developed. They deactivate more slowly than the catalysts containing platinum alone. This permits them to be used for longer intervals between regenerations or to be used to give higher severity.

It is important to prepare a catalyst with the proper proportion of hydrogenating activity and of strength and extent of acidity. By maximizing these characteristics in conjunction with the proper choice of operating conditions, one can operate other processes which are related to platinum reforming. For example, one might wish to optimize the isomerization reaction and to isomerize pentane to isopentane. Because of the variation of equilibrium constant for this reaction with temperature, one should employ as low a temperature as possible. In practice one can operate at about 150°C using about the same pressure of hydrogen as in reforming. The platinum keeps the catalyst clean and intervals between regenerations are months. At 150°C, the equilibrium quantity of olefin is very low and the method by which initial carbonium ions are formed is not clear.

Reaction (5) can be augmented, and catalysts which promote *hydrocracking* have been developed. High pressures of hydrogen are needed to diminish deactivation but catalyst lives of one to three years can be obtained. The acidic component can be acidic molecular sieves, the hydrogenating component can be Pd, NiS_x or NiS_x-WS_x. Operating temperatures are about 400°C. In hydrocracking, much the same result is obtained as in cracking but the olefin fragments are hydrogenated at the hydrogenating component (with consequent substantial consumption of hydrogen) and the life of the catalyst is enormously extended over that in catalytic cracking.

E. Heterogeneous Catalytic Oxidation

1. *General*

Some examples of oxidation reactions have been listed in Table II. The area of oxidation catalysis has received less academic attention than that of reactions involving hydrogen. It is a more difficult area. It has been more difficult to apply some of the mechanistic probes described in Section IV,C. Most of the reactions require rather high temperatures and the reactions usually appear to involve a rather complicated set of elementary processes. The best practical catalysts are often complicated mixtures. However, the area is of considerable practical importance and it is almost sure to increase in importance. It must be studied.

The reaction

$$2CH_4 + O_2 \rightarrow CH_3{-}O{-}CH_3 + H_2O$$

would seem to be a very simple reaction and it is one which is strongly favored thermodynamically. However, at present, we have no notion whatever as to how the reaction might be effected in any significant yield.

The overriding problem in heterogeneous catalytic oxidation is selectivity. It is easy to convert CH_4 to CO_2 but one cannot stop with any selectivity en route. All intermediates appear to be more reactive than the reactant. There is currently no useful heterogeneous catalytic process which involves the oxidation of alkanes.

There has been a large amount of work dealing with the oxidation of simple hydrocarbons, of carbon monoxide to carbon dioxide, and of hydrogen to water. In general the ease of oxidation increases in the sequence alkanes, aromatic hydrocarbons, olefins. Further, ethane is more easily oxidized than methane, and propane, more than ethane. The controls imposed by the Government of the United States upon motor vehicle emissions have led to a burst of practical work in this area. The catalysts described in Section II,B,1 reduce concentrations of hydrocarbons and carbon monoxide to mere traces. They effect this under what would earlier have been thought conditions extremely unfavorable to heterogeneous catalysts: temperatures which change rapidly from 0° to 1000°C, flow rates which change rapidly, and a requirement that the catalyst have a very low thermal mass so that it will warm very rapidly. On top of this, the catalyst must resist being shaken unmercifully and last for 25,000 miles with no attention.

Myriads of different catalysts effect the oxidation of carbon monoxide and hydrocarbons: Group VIII metals, CuO, Cr_2O_3, $CuO \cdot Cr_2O_3$, rare earth oxides, and many complicated oxides like perovskites. A number of studies have attempted to characterize the activity sequence for simple oxides. In the oxidation of ethylene at 262°C, the sequence of oxides is Co > Cr > Ag > Mn > Cu > Ni > V > Cd > Fe > Mo > W > Ti > Zn (Germain, 1969, Section V,2). Other reactants and conditions give other series. Despite extensive study of oxides, only platinum-containing catalysts have proved effective in converting motor vehicle emissions. One may wonder what the future will think of our using up this rare and valuable metal for this purpose. Little of the platinum will be recovered.

One may imagine two possible extreme cases of oxidation catalysis.

(a) The oxygen used to produce H_2O, CO_2, or other products comes from adsorbed oxygen.

(b) The oxygen comes from lattice O^2 which is replaced by adsorption of oxygen in a subsequent step.

For a simple example, consider the oxidation of carbon monoxide. In case (a), the sequence might be

$$M^{n+}(\text{cus}) + \tfrac{1}{2}O_2 \rightarrow M^{(n+1)+}O^-$$

$$M^{(n+1)+}O^- + CO \rightarrow M^{n+}(\text{cus}) + CO_2$$

In case (b), the sequence could be

$$M^{n+}O^{2-} + CO \rightarrow M^{(n-2)+} + CO_2$$

$$M^{(n-2)+} + \tfrac{1}{2}O_2 \rightarrow M^{n+}O^{2-}$$

Overall, $O_2(g)$ is the oxidizing agent in both cases, but in the step which actually gives the oxidation of CO to CO_2, the oxidizing agent is both O^- and $M^{(n+1)+}$ in case (a) whereas in case (b) only M^{n+} is the oxidizing agent since the oxidation number of oxygen is the same in O^{2-} and CO_2. Where adsorbed oxygen is O^{2-} rather than O^-, then only the metal ion is the effective oxidizing agent in both cases. Since platinum does not form bulk oxide but only a surface monolayer of oxygen, oxidations on platinum must constitute an example of case (a).

One would also expect to observe case (a) with an oxide which cannot easily be reduced and which is relatively slowly oxidized to higher oxides, possibly, for example, Cr_2O_3. On the contrary, one might expect to observe case (b) for an oxide in which the metal ion is reoxidized rapidly to the highest oxidation number, possibly, for example, MoO_3.

2. *Adsorption of Oxygen*

It is curious that adsorbed oxygen can be more reactive than elemental oxygen. After all, the energy of the system falls when oxygen is adsorbed yet one must assume in some cases that the reactivity of oxygen rises. However, this situation is obtained in all of catalysis. Clearly, though, one can go too far in lowering the energy of the system. Metallic titanium is not going to be a good oxidation catalyst. It adsorbs oxygen very readily in the first step of case (a) but the second step will have rate of zero. On the contrary, oxygen adsorbed on the surface of metallic platinum will oxidize hydrogen and carbon monoxide at room temperatures. The idea that effective catalytic action requires that a reactant be adsorbed strongly enough to give some reasonable value of the quantity adsorbed but not so strongly that the adsorbed species is of low reactivity is an old one going back to Sabatier, one of the founders of modern heterogeneous catalysis.

The adsorption of oxygen on oxides in what might be considered the first step of case (a) presents rather complicated possibilities since the following adsorbed species are possible: O^-, O_2^- (the superoxide ion known in bulk in KO_2), O_3^- (the ozonide ion known in bulk in KO_3), O^{2-} (the oxide ion), and O_2^{2-} (the peroxide ion). However, O^-, O_2^-, and O_3^- have an odd electron and are paramagnetic. Therefore, studies using electron paramagnetic resonance (EPR) have been fruitful (Lunsford, 1973). Use of ^{17}O which has a nuclear spin of 5/2 (unlike ^{16}O which has zero nuclear spin) has greatly extended the information coming from EPR studies because by the examination of the hyperfine splitting, one can unequivocally tell how many oxygen atoms are involved in the adsorbed species and in the case of O_2^- the hyperfine splitting indicates whether the two oxygen atoms are equivalent. In some instances, the hyperfine splitting identifies the surface ion to which the adsorbed oxygen is bonded. Thus, a reduced MoO_3 exhibits the EPR spectrum of Mo^{5+}. As oxygen adsorbs, the signal of O_2^- appears. The hyperfine splitting indicates that there is substantial covalent bonding of the ion and that the two oxygen atoms are not equivalent. The process can probably best be represented as

$$Mo^{5+}(cus) + O_2(g) \rightarrow Mo^{6+}O_2^-$$

Treatment of the same material with N_2O leads to the signal of O^-. In the few cases in which the matter has been studied, the reactivity of O^- with carbon monoxide and simple hydrocarbons appears to be much greater than that of O_2^-.

Cimino and Stone, separately and jointly, have reported a series of important studies on solid solutions of oxides of transition elements in insulating oxides like MgO and Al_2O_3 (Cimino *et al.*, 1973; Egerton *et al.*, 1974; Cimino, 1974). Information on two items has come from this work: the adsorptive and catalytic behavior of ions of different transition elements in the same environment, and the effect of the dilution of a given ion upon these properties. Ions of transition elements have adsorptive and catalytic properties even in high dilution in insulating oxides where the binding of adsorbed species in a purely localized fashion to the ion of the transition element seems certain. However, there is an effect of concentration. In particular, Egerton *et al.* (1974) conclude that electronic interaction exists among neighboring Cr^{3+} ions in solutions of higher concentrations of Cr^{3+} in $(Cr^{3+}, Al^{3+})_2O_3$. At least under oxidizing conditions, the purely localized representation of adsorption used in Section V,B would appear to be an oversimplification.

3. *Selective Heterogeneous Catalytic Oxidations*

In recent years there have been discovered a number of reactions involving hydrocarbons and hydrogen which are catalyzed by organometallic complexes (J. Chatt and J. Halpern, in Basolo and Burwell, 1973). One case in which both homogeneous and heterogeneous catalysts are known, the metathesis of olefins, was discussed in Section I and another in Section V,B,2 (allylic double bond migration). Many others which could be fitted into Sections V,A and V,B are known. The molecular analogs have been useful in guiding mechanistic thinking in heterogeneous catalysts (Ugo, 1975). Many fewer homogeneous analogs are known for heterogeneous catalytic oxidation. The high temperatures needed in most heterogeneous catalytic oxidations militates against this. A low temperature example, however, follows.

The catalyzed reaction between *tert*-butyl hydroperoxide and cyclohexene at about 80°C produces cyclohexene oxide and *tert*-butyl alcohol.

MoO_3 is a heterogeneous catalyst for the reaction and $MoO_2(acac)_2$ (acac is a bidentate, negatively-charged ligand, the conjugate base of acetylacetone) and some other, but far from all, molybdenum complexes are homogeneous catalysts (Triforo *et al.*, 1975). The system presents the advantage that the oxidation number of molybdenum during the catalytic reaction can be monitored by optical spectroscopy. Coordinatively unsaturated Mo species appear to be involved in both the heterogeneous and homogeneous catalyses.

The reaction just described is rather selective, but it does not involve elemental oxygen. Metallic silver catalyzes the oxidation of ethylene to ethylene oxide at about 250°C in yields exceeding 70%. Adsorbed oxygen must be the reactant and the reaction must be an example of case (a) of Section V,E,1. Probably the reaction involves abstraction of one oxygen atom from oxygen adsorbed as superoxide, $O—O^-$. This leaves an adsorbed atom of oxygen which reacts to burn ethylene completely to carbon dioxide (Kilty and Sachtler, 1974). The overall equation may be represented

$$7C_2H_4 + 6O_{2ads} \rightarrow 6C_2H_4O + 2CO_2 + 2H_2O$$

However, silver is the only known catalyst which gives significant yields in this reaction and the reason for this is as yet unknown. Further, the reaction is not selective with propylene on silver or on any other known heterogeneous catalyst. Very probably the reason silver does not lead to good yields of propylene oxide is that propylene, unlike ethylene, possesses allylic hydrogen atoms and attack starts at the reactive allylic C—H bond.

Selectivity can sometimes be achieved in cases in which some intermediate between hydrocarbon and carbon dioxide is relatively stable and, in vapor-phase oxidations, relatively volatile. The oxidation of naphthalene to phthalic anhydride listed in Table II and the related oxidation of benzene to maleic anhydride are of the latter type [for a description of the industrial aspects of reactions of this section, see Thomas (1970, Chapter 20); for the chemistry, see Germain (1969, Chapter V) and Hucknall (1974)].

In general it is unlikely that case (a) will usually lead to selective oxidation. The active adsorbed forms of oxygen are even more apt to attack the intermediates than the original reactant. Case (b) presents more favorable opportunities for selectivity. The oxidation potential can be widely varied and there are some possibilities for variety in the natures of the initial adsorption and of the adsorptions of the intermediate species.

Some oxidations of several olefins on Bi_2MoO_6 are among the most remarkable examples of selective oxidation. These reactions occur at about 450°C. Further details are available in Schuit (1974) and Hucknall (1974). The reactions are:

$$CH_2{=}CH{-}CH_3 + O_2 \rightarrow CH_2{=}CH{-}CHO + H_2O$$

$$CH_2{=}CH{-}CH_3 + NH_3 + \tfrac{3}{2}O_2 \rightarrow CH_2{=}CH{-}C{\equiv}N + 3H_2O$$

$$CH_3{-}CH{=}CH{-}CH_3 + \tfrac{1}{2}O_2 \rightarrow CH_2{=}CH{-}CH{=}CH_2 + H_2O$$

We shall discuss only the first reaction. Isotopic-labeling experiments using deuterium and carbon isotopes establish that propylene is dissociatively adsorbed in the first step as some kind of allylic species and that this step is rate limiting. Thus, $^{14}CH_2{=}CH{-}CH_3$ and $^{14}CH_3{-}CH{=}CH_2$ give the same product (i.e., $CH_2 \overset{\cdots}{-} CH \overset{\cdots}{-} CH_2$ is the first adsorbed species and it can react subsequently at either end). Further there is a kinetic isotope effect of 0.55 disfavoring D−C vs H−C at 450°C. This number results from measuring the rates of reaction of C_3H_6 and C_3D_6. This kinetic isotope effect indicates that the initial cleavage of allylic C−H bond is rate limiting.

A remarkable feature of the reaction is that one may omit oxygen and find the reaction to proceed with the usual selectivity until the catalyst composition falls to $Bi_2MoO_{5.5}$. During this period the electrical conductivity increases and the EPR signal of Mo^{5+} appears. Treatment with oxygen restores the original properties and composition. Keulks has shown that the acrolein formed initially using $^{18}O_2$ is nearly devoid of ^{18}O. It appears very probable then that the oxygen which is incorporated comes from the lattice. The reaction is first order in propylene and zero order in oxygen. These orders are compatible with initial adsorption of propylene being rate limiting and with reoxidation by oxygen being fast.

The crystal structure of Bi_2MoO_6 consists of sheets of O^{2-}, $(Bi_2O_2)^{2+}$, and $(MoO_2)^{2+}$ alternating in the following manner

$$(Bi_2O_2)^{2+} \quad O^{2-} \quad (MoO_2)^{2+} \quad O^{2-} \quad (BiO_2)^{2+} \quad O^{2-} \ldots$$

The oxygen between the layers containing the metal ions is clearly mobile, and it has been suggested that the oxygen which replaces the O^{2-} consumed in the first step of case (a) may be added at some distance from the point of loss.

The initial reaction may be

$$C_3H_6 + Mo^{6+} + O^{2-} \rightarrow (MoC_3H_5)^{5+} + OH^-$$

During subsequent stages, it appears likely that the adsorbed organic intermediates are not exposed to adsorbed oxygen and this may be the key to the selectivity.

A number of variants of the Bi_2MoO_6 catalyst have been reported and some appear to be superior in selectivity for some of these reactions. Many consist of complicated combinations of some of the oxides of Mo, Co, W, V, or U plus some of the oxides of Bi, Pb, As, Sb, P, Se, or Te.

One final comment must be made with respect to heterogeneous catalytic oxidation. Free radicals may propagate from the surface of the catalyst into the fluid (gas or liquid) phase and initiate reaction sequences there. This is common in the oxidation of hydrocarbons in the liquid phase on oxides of transition elements (Margolis, 1973), and the same phenomenon appears to be involved as a side reaction in oxidations on Bi_2MoO_6.

VI. Current Problems in Heterogeneous Catalysis

This section will cover some of the current problems in heterogeneous catalysis omitting, however, oxidation since such matters were treated even if briefly in Section V,E.

A. Structure Sensitivity

This subject has been mentioned at the end of Section V,A. The question is just what effect the exact nature of the surface (concentration of defects, edges, steps, and the identities and concentrations of various crystal planes), the particle size, and the nature of the support has upon catalytic reactions. If the effect is very small for a particular reaction on a particular catalyst, the reaction is said to be *structure insensitive* on that catalyst; if the effect is substantial, the reaction is said to be *structure sensitive*. The basic question is old, it goes back to the work of H. S. Taylor in the 1920s. However, interest in the matter resumed recently because methods of measuring the percentage exposed of supported Group VIII metal catalysts had become available. The percentage exposed should have some relation to particle size. One could now ask how the turnover number of a particular reaction (the number of molecules of the reactant converted per unit time per surface atom of the Group VIII metal) depends upon the percentage exposed and upon the identity of the support. The rates of some reactions, for example, the hydrogenation of ethylene on platinum catalysts and the isomerization of *neo*pentane to isopentane at 300°C on platinum, appear to be nearly structure insensitive, whereas those of others, for example, the hydrogenolysis of *neo*pentane to isobutane and methane on platinum at 300°C, appear to be substantially structure sensitive (Boudart, 1969).

However, mere measurement of rate does not fully test the question of structure sensitivity. Selectivity data provides additional information and a reaction may appear structure insensitive as judged merely by rate data but structure sensitive as judged by selectivity data.

A number of investigations in this area are under way and it is as yet too early to draw general conclusions.

B. Localized Versus Nonlocalized Adsorption

Throughout the preceding sections, we have treated the bond between the chemisorbed species and the surface as localized, i.e., we have assumed that the electronic effects of the bond are confined to the immediate vicinity of the adsorption site. We have frequently likened the adsorption complex to organometallic or coordination complexes in which the bonding is necessarily localized. This assumption must often be an oversimplification, particularly for metals but also for oxides of transition metals.

During the 1950s a number of attempts were made to treat chemisorption on metals and semiconductors as involving the delocalized electrons of the metal or semiconductor. Such attempts have not turned out to be very useful. The current interest of theoreticians in surfaces and chemisorption provides the hope that a more accurate interpretation of chemisorption will be forthcoming. At the moment, it is the opinion of the writer that localized adsorption provides the most useful zero-order approximation.

C. Alloy Catalysts

There is considerable current interest in the question of the chemisorptive and catalytic characteristics of metallic alloys. Will the surface of a palladium–gold alloy exhibit properties like the sum of the properties of the two separate metals, or will the properties be different from that of the mere sum? That is, do the palladium and gold atoms at the surface of the alloy behave independently or not? This question is related to that of Section VI,B. It is still not clear what the final answer will be. For recent reviews of this areas, see Sinfelt (1974) and Ponec (1975).

D. Stereospecific Catalysts

If an enzyme catalyzes the hydrogenation of a prochiral unsaturated molecule, the product is usually just one of two possible optically active forms. Some homogeneous catalysts have been discovered which give high optical purities (although less than 100%). Stereospecific heterogeneous catalysts are known (Burwell, 1972b) but they give rather low optical purities. What are probably the most successful of such heterogeneous catalysts have been made by causing the adsorption on the surface of Group VIII metals of optically active acids or aminoacids. It would be important to discover heterogeneous catalysts giving high optical purities. For example, one could then hydrogenate such molecules as $CH_3C(COOH){=}NH$ to optically pure amino acids useful as nutritional supplements in protein-poor areas.

E. "Energy"

As natural gas and petroleum become exhausted, one may expect the development of a very large new industry concerned with converting coal, oil shale, and tar sands into liquid hydrocarbon products and into methane as a substitute for natural gas. Heterogeneous catalytic processes will figure very prominently in this industry. Many of the needed processes can probably be developed from existing technology. It is not yet clear to just what degree new processes will be needed. The so-called *hydrogen economy* would involve the electrolysis of water at sea coasts using nuclear power. The hydrogen would then be shipped inland by pipeline. The availability of large amounts of cheap hydrogen would potentiate the development of a number of new catalytic processes as well as the more extensive use of present ones. Large-scale production of methanol as a fuel would involve a number of catalytic problems, but they would probably be of a developmental nature (see G. C. Bond, in Basolo and Burwell, 1973).

F. Heterogenizing Homogeneous Catalysts

A substantial number of homogeneous reactions catalyzed by organometallic complexes have been discovered in recent years (J. Chatt and J. Halpern, in Basolo and Burwell, 1973). Homogeneous catalysts are available for reactions for which there are no conventional heterogeneous catalysts, for example, the hydroformylation reaction, $RHC{=}CH_2 + CO \rightarrow RH_2C{-}CHO$, and the formation of acetic acid from methanol, $CH_3OH + CO \rightarrow CH_3COOH$. Homogeneous catalytic reactions may in some cases be more selective than heterogeneous ones. On the other hand, heterogeneous catalytic processes are simpler. In a batch reactor, one can easily recover the catalyst at the end of a reaction; in a flow reactor, the problem of catalyst recovery does not even arise. In homogeneous catalytic reactions, recovery of the catalyst is a separate process. Further, in most cases the recovery needs to be done very well since many of the catalysts are complexes of expensive metals (rhodium, for example, in the synthesis of acetic acid). Thus, other things being equal a heterogeneous catalytic reaction will be preferred to a homogeneous one and, for example, it is highly unlikely that any homogeneous catalytic process for the hydrogenation of nitrogen will replace the Haber process in the foreseeable future.

Things are not, however, always equal and in a substantial number of cases the homogeneous catalytic process presents advantages. One might hope to achieve the best of both worlds by attaching the homogeneous catalyst to some kind of a high-area, porous, inert support. There is considerable current work in this area (see Bailar, 1974; J. Manassen, in Basolo and Burwell, 1973). Generally speaking, the homogeneous catalysts are complexes of transition elements in lower oxidation numbers with such ligands as phosphines, hydrides, and carbonyl. A common technique for heterogenizing a homogeneous catalyst has been to attach a $(C_6H_5)_2P$-group to a high polymer or to silica gel. The modified support can then be exposed to a suitable complex of the metal which, upon binding to the surface ligand, generates the desired catalyst. Examples of reactions for which heterogenized homogeneous catalysts have been described are olefin hydrogenation, hydroformylation, and hydrolysis. It may well prove possible to make highly stereospecific catalysts of this type.

The heterogenized homogeneous catalyst has one potential disadvantage over the homogeneous catalyst which will probably limit its applicability. Consider a reaction like hydroformylation which is rather strongly exothermic. When the reaction is run in solution using homogeneous catalysts, the heat evolved can readily be removed by stirring and provision for cooling. However, the porous supports of heterogenized homogeneous catalysts are poor conductors of heat. It would be difficult in a fixed bed reactor of any reasonable size to keep the temperature from rising above desired operating temperatures. One can imagine using a fluidized reactor with cooling, but unless the catalyst is remarkably attrition resistant, one would still face the problem of catalyst recovery.

VII. Historical Note

The references which follow this section primarily consist of reviews with only a few research papers. This article makes no attempt to indicate priorities but that matter will generally appear in the references. Although the choice of reviews as references seems likely to be most useful to the reader, it has the disadvantage that there are no references made to the giants who forged the basis of modern heterogeneous catalysis, Ipatieff, Sabatier, Haber, Langmuir, H. S. Taylor, Rideal, Schwab, and Emmett.

It is not easy to assign a formal beginning to heterogeneous catalysis, but Faraday was one of the first who clearly realized that he had observed it. A number of heterogeneous catalytic reactions of inorganic compounds were studied descriptively during the nineteenth century, the oxidations of sulfur dioxide, hydrogen, and ammonia, for example. The greatest achievement in this area did not come until the early twentieth century when Haber developed the hydrogenation of nitrogen to ammonia, one of the most important processes in the chemical industry and the one which, for good or evil, has permitted the population of the earth to continue to increase so rapidly. That is, ammonia provides the nitrogeneous fertilizer without which drastic famines would have resulted.

Modern heterogeneous catalysis may be considered to have started with the development of applications of heterogeneous catalysis to organic reactions by Sabatier and Ipatieff at the very end of the nineteenth century: the hydrogenations of olefins, alkynes, and benzenes and the dehydration of alcohols. Such largely preparative work dominated heterogeneous catalysis until about 1920.

The period extending to about 1935 might be characterized as that of adsorption. It began with the appearance of the Langmuir adsorption isotherm, extended through the work of Taylor in studies of chemisorption and the recognition of activated adsorption, and might be considered to close with the studies of physical adsorption which led to the Brunauer–Emmett–Teller isotherm. This latter discovery provided the first useful method for measuring surface areas and this work led also to the work of others which permitted the characterization of texture. Both were key developments in the characterization of catalysts. The period was further characterized by development of the role of chemisorption in heterogeneous catalysis.

The next period was characterized by the beginning of the development of mechanism in heterogeneous catalysis by such men as Rideal, Horiuti and Polanyi, and Bonhoeffer and A. Farkas. This work was stimulated by the new availability of isotopes, particularly deuterium.

The next period, which might be taken to start about 1950, was characterized by a much more penetrating development of mechanism and by the applications of such new instruments as infrared spectrometers, EPR spectrometers, mass spectrometers, gas chromatographs, and others, to studies of mechanism, chemisorption, and catalyst structure. This is the period which is primarily treated in this article.

Finally, of course, starting with the period of Ipatieff and Sabatier, there has been a series of discoveries of major industrial heterogeneous catalytic processes which revolutionized the petroleum industry and forms much of the basis for the chemical industry.

References

Anderson, J. R. (1975). "Structure of Metallic Catalysts." Academic Press, London.

Anderson, R. B., ed. (1968). "Experimental Methods in Catalytic Research." Academic Press, New York.

Bailar, J. C., Jr., (1974). *Catal. Rev.* **10**, 17.

Basolo, F., and Burwell, R. L., Jr., eds. (1973). "Catalysis Progress in Research." Plenum, New York.

Boudart, M. (1968). "Kinetics of Chemical Processes." Prentice-Hall, Englewood Cliffs, New Jersey.

Boudart, M. (1969). *Adv. Catal.* **20**, 153.

Boudart, M., and Burwell, R. L., Jr. (1974). *Tech. Chem.* (*N.Y.*) **6**, Part I, 693.

Boudart, M., and Schlatter, J. C. (1972). *J. Catal.* **24**, 482.

Burwell, R. L., Jr. (1969). *Acc. Chem. Res.* **2**, 289.

Burwell, R. L., Jr. (1972a). *Catal. Rev.* **7**, 25.

Burwell, R. L., Jr. (1972b). *Intra-Sci. Chem. Rep.* **6**, 135.

Burwell, R. L., Jr., Haller, G. L., Taylor, K. C., and Read, J. F. (1969). *Adv. Catal.* **20**, 1.

Calderon, N., Ofstead, E. A., and Judy, W. A. (1976). *Angew. Chem. Int. Ed. Engl.* **15**, 401.

Cimino, A. (1974). *Chim. Ind.* (*Milan*) **56**, 27.

Cimino, A., Indovina, V., Pepe, F., and Stone, F. S. (1973). *Gazz. Chim. Ital.* **103**, 935.

Drauglis, E., and Jaffee, R. I., eds. (1975). "The Physical Basis for Heterogeneous Catalysis." Plenum, New York.

Egerton, T. A., Stone, F. S., and Vickerman, J. C. (1974). *J. Catal.* **33**, 299.

Emmett, P. H. (1972). *Catal. Rev.* **7**, 1.

Gager, H. M., and Hobson, M. C., Jr. (1975). *Catal. Rev.* **11**, 117.

Germain, J. E. (1969). "Catalytic Conversion of Hydrocarbons." Academic Press, New York.

Gomer, R. (1975). *Acc. Chem. Res.* **8**, 420.

Gregg, S. J., and Sing, K. S. W. (1967). "Adsorption, Surface Area and Porosity." Academic Press, New York.

Haines, R. J., and Leigh, G. J. (1975). *Chem. Soc. Rev.* **4**, 155.

Hall, W. K. (1975). *Acc. Chem. Res.* **8**, 257.

Hucknall, D. J. (1974). "Selective Oxidation of Hydrocarbons." Academic Press, New York.

IUPAC (1976). "Definitions, Terminology and Symbols in Colloid and Surface Chemistry—Part II. Heterogeneous Catalysis." *Pure Appl. Chem.* **45**, 71 (1976); *Adv. Catal.* **26**, 351 (1977).

Kemball, C. (1971). *Catal. Rev.* **5**, 33.

Kilty, P. A., and Sachtler, W. M. H. (1974). *Catal. Rev.* **10**, 1.

Kokes, R. J. (1973). *Acc. Chem. Res.* **6**, 226.
Lunsford, J. H. (1973). *Catal. Rev.* **8**, 135.
Margolis, L. Ya. (1973). *Catal. Rev.* **8**, 241.
Pearson, R. G., ed. (1973). "Hard and Soft Acids and Bases." Dowden, Hutchinson, and Ross, Stroudsburg, Pennsylvania.
Poneč, V. (1975). *Catal. Rev.* **11**, 41.
Schuit, G. C. A. (1974). *J. Less-Common Met.* **36**, 329.
Siegel, S. (1966). *Adv. Catal.* **16**, 123.
Sinfelt, J. H. (1974). *Catal. Rev.* **9**, 147.
Somorjai, G. A. (1972). "Principles of Surface Chemistry." Prentice-Hall, Englewood Cliffs, New Jersey.
Tanabe, K. (1970). "Solid Acids and Bases." Academic Press, New York.
Thomas, C. L. (1970). "Catalytic Processes and Proven Catalysts." Academic Press, New York.
Triforo, F., Forzatti, P., and Pasquon, I. (1975). *In* "Catalysis, Heterogeneous and Homogeneous" (B. Delmon and G. Jannes, eds.), p. 509. Elsevier, Amsterdam.
Ugo, R. (1975). *Catal. Rev.* **11**, 225.

Recent Developments in the Chemistry of Transition Metal Carbides and Nitrides

H. A. JOHANSEN*

Westinghouse Research Laboratories, Pittsburgh, Pennsylvania

I. INTRODUCTION

Under the impetus of the demands of military, nuclear, space age, and commercial technology over the past two decades, the fields of transition metal chemistry, physics, and metallurgy have achieved great advances. In particular, the elements of the fourth, fifth, and sixth groups received much attention, and the pure metals and many of their compounds have moved in that time span from laboratory curiosities to important industrial

* Present address: Rt. 1, Box 89, McMinnville, Oregon 97128.

commodities. Compounds such as tungsten carbide, now produced in hundreds of tons annually, and titanium metal, in thousands of tons, illustrate how great the advance has been.

The carbides and nitrides of the fourth-, fifth-, and sixth-group transition metals show most unusual properties. They exhibit great hardness and high melting points. For example, TiC with a hardness of ~ 3000 kg/mm^2 is exceeded only by some borides and diamond. TaC, at $\sim 4000°C$ has the highest melting point of any substance. Superconductivity is exhibited by some of these materials—until fairly recently, the record for the highest superconducting transition temperature was held by $NbC_{0.3}N_{0.7}$ at $\sim$ 18°K. The high hardness, particularly of the carbides, goes along with great strength. At low temperatures (i.e., below 1000°C), the materials are subject to brittle failure in stress, but at higher temperatures, are ductile while retaining high strength.

The high electrical and thermal conductivities, as well as optical and mechanical properties, show the carbides and nitrides to behave much like metals, e.g., positive temperature coefficient of resistivity, spectral emissivity similar to the parent metal, and some ductility, at least at elevated temperature. The high hardnesses and brittleness indicate some covalent bonding, while the high electrical and thermal conductivity indicate some metallic bonding.

The formation of the carbide and nitride takes place by either the C or N entering the metal lattice to occupy an interstitial site. Since the C and N atoms are quite similar in size, this results in many similarities in the preparation and properties of the carbides and nitrides: hence, it is useful to consider the carbides and nitrides together.

Because of the unusual properties and potential commercial applications of the fourth-, fifth-, and sixth-group transition metal carbides and nitrides, large research efforts were expended in many laboratories, and the literature has consequently become voluminous. Authoritative and critical reviews and descriptions of the field of transition metal carbides and nitrides may be found in the early work of Schwarzkopf and Kieffer (1953), the later work of Kieffer and Benesovsky (1963), Schaffer (1964), Storms (1967), Hausner and Bowman (1968), and Toth (1971). In addition there is the guide to the earlier extensive Russian work by Samsonov (1964). A bibliography of the older literature is given by Haughton and Prince (1956).

It is the purpose of the present discussion to describe the principal scientific and industrial developments of the last five to ten years in this field, primarily from the chemical standpoint. Older work and specialized topics in the associated disciplines of physics and metallurgy will be cited on occasion for illustration and explanation. The reader will find detailed treatment of the earlier work and of special topics in the sources mentioned above.

II. Preparation and Characterization

A. Preparative Methods

The preparation of carbides and nitrides of the transition group metals is accomplished most directly by reaction of the elements. The carbides may be produced by intimate mixing of the metal with appropriate amount of carbon in the form of graphite, and heating in graphite crucibles under inert atmosphere. For nitrides, the metal is heated in a stream of nitrogen. The sintered powder may be hot-pressed, arc-melted, or, in certain cases, levitation-melted or extruded to obtain the necessary high-density shapes required.

Very high temperatures are needed in the preparation and processing of many of the materials, and this requirement has given rise to a whole new field of high temperature chemistry, which includes, as well, borides, silicides, and various ceramics and intermetallics.

Depending on purity desired and end-use form, carbides may also be produced by reduction of the metal oxide with carbon,

$$ZrO_2 + 2C \rightarrow ZrC + 2CO \qquad (1)$$

by reaction of the metal with acetylene or CO,

$$2Ta + C_2H_2 \rightarrow 2TaC + H_2 \qquad (2)$$

$$Ta + 2CO \rightarrow TaC + CO_2 \qquad (3)$$

More exotic techniques have also been used, such as the so-called menstruum method, where the elements are allowed to react in a molten metal bath such as cobalt (Warren, 1969). This is a method particularly suitable for producing high-stoichiometry single crystals.

Another method, reported by Zainulin and co-workers (1972), shows that ZrC can be synthesized in molten NaCl.

For the nitrides, the preparative methods, in addition to direct reaction of the metal with N_2, include the reaction of the anhydrous halide with ammonia,

$$2TiCl_4 + 10NH_3 + H_2 \rightarrow 2TiN + 8NH_4Cl \qquad (4)$$

or the reaction of the halide with N_2 and H_2,

$$2TiCl_4 + 5N_2 + 16H_2 \rightarrow 2TiN + 8NH_4Cl \qquad (5)$$

or carbon reduction and nitriding simultaneously,

$$2TiO_2 + 4C + N_2 \rightarrow 2TiN + 4CO \qquad (6)$$

or the ammonolysis of TiO_2 with NH_3

$$2TiO_2 + 2NH_3 + H_2 \rightarrow 2TiN + 4H_2O \tag{7}$$

The direct carbiding or nitriding of the transition metals by means of gases finds special interest because the properties of the carbide or nitride can be studied as formed and *in situ* without further processing. It is also possible to form complex shapes from the parent ductile metal, e.g., tantalum, and then carbide, in order to obtain shapes impossible to form from the brittle TaC after carbiding. Crucibles and other simple shapes may be made from pressed and sintered carbide powders, but complex shapes, e.g., coiled filaments for incandescent lamps, are made most expeditiously by carbiding the metal after forming it into coils (Johansen, 1970a).

Examples of the usefulness of *in situ* carbiding on wires are shown by Chermant *et al.* (1969), who carbided titanium wires and found the rates of carbiding to be faster in methane than in toluene and benzene. Observations were made of crystal growth and diffusion rates. The formation of carbides on tungsten wires is of some practical importance in the tungsten

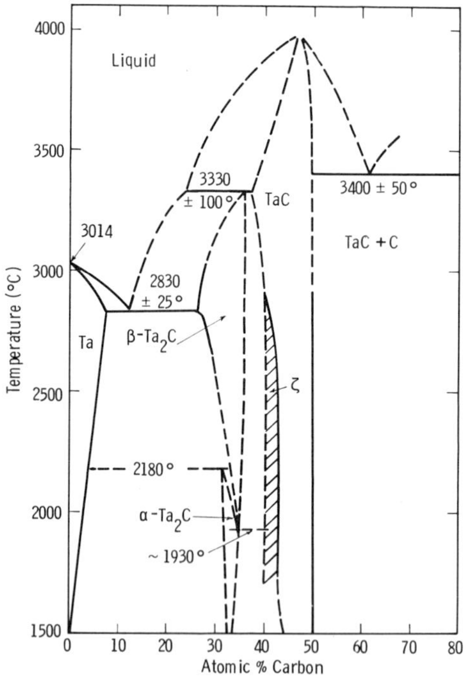

FIG. 1. Proposed phase diagram for tantalum carbon. [After Storms (1967) and Toth (1971).]

incandescent lamp. The carbide (W_2C) on tungsten wire was found to originate from the carbon impurities in the metal. If the tungsten wire were heated in a *poor* vacuum (10^{-4} torr), no carbide was formed because the traces of oxygen oxidized the carbon. If heated in a *good* vacuum (10^{-9} torr), the carbide was found as a surface layer (Croft and Nygaard, 1969).

The typical reactions illustrated in the preceding section appear simple, but this is deceptive. Many of the reactions involving halides, oxides, or sulfides have intermediates and subcompounds, so that their chemistry is more complex than it appears. More significant, however, is the realization of the wide range of nonstoichiometry that these carbides and nitrides may exhibit. Phase boundaries may be wide, as can be seen in Fig. 1, where the Ta–C phase diagram (Storms, 1967) is illustrated. The homogeneity range for TaC is $TaC_{0.75}$ to $TaC_{0.99}$, with the lower limit still subject to some uncertainty. It is frequently more useful to write such compounds as TaC_{1-x}, with $0.01 \leqq x \leqq 0.25$ in this case.

Accordingly, the problem of describing such widely varying compositions within the same phase arose.

B. Characterization

The result of the realization of nonstoichiometry, and of the presence of wide homogeneity ranges in these compounds was a reexamination of various properties as a function of the carbon or nitrogen content. To do this required the more careful characterization of the sample being tested. Conventional chemical analytical techniques, particularly for C, O, N, and H, in the carbides and nitrides are still very important, but must be augmented by tests on crystal structure and lattice parameters, grain size, porosity, and homogeneity. The particular characterization depends on the particular property being measured. In all, however, the chemical characterization is fundamental, and unfortunately is frequently neglected or poorly done. A description of characterization techniques may be found in Toth (1971) and Hannay (1973).

The wide discrepancies in properties from various sources in the literature may thus be sometimes explained on the basis of insufficient characterization. Unfortunately, also, one laboratory's characterization may not agree with that of another. Only time-consuming and careful comparison of methods and results can allow the true state of affairs to appear. This requires mutual patience and helpfulness among investigators that is at present not common.

An example of the results of insufficient characterization (coupled with a property difficult to measure) is shown in Fig. 2, where the spectral (at 6500 Å) emittance of "TaC" is plotted *vs* temperature for a variety of literature

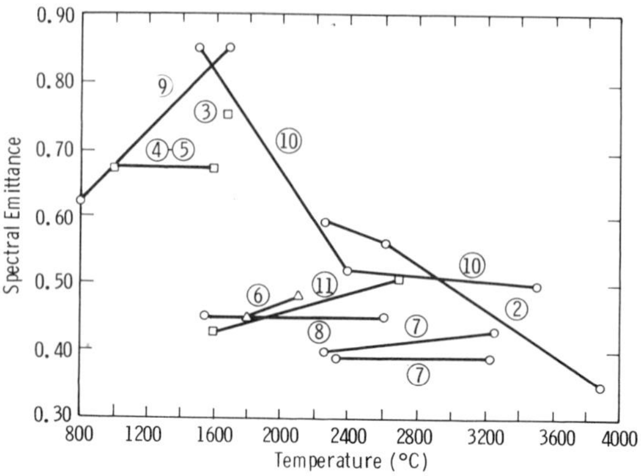

FIG. 2. Literature values of spectral emittance of TaC. (Numerals and symbols represent independent investigations.)

values. The numerals and symbols indicate different independent investigations. The results of Johansen and Comenetz (1969) for $TaC_{1-x}(0.1 \leqq x \leqq 0.3)$ are shown in Fig. 3 as the shaded portion superimposed on the existing literature values. (The shaded portion represents the range of values obtained for composition and other variations.) The fit of the shaded portion with the literature values is seen to be quite good; but what this shows is that the literature values of the emittance were measured on material (with at least surface compositions) quite deficient in carbon, perhaps with x as much as 0.2. The accepted value in the literature of 0.45 for the spectral emittance of TaC at temperatures above 2500°C is probably therefore too low. Careful measurements on TaC_{1-x} with $0.02 \leqq x \leqq 0.1$ showed that at temperatures above 2500°C, the value of the spectral emittance is 0.55.

Besides the intrinsic difficulty of measuring the emittance property at

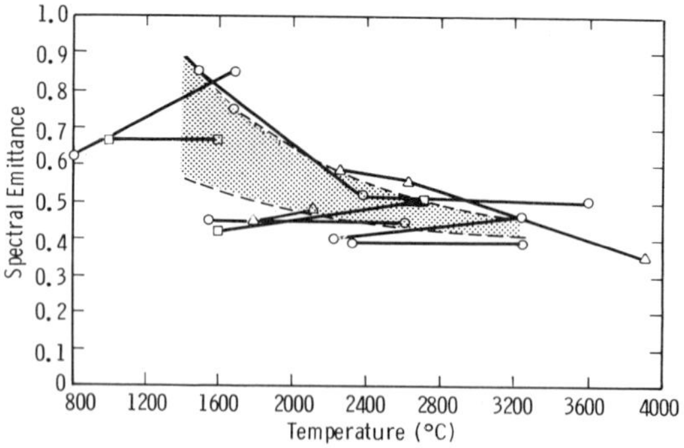

FIG. 3. Spectral emittance of TaC_x.

high temperatures, factors such as surface alteration, gas contamination, surface films, and smoothness contributed to the variation as well as the stoichiometry. Knowing that spectral emittances depend on composition, and that composition and physical state may change with time at high temperature, serves to alert the investigator to the difficulties of accurate measurements at high temperatures.

C. Interstitial Compounds

The transition metal carbides and nitrides form relatively simple crystal structures. All the compounds exist in a cubic form, principally the NaCl (B1) structure, and, except for the Group IV carbides and nitrides, form hexagonal crystal structures, as well.

The crystal chemistry of these materials is highly developed and thoroughly reported. While the field is still quite active, the emphasis appears to be mostly in the direction of unraveling the complex structures of Hägg compounds and the interactions of the Nowotny octahedral phases. Ordering in the structures was reported by Nowotny and Benesovsky (1968) and ordering of vacancies, by Hiraga (1973). A high temperature phase with a complex ζ–carbide structure in the Mo–Cr–C system was reported by Rudy (1973), and a high pressure γ-TaN compound was reported by Brauer and co-workers (1972). These few references will serve for orientation to the direction of present investigations and will indicate the sophistication and the complex nature of some of the studies. For further treatment, the reader is referred to a detailed review of the crystal chemistry of complex carbides and related compounds by Nowotny (1972).

D. Phase Relationships

Since phase diagrams are fundamental to any studies involving preparations or uses of the transition metal carbides and nitrides, these systems have been widely investigated. Unfortunately, the early work did not take into account the powerful effect of small amounts of stray interstitials, such as O and H, on the melting points and crystal structures of the carbides and nitrides. The binary systems were therefore, in many cases, ternary systems. The unraveling of many of these problems has filled a large number of books. Elucidating phase diagrams requires sophisticated high-temperature equipment and analytical techniques. The time of the lone investigator appears to have passed in this field; large, well-equipped laboratories with abundant resources appear to be the only active participants.

Disputes over even simple binary systems are still not entirely resolved, particularly in the metal–nitrogen systems, and may never be. The Hf–N system was recently studied and reported by Billy and Teyssedre (1973), as was the V–N system by Hörz (1971) and the Mo–N system by Ettmayer (1970a).

Of late, most efforts have been in the direction of investigating special portions of a phase diagram such as the studies of phase equilibria on the metal-rich side of the Ta–N system reported by Geils and Potter (1973), and the Ti-TiN system reported by Wood (1974).

Systems of a carbide or nitride with an added transition metal have also been receiving increased attention, for example, systems of NbC with other transition metals were reported by Guha and Kolar (1972), and the composition of phases in Nb–Zr–N alloys has been reported by Savage and Alstetter (1970). The Co–Ti–C system at 1100°C has been studied by Kossowsky and Harvey (1974), and a phase study of a transition metal with both C and N has been reported by Ettmayer (1970b).

Although many of these studies are related to problems of commercial significance, they also illuminate the scientific problems of transition metal carbides and nitrides.

Recent critical reviews of phase relationships of transition metal carbides and nitrides are furnished by Toth (1971), Storms (1967), and Rudy (1969).

III. Thermochemical Considerations

A. Thermodynamics

Thermodynamic data may be used to predict the course of reactions, and thus the behavior of a substance in reaction with other materials, or the decomposition of a substance to various species.

The use of such thermodynamic data and functions is particularly attractive for the high temperature systems involved in preparing and using the systems of the transition metal carbides and nitrides because the difficulties of carrying out high temperature experiments may thus be partly anticipated and avoided.

Unfortunately, to make accurate predictions from thermodynamic data requires reliable thermodynamic data, and here is where the dispute begins. Probably in no part of the transition metal carbide and nitride chemistry field is the literature more voluminous and controversial, and the activity continues. The field has had extensive critical review, and it is now clear that the unreliability in the thermodynamic data for these materials results in part from poor characterization of the compounds (for example nonstoichiometry), inherent difficulties in high temperature measurements, conflicting results from different investigators, and in some cases, no measurements at all.

Despite these difficulties, attempts to rationalize the data in the literature and collect them in usable form abound. Of particular importance are the tables compiled by Stull and Prophet (1971), Schick (1966), Chang (1965), Storms (1967), Samsonov (1964), Kelley and King (1961), Kelley (1963), and Wicks and Block (1963). A discussion of methods of use of thermodynamic functions in high temperature chemistry involving carbides and nitrides (as well as other compounds) is given by Searcy (1963). Tables of thermal functions, as well as a discussion of thermodynamic properties, are given by Storms (1968). The information given by Toth (1971) is particularly useful. A discussion of thermochemical stabilities and structures of many compounds including some carbides at high temperatures from the recent literature is presented by Hastie and co-workers (1970). The authors state that they had to be selective in quoting the recent literature, but list over 230 references!

Thermodynamic analyses of the transition metal carbides appear to be fairly well developed but the situation for the nitrides is much less satisfactory. The standard enthalpy of formation for ZrC was reported by Baker and co-workers (1969) to be $\Delta H^\circ_{f,298\,^\circ K} = 307.7 \pm 0.4$ kcal, a correction of an earlier value by less than one kcal, and due mainly to careful assessment of impurities. By contrast, the case for ZrN, and for other nitrides as well, finds mainly estimates existing. The JANAF-Tables (Stull and Propher, 1971) value for ZrN was $\Delta H^\circ_{f,298\,^\circ K} = 170.5$ kcal (estimated), and an experimental determination reported by Gingerich (1968) gave 123. $\pm$ 7.0 kcal. The situation with the thermodynamics of the nitrides plainly requires much further experimental work.

Metallurgists and some chemists have used a thermodynamic data display technique called thermochemical diagrams which can help in evaluating chemical reactions for practical systems. Such diagrams show the composition and pressure of gaseous species in equilibrium with condensed phases for a particular temperature. A report by Jansson (1970) describes the technique and some results. A typical diagram for Ti–O–C at 1800°K is shown in Fig. 4. Again, the success of the method depends heavily on more accurate experimental evidence and better theoretical treatment.

B. Kinetics

Almost as active as the thermodynamics field for transition metal carbides and nitrides is the field of chemical reaction kinetics. Questions relating to reaction rates, rates of solution, diffusivity, and electro-transport, are all being energetically explored. The answers are frequently sought, of course,

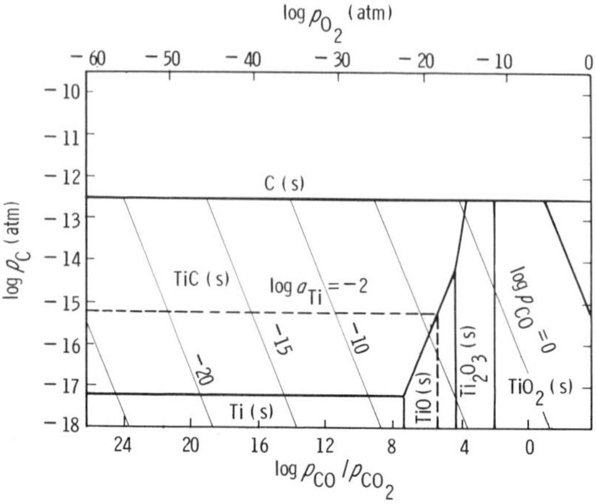

FIG. 4. Thermochemistry of condensed phases in the Ti—O—C system at 1800°K. [After Jansson (1970).]

because of the bearing they may have on stabilities, corrosion resistance, and preparative methods; and these, in turn, have a bearing on commercial and scientific applications.

Kinetic studies may be made by a variety of methods; layer growth, weight gain (or loss), diffusion measurements, or electro-transport. The questions of just how the transition metal carbides and nitrides form and at what rates are answered by such studies.

If a high density electric current is passed through a metal bar, interstitial atoms may be swept along the bar. From such electrotransport studies, the electric mobility, diffusion coefficient, and effective valencies may be found. The method also offers a purification of the metal from O_2 and N_2 (but apparently not C). Such studies on Hf have been reported by Carlson and co-workers (1973), and for electro-transport of C in Nb and Ta by Schmidt and Carlson (1972) who also list recent reviews on the technique.

The carburization of niobium wire in acetylene or methane was reported by Hörz and Lindenmaier (1972a) using a method whereby the reaction rate was monitored continuously by following the change in resistivity with carburization. Decarburization of niobium carbide in flowing oxygen was followed in the same manner (Hörz and Lindenmaier, 1972b).

Kinetic studies of the carbiding of tantalum using a gravimetric method was reported by Johansen (1970b). The Ta–C system (Fig. 1) shows two carbides, Ta_2C and TaC. Initially, in any carbiding process, a carbon-rich layer on the surface of the Ta is formed, and a carbon concentration gradient diminishing toward the interior is created. The diffusion of the carbon then must proceed through three separate zones; the outermost layer TaC, the inner Ta_2C, and the solid solubility zone in the bulk metal. By careful analysis of the data, it was possible to deduce the two diffusion coefficients for carbon through TaC and through Ta_2C. The two diffusion constants at 2600°C were 4.75×10^{-8} cm²/sec and 2.97×10^{-7} cm²/sec, respectively, showing that the diffusion was more rapid through the subcarbide Ta_2C than the carbide TaC. This is reasonable since it is to be expected that the diffusion pathway should be through the interstitial sites in the metal, and more empty sites would be available in the lower carbide.

Diffusion studies using isotopic methods have been reported by numerous investigators. The method lends itself to relatively simple determinations of diffusion coefficients. From such studies the activation energies and preexponential factors for diffusion are now known for most of the metals and interstitials over a wide range of temperatures. The diffusion of carbon 14 in tungsten (Shapela, 1972) and Ti-44 in TiC_x (Sarian, 1969) and measurements of carbon monoxide isotope exchange with tantalum carbide (Gasser *et al.*, 1968) are reports illustrative of the modern development of this technique.

Of increasing interest in recent years has been the interaction of carbon and nitrogen in the carbiding or nitriding processes. It is apparent that in certain cases the presence or absence of the other interstitial can either enhance or suppress the processes. Carbon diffusion has been measured in zirconium nitride as reported by Khromov and co-workers (1972), and the diffusivity and solubility of nitrogen in molybdenum and the trapping of nitrogen by carbon in molybdenum has been reported by Weaver (1972). Carbiding processes using carbon vapor are greatly enhanced by the presence of nitrogen at 2500°C, with rates comparable to the use of acetylene (Johansen, 1970a).

It is to be expected that work on the kinetics and gas reactions, as well as the thermodynamics of the transition metal carbides, will continue to be of scientific and technical interest.

IV. Properties from the Chemical Standpoint

A. High Temperature Species

Early research in discovering the properties of transition metal carbides was aimed primarily at processes that could utilize their great strength and hardness. To some extent this was successful, particularly for materials for cutting tools, dies, abrasion resistant surfaces, and armor piercing projectiles. Mostly, however, the full use of the carbides and nitrides has been thwarted by their brittleness at room temperature and the high temperature processes required for forming shapes. Except for small simple shapes, the high melting points and reactivity at high temperature preclude casting; hence, the only alternatives are powder metallurgy techniques, or else bonding of the powders with lower melting metals such as nickel or cobalt.

The powder metallurgy techniques that are used rarely enable fully dense specimens to be produced for testing of properties, such as optical, mechanical, electrical, magnetic, or interatomic bonding. The residual porosity always affected the values to greater or lesser degree. Nevertheless, measurements have been made and reported, as a start had to be made somewhere. The values of most properties, as a consequence of the porosity differences alone, cover a range sometimes quite wide.

The problem of the brittleness at room temperature of the transition metal carbides and nitrides has been even more serious. While extremely hard and strong, no toughness is apparent until fairly high temperatures are reached for the carbides. (The data for nitrides are scarce or nonexistent.) The transition temperature T_{db} for the ductile–brittle behavior for TaC_x on hot-pressed powder specimens bonded with cobalt and then de-cobalted was found to vary as shown in Fig. 5 (Johansen and Cleary, 1966). The T_{db} for tantalum carbide with x near 1 was found to be about 1850°C. A value of T_{db} for tantalum carbide of 1730°C was reported by Kelly and Rowcliffe (1967) also on polycrystalline specimens. The T_{db} values for single crystals of various carbides are several hundred degrees lower than for pressed and sintered polycrystalline specimens, thus indicating that the voids in the polycrystalline samples act as notches and stress risers, thus raising the T_{db}. In general, the T_{db} appears to be lower for increasing deviations from stoichiometry below a carbon to metal ratio of 1. Plainly, the T_{db} for the carbides, which range from 1000° to 2000°C, must be lowered a very great deal to allow any hope for widespread structural use.

In the section that follows, the results of studies of various properties that are dependant on the stoichiometry of various selected carbides and

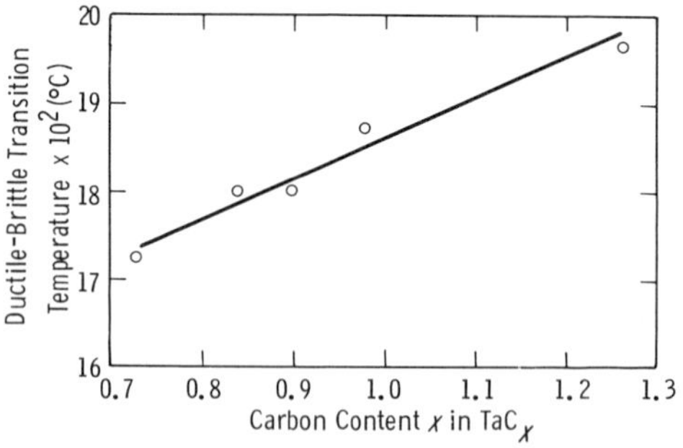

FIG. 5. Ductile–brittle transition temperature *vs* carbon content x in TaC_x.

nitrides are summarized. A more detailed treatment for some of the properties is available from Toth (1971).

1. *Hardness Properties*

The hardness of TiC_{1-x} increases strongly with increasing carbon content, rising from about 1600 Vickers hardness number (VHN) at $x = 0.4$ to a hardness of 2700 VHN at $x = 0.01$ stoichiometry (Cadoff *et al.*, 1956). The results for TaC_{1-x}, however, showed an anomalous behavior with a maximum in the hardness at about $TaC_{0.85}$ decreasing in both directions, to higher and lower stoichiometry (Santoro, 1963). The hot hardness of NbC–C alloys was found by Asai *et al.* (1971) to be independent of carbon content.

2. *Electrical and Magnetic*

In general, the electrical and magnetic properties of the transition metal carbides and nitrides appear to follow those of the parent metals. The data are, however, quite fragmentary, and the poorly characterized materials have contributed a wide scatter to the recorded values. The compounds are quite metallic, and, in fact, the electrical conductivities may approach or equal those of the parent metals. The room temperature resistivities range from 7 $\mu\Omega$-cm for ZrN (Pessall *et al.*, 1967) to about 110 $\mu\Omega$-cm for TiC (Rudy and Benesovsky, 1960). Most values for the other carbides and nitrides lie between 7 and 110. There are numerous anomalies resulting from influence of impurities and effects of stoichiometry.

3. *Magnetic Susceptibility*

This is a property for which relatively little information is available. Toth (1971) has listed some values of fourth and fifth-group carbides and nitrides which range from about -30 for the fourth-group carbides to $+30 \times 10^6$ emu/mole for other carbides and nitrides. Large variation of values for nominally the same compositions indicates a strong dependence on stoichiometry or impurity levels, or both. Accurate data for this property would be useful in understanding the band structure of the carbides and nitrides. Recent magnetic susceptibility measurements on sixth-group compounds appear to be largely missing. Chromium nitride is reported to be ferromagnetic by Corliss and co-workers (1961).

4. *Hall Coefficient*

A fair amount of information is available on Hall coefficients for fourth- to sixth-group carbides and nitrides. The values at room temperature range from about -0.38×10^{-4} cm^3/C for Cr_7C_3 to about -21.8×10^{-4} cm^3/C for WC, with the other compounds showing intermediate values. One study (Santoro and Dolloff, 1968) showed that the Hall coefficient for TaC_{1-x} did vary with x, the coefficient becoming more negative as stoichiometry is approached.

5. *Thermoelectric Power and Thermionic Emission*

In the 1960s, a great deal of interest was raised in the possible uses of the transition metal carbides and nitrides for direct conversion of heat to electricity via thermoelectric and thermionic generators. Unfortunately, despite the high temperature stabilities, the efficiencies were disappointingly low and the great expectations fell through. For fourth- and fifth-group carbides and nitrides, Toth (1971) has tabulated values from -1 to -15 μV/°K at room temperature. Values (in μV/°K) for the sixth-group carbides, as determined by H. A. Johansen (private communication, 1960), are Cr_7C_3 – 10.8, MoC – 3.1, WC – 16.6, and for CrN – 11.6; thus, the Group VI values are seen to have the same range as fourth- and fifth-group compounds.

The work function of carbides and nitrides derived from thermionic emission measurements was also the subject of an intensive but brief period of study in the last decade. Values of work functions for transition metal carbides and nitrides as well as other compounds have been tabulated by Fomenko (1966). A survey of more recent work is given by Matskevich (1968), particularly of the extensive Russian literature. Yamamoto and co-workers (1974) using data from Fomenko (1966) have developed a semi-empirical model for the work function of binary compounds including transition metal carbides and nitrides. They concluded that the model gave a satisfactory prediction of work functions for the compounds. Studies of the work function of TiC_x as dependent on stoichiometry have been reported by Samsonov and co-workers (1972), who observed that the work function depended on the number of vacancies in the carbon sublattice.

6. *Thermal Conductivity*

Heat conductivities for the various carbides and nitrides range from about 0.01 cal/cm-sec-°C for niobium nitride to 0.07 cal/cm-sec-°C for tungsten carbide. This property received early attention in a variety of studies because of its bearing on efficiencies in the many thermoelectric and thermionic programs of the 1960s. More recent work has focused on the property at higher temperatures, for example, the thermal conductivity of $TiC_{0.995}$ and $ZrC_{0.97}$ in the range of 1300–2500°K was reported by Korshunov and co-workers (1973). The values were calculated using thermal diffusivity and heat capacity with a novel acoustic method. This work and the heat conduction of $VN_{0.95}$ and $CrN_{0.98}$ reported by Aivazov and co-workers (1973), who found a pronounced ionic character to the Cr—N bond from i.r. studies, show the degree of sophistication and effort of the work carried out in many of the Russian institutes.

7. *Thermal Expansivity*

The coefficient of thermal expansion is a property of importance when a material begins to be used. As soon as components are designed and fits have to be made, then suddenly the thermal expansivity becomes very important. For the transition metal carbides and nitrides whose actual and potential uses may involve extreme temperature changes, the property is of more than academic interest. Toth (1971) has listed some selected values of room temperature coefficients of thermal expansion which vary from $2.3 \times 10^{-6}/°C$ for chromium nitride to $10.3 \times 10^{-6}/°C$ for Cr_3C_2. The other carbides and nitrides are intermediate. Further data for other temperature ranges are given by Samsonov (1964).

Recent expansivity determinations for niobium carbide, hafnium carbides, and tantalum carbide were reported by Jun and Shaffer (1971) at high temperature. They found the linear expansion coefficient to increase with increasing carbon content as well as with temperature.

Thermal expansion determination by means of neutron diffraction of a complex carbide $UMoC_2$ was reported by Bowman and co-workers (1970) who found a value of $8.6 \times 10^{-6}/°C$, i.e., roughly the reported value for Mo_2C.

In part, the present direction has been to emphasize work on single crystals, or fully dense polycrystalline carbides. The study of the mechanical behavior of the nitrides has been almost entirely neglected. Single-crystal carbide techniques are well advanced, however, and are reviewed by Toth (1971). Especially successful in producing large, flaw-free single crystals has been the solid state, grain growth method reported by Fleischer and Tobin (1969, 1971). Their method involves heating the metal, in rod form in a graphite crucible, to high temperatures for prolonged periods. The metal reacts with the carbon, and the unreacted core metal distorts the carbide outer layer. As carbiding progresses, the temperature is gradually raised. The strained carbide is annealed and grain growth results. This is the classic metallurgical technique of strain-anneal but transmuted to the 2000–3000°C range. Large centimeter-size single crystals can be produced by this method.

By contrast, there are no reports of single-crystal preparations of the nitrides in the recent literature. Undoubtedly, this should be a fertile field for future chemical and physical investigations.

B. Superconductors

The preceding section dealt with the properties of the transition metal carbides and nitrides from the high temperature standpoint: but these materials have equally interesting and important low temperature properties. They are high transition-temperature (T_c) superconductors and many are high current density, and high magnetic-field superconductors as well.

While recent discoveries in superconductor research have focused on the β–tungsten-type superconductor (A15 structure, transition metal compounds such as Nb_3Sn, $T_c = 18°K$, and $Nb_3(Al,Ge)$, $T_c = 21°K$), the B1 (NaCl) structure compounds of the transition metal carbides and nitrides, and particularly the carbonitrides ($NbC_{0.3}N_{0.7}$ with $T_c \sim 18°K$) still are of great technical and scientific interest.

Until the fact of the nonstoichiometry of the carbides and nitrides became understood, there was nothing but contradiction in the literature for superconducting transitions, and there are still frequent discrepancies due to the difficulties of the chemical and other analyses necessary for characterization.

In general, there is a strong dependence of T_c on the transition metal to nonmetal ratio. For example, the system TaC_{1-x} is shown in Fig. 6 (Toth, 1971). Within the homogeneity range for the cubic NaCl structure, T_c goes from about 1° to 10°K. The superconducting transition therefore is a function of the number of vacancies and the vacancy structure of the lattice in the Bl-type carbides and nitrides. As the octahedral sites are filled, T_c increases.

The use of a linear chain model for the A15 structure (A_3B compounds, V_3Si for example) has been proposed by Weger (1964) wherein the A or metal

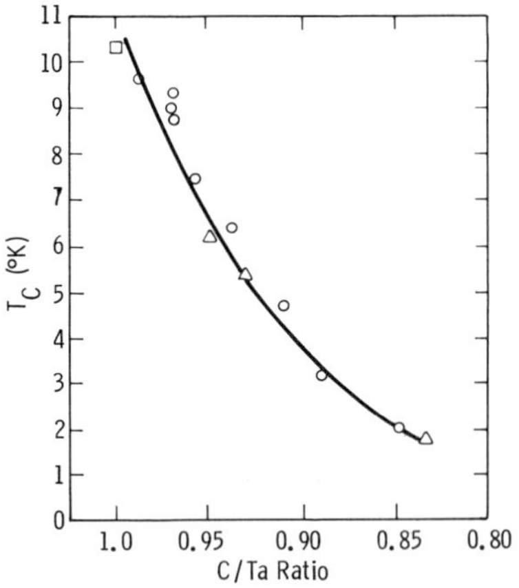

FIG. 6. Superconducting transition temperatures in the TaC_{1-x} binary system. [After Toth (1971).]

atoms in the lattice form three mutually orthogonal linear chains. The B atoms occupy isolated pockets within the chain system. In the B1(NaCl)-structure system of the carbides and nitrides, only two linear orthogonal chains of the transition metal are present and these are "balanced" by two linear orthogonal chains of the nonmetal nitrogen or carbon, so it is problematical whether this model of linear orthogonality can be applied to the carbide and nitride superconductors.

Besides the stoichiometry of the carbides and nitrides, various impurities can affect the superconducting T_c. Since oxygen can also enter the octahedral lattice sites, as well as carbon and nitrogen, it thereby interferes with the processes producing the superconduction, and a lower T_c results. However, in the B1-structure compound TiO, which has a 15% lattice deficiency of both Ti and O, added pressure (90 kbar) and temperature (1650°C) reduced the vacancy concentration and the T_c increased from 0.7° to 2.3°K (Doyle *et al.*, 1968). Certain magnetic impurities (Cr, Fe, Co, Ni) have been reported to lower drastically the T_c of several superconducting metal alloy systems such as $Mo_{0.8}Re_{0.2}$ (Mathias, 1962), but such effects have not been studied in the B1 carbides and nitrides. As a practical matter it is difficult to exclude at least traces of iron from these materials and the likelihood of more than traces appearing is frequently quite high.

The superconducting T_cs of the binary systems of the transition metals with carbon or nitrogen have all been thoroughly studied and reported (for a review, see Toth, 1971). Of greater interest, both technically and scientifically, however, is the work on ternary systems such as Nb–C–N. This system can also be considered a pseudobinary system of NbC and NbN. Table I shows a listing of pseudo binaries, pseudo ternaries, and one pseudo quaternary investigated by Pessall and co-workers (1968). The addition of NbC to NbN raises the T_c of NbN from about 14°K to about 18°K at 30 mol % addition of NbC. Also, the addition of TaC to NbN has a similar effect but the T_c only reaches 16°K. The addition of VC to NbN drastically reduces T_c as does HfN and ZrN to NbN. The effects of the various additions followed no apparent set relationships, although it was found that a variety of pseudo ternaries could be made with T_c of 17.5°K. However, none of the pseudo ternaries exceeded the maximum values of the pseudo binaries.

For a superconductor to be of technical and commercial interest, it must possess a high T_c, preferably above 21°K (bp hydrogen). It must possess, as well, high values for upper critical magnetic fields (H_{c2}) and critical current densities (J_c). The most likely uses for superconducting devices are those such as superconducting magnets and solenoids in which there may be high magnetic fields, and thus a superconductor with high H_{c2} is required. Similarly for devices such as superconducting motors and transmissions lines where high currents may be encountered, a superconductor with high J_c is required.

TABLE I

CHEMICAL ANALYSES OF PSEUDO-BINARY AND PSEUDO-TERNARY ALLOYS

Nominal composition	N:metal	C:metal	O:metal	Start of superconducting transition T_c(°K)
$NbN_{0.95}ZrN_{0.05}$	0.82[a]	0.050	—	14.2
$NbN_{0.10}ZrN_{0.90}$	0.88[a]	—	—	10.0
$NbN_{0.85}TiN_{0.15}$	0.82[a]	0.028	—	16.6
$NbN_{0.75}TiN_{0.25}$	0.83[a]	0.020	—	16.9
$NbN_{0.70}TiN_{0.30}$	0.85[a]	0.002	—	17.0
$NbN_{0.65}TiN_{0.35}$	0.87[a]	0.011	—	17.0
$NbN_{0.50}TiN_{0.50}$	0.77[a]	0.081	—	17.5
$NbN_{0.50}VN_{0.50}$	0.85[a]	—	—	3.8
$NbN_{0.70}ZrN_{0.24}TiN_{0.06}$	0.87[a]	—	—	14.3
$NbN_{0.70}ZrN_{0.17}TiN_{0.13}$	0.91[a]	0.037	—	15.7
$NbN_{0.70}ZrN_{0.10}TiN_{0.20}$	0.89[a]	—	—	16.2
$NbN_{0.675}NbC_{0.225}NbO_{0.10}$	0.82[b]	0.11	0.0003	17.2
$NbN_{0.63}NbC_{0.27}NbO_{0.10}$	0.86[b]	0.08	0.0002	16.8
$NbN_{0.56}NbC_{0.24}NbO_{0.20}$	0.72[b]	0.22	0.0006	17.8
$NbN_{0.90}NbC_{0.10}$	0.78[b]	0.19	0.003	17.7
$NbN_{0.10}NbC_{0.90}$	0.11[b]	0.87	0.008	12.8
$NbN_{0.45}NbC_{0.45}NbO_{0.10}$	0.54[b]	0.39	0.0002	17.9
$NbN_{0.54}NbC_{0.36}NbO_{0.10}$	0.65[b]	0.24	0.001	17.6
$NbN_{0.495}NbC_{0.405}NbO_{0.10}$	0.67[b]	0.24	0.001	17.6
$NbN_{0.75}NbC_{0.25}$	0.71[b]	0.23	—	17.8
$NbN_{0.7}NbC_{0.3}$	0.63[b]	0.27	—	17.8
$NbN_{0.95}TiC_{0.05}$	0.83[b]	0.04	—	17.3
$NbN_{0.9}TiC_{0.1}$	0.83[b]	0.16	—	17.7
$NbN_{0.8}TiC_{0.2}$	0.74[b] $\pm$ 0.03	0.15 $\pm$ 0.01	—	18.0
$NbN_{0.75}TiC_{0.25}$	0.76[b] $\pm$ 0.03	0.21 $\pm$ 0.01	—	17.9
$NbN_{0.75}TiC_{0.25}$	0.69[b]	0.18	—	17.9
$NbN_{0.7}TiC_{0.3}$	0.64[b]	0.19	—	17.8
$NbN_{0.60}TiC_{0.4}$	0.60[b]	0.33	—	15.4

[a]Nitrogen analyzed using Kjeldahl method.
[b]Nitrogen analyzed using Dumas method.

Both properties H_{c2} and J_c have been extensively studied for the B1 carbides and nitrides and found to depend on metal to nonmetal ratio, impurities, method of preparation, porosity, sample form (bulk or thin film), and method of testing (Pessall *et al.*, 1965; Toth, 1971).

A variety of attempts to correlate the T_cs, H_{c2}s, and J_cs with various factors in the transition metal carbides and nitrides have been made. The aim of course has been to find new and higher T_c superconductors. Such correlations have had only limited success, but the search goes on.

A good correlation of the superconducting T_c with the optimized distance of the transition metal atoms was cited by Pessall and co-workers (1968), and Pessall and Hulm (1966). Another approach was suggested by Toth and co-workers (1965) in which the valence electron concentration controls the T_c through the total number of valence electrons of both transition metal and the carbon or nitrogen. This correlation was found to be quite good. A discussion of the valence electron concentration approach is found in Toth (1971).

The superconducting properties of the B1-structure carbides and nitrides have briefly been thrust out of the limelight by the A15 compounds. Recently, interest in the B1 compounds has been renewed. A theoretical T_c of 22°K has been predicted by Phillips (1972). A recent review of the A15 materials and their relation to the superconducting alloys and the B1 compounds is furnished by Hulm and Blaugher (1972). For a complete review of recent work in this burgeoning field of superconducting research and development see the work by Parks (1969) and Douglass (1972).

C. Special Topics

1. *Bonding*

For the transition metal carbides and nitrides with their unusual properties of high melting points, great hardness, mixed metallic character, and superconduction, there has to be some unusual bonding within the compounds. Unfortunately, despite attempts to explain the various different properties by several theories, the understanding is still quite incomplete.

The contributions to the bonding in the compounds may be metallic, covalent, or ionic, and the various theories emphasize one or the other contribution. The metal–metal bonding theories tend to explain the metal-like behavior of the carbides and nitrides, especially the electrical properties. For the metal–nonmetal bonds, the covalent bonding theories may be applied to explain the hardness, strength, and high melting-point behavior.

The metal–nonmetal bonding theories postulate bonding of the *sp* type, so that the nonmetal orbitals are directed toward the metal atoms in the lattice.

An opposite theory postulates the formation of metal–metal bonds which are strengthened by donations of electrons from the carbon or nitrogen interstitials. The *d* bands of the metals become the receptors of the electrons.

It is probable that each of the (metallic, covalent, and ionic) bonding mechanisms has an effect, and the newer theories take this into account.

The detailed consideration of the chemistry and theories of bonding and band structures in the transition metal carbides and nitrides while fascinating is beyond the scope of this article. For a discussion of the topic see Toth (1971), and for a recent review of bonding theories see Phillips (1973).

2. *The Chromium Enigma*

Throughout the foregoing treatment, the reader may have noticed that while the subject matter is concerned with the transition metal carbides and nitrides of the fourth, fifth, and sixth groups of the periodic table, references to the compounds of the sixth group, chromium, molybdenum, and tungsten, have been quite scanty or missing entirely. The reason for this is that the sixth-group carbides and nitrides are significantly less stable than their counterparts in the fourth and fifth groups, and thus less readily prepared and studied. The carbides can be formed by conventional powder metallurgy techniques or by arc melting in contact with graphite, but the nitrides cannot be formed successfully by the reaction of the metal with gaseous nitrogen. Only 2–3% nitrogen uptake is obtained for example by melting chromium in a nitrogen atmosphere. Ammonolysis is the usual method of preparation at fairly low temperatures,

$$2Cr + 2NH_3 \xrightarrow{800°C} 2CrN + 3H_2 \tag{8}$$

for example as shown in equation (8). Two compounds, Cr_2N (11.87 wt.% N_2) and a cubic CrN (21.22 wt.% N_2) have been identified. The Cr_2N phase has an hexagonal close packed lattice and the CrN phase has the B1 rock-salt structure. There are at least three phases identified in the chromium–carbon system: $Cr_{23}C_6$, Cr_7C_3, and Cr_3C_2, with the structures complex face-centered cubic, hexagonal, and orthorhombic, respectively.

Similarly with the molybdenum and tungsten carbides and nitrides, which are more difficult to prepare, characterize, and investigate at high temperatures without thermal instabilities arising. Therefore, the state of the knowledge concerning the sixth-group carbides and nitrides is relatively not in as satisfactory a position as the fourth- and fifth-group compounds. This is not for lack of interest or effort, but due primarily to the difficulties of behavior of these most unusual metals.

Chromium, in particular, deserves a special comment at this point not only because of the great efforts that have been made to prepare and study the properties of the carbides and nitrides but also because of the efforts undertaken to *remove* carbon and nitrogen as well as other interstitials in the quest for means to produce room-temperature ductile chromium.

Chromium metal prepared to highest purities (99.99% or better) still exhibits no ductility whatsoever at room temperature. This is unexpected, for as a body-centered-cubic metal, Cr should be as ductile as Fe. Such puzzling behavior was originally thought to be due to the presence of oxide inclusions (as Cr_2O_3) in the microstructure of the chromium metal (Kroll *et al.*, 1950) and in fact chromium metal with oxygen content reduced to $< 0.05\%$ did show some degree of room temperature ductility (Johansen and Asai, 1954). In addition to purity, other factors were discovered to be important: degree of cold work in the metal, and kind and extent of surface treatment. It was as a result of surface removal procedures preparatory for wire drawing that it was discovered that high-purity Cr did exhibit some degree of room temperature ductility. While no single cause can yet be assigned to the lack of room temperature ductility in chromium, the evidence began to point to nitrogen as a main cause (Wain *et al.*, 1955).

By heating chromium in air or nitrogen, the metal could be embrittled. By electropolishing the nitrided layer away, the metal became ductile again. The ductile–brittle transition temperature (T_{db}) for rolled chromium with 20 ppm nitrogen was $-60°C$, and increased to 137°C with 260 ppm nitrogen, and 260°C with 1000 ppm.

That the damaging effect of nitrogen to the ductility is due to nitrogen in solution was shown by Cairns and Grant (1964) who found that quenched samples of chromium had a higher T_{db} than do slow cooled specimens. They found also that oxygen in solution had an effect on raising the T_{db}, but to smaller degree than nitrogen.

The mechanism whereby the nitrogen in the surface causes the brittleness is thought to be due to a tightly bound layer of nitride which when stressed cracks, and these nitride cracks propagate through the underlying chromium metal (Sully and Brandes, 1967). This does not explain why the chromium metal is so extraordinarily notch sensitive to notches originating in such a thin layer; the question of nitrogen involvement in the basis metal is therefore still alive. A comprehensive review of the subject of chromium ductility and the nitrogen influence can be found in Sully and Brandes (1967).

Studies in the Cr—N system continue to be made. The nitriding of chromium in nitrogen between 800° and 1400°C was studied by Cadiou and Paidassi (1969) who found a parabolic rate and an energy of activation of 52.2 Kcal/mole. The solubility of nitrogen in solid chromium over the temperature range 1050° to 1350°C was measured by Mills (1971). The solubilities at the two extrema were found to be 0.032 wt.% and 0.29%, respectively. The enthalpy of the reaction

$$\tfrac{1}{2}N_2 \rightarrow N_{Cr} \qquad \text{(nitrogen in solution)} \tag{9}$$

was calculated to be 4220 cal/g-atom nitrogen which was somewhat lower than earlier values. Mills (1970) also has reported new values for the equilibrium pressures and homogeneity limits for Cr_2N and CrN, and Mills (1972) reported partial and integral values of thermodynamic relations in the

Cr—N system. Mills' value for the heat of formation of CrN of −25.5 kcal/mole may be compared with −28.4 ± 1.5 kcal/mole calculated by DeLuca and Leitnaker (1973) who reviewed the thermodynamic properties of the Cr—N system.

Unfortunately, no clear cut picture yet emerges for the role of nitrogen and other interstitials in the behavior of the various properties of chromium, particularly the ductile–brittle transition and the "room temperature" anomalies. In the latter, nine or more properties, such as thermal expansion, elastic modulus, internal friction, electrical resistance, specific heat, and others, all show anomalous inflections with temperature, all at or near 38°C.

3. *Catalytic Behavior of Interstitial Compounds*

While platinum is the catalyst par excellence, its cost has caused various industries to seek alternative materials. Investigations carried out on hydrocarbons using various metals (such as Fe, Co, Ni, and W) as catalysts showed that the pure metal had little effect for hydrogenation or isomerization. However, after a suitable induction period some of the metals were then found to exhibit the desired catalytic activity.

What had been discovered was that in the "induction period" the metal was becoming carbided; and the more carbiding the better the catalytic activity. The next step, of course, was to use the carbide deliberately, instead of the metal.

Such experiments for ethane hydrogenolysis using molybdenum were reported by Sinfelt and Yates (1971). They had noted the results of the decomposition of ammonia experiments of Tsuchiya and Ozaki (1969), wherein molybdenum catalysts were found to be nitrided and the catalytic activity suspected to be due to the nitride layer.

In addition, it had been noted by Muller and Gault (1970) that rapid, extensive cracking of 1,1,3-trimethylcyclopentane occurred on tungsten at 350°C, followed by slower reactions including the formation of xylenes.

The anodic oxidation of hydrogen on a tungsten carbide electrode was reported by Böhm (1970). The kinetics of the process showed the activity of WC to compare favorably with that of platinum.

All these results and others led to the conclusion by Levy and Boudart (1973) that the addition of carbon to the tungsten lattice gave a platinum-like catalytic activity and was greatest with actual bulk carbide formation. Levy and Boudart speculated that some change in the electron distribution in the tungsten as a result of the carbiding conferred the catalytic behavior.

Such results show exciting promise both for the field of transition metal carbides and nitrides and for the field of catalysis, where the exact mechanisms may be further clarified.

V. Conclusion

As a field, the chemistry of the transition metal carbides and nitrides appears to have matured. While fairly large gaps in knowledge still remain, the general behavior and structure of the field has become more clear. A fair number of well-equipped laboratories with competent investigators around the world continue to study the various problems and refine the data. As before, the emphasis appears to be mainly on the high temperature properties with potential for uses in conventional chemical and metallurgical processes, as well as in new applications, such as high-power circuit breakers, fuel cell technology, magnetohydrodynamic devices, and possibly fusion power reactors. There is certainly renewed interest in superconducting properties, and brand new interest in catalytic properties.

Future work in these fields should be most interesting and rewarding.

References

Asai, G., Paulson, D. L., and Kelly, H. J. (1971). *U.S. Bur. Mines, Rep. Inrest.* **7479**.

Aivazov, M. I., Muranevich, A. Kh., and Domashnev, I. A. (1973). *Teplofiz. Vys. Temp.* **11**, 768; *High Temp. (Engl. Transl.)* **11**, 690 (1973).

Baker, F. B., Storms, E. K., and Holley, C. E., Jr. (1969). *J. Chem. Eng. Data* **14**, 244.

Billy, M., and Teyssedre, B. (1973). *Bull. Soc. Chim. Fr.* **5**, Part 1, 1537.

Böhm, H. (1970). *Electrochim. Acta* **15**, 1273.

Bowman, A. L., Arnold, G. P., and Krikorian, N. H. (1970). *J. Appl. Phys.* **41**, 5080.

Brauer, G., Mohr, E., Neuhaus, A., and Skokan, A. (1972). *Monatsh. Chem.* **103**, 794.

Cadiou, L., and Paidassi, J. (1969). *C.R. Hebd. Seances Acad. Sci., Ser. C* **268**, 743.

Cadoff, I., Neisen, J. P., and Miller, E. (1956). *Plansee Proc., Pap. Plansee Semin.*, "*De Re Met.*," *2nd, 1955* p. 50.

Cairns, R. E., and Grant, N. J. (1964). *Trans. AIME* **230**, 1150.

Carlson, O. N., Schmidt, F. A., and Sever, J. C. (1973). *Met. Trans.* **4**, 2407.

Chang, Y. A. (1965). "Thermodynamic Properties of Group 4, 5, and 6 Binary Transition Metal Carbides," AFML-TR-65-2, Part IV, Vol. I.

Chermant, J. L., Deschanvres, A., Manier, G., and Verge, G. (1969). *C.R. Hebd. Seances Acad. Sci., Ser. C* **268**, 2291.

Corliss, L., Eliot, N., and Hastings, J. (1961). *Phys. Rev.* **117**, 929.

Croft, W. J., and Nygaard, K. J. (1969). *J. Sci. Instrum.* **2**, 1012.

DeLuca, J. P., and Leitnaker, J. M. (1973). *J. Am. Ceram. Soc.* **56**, 126.

Douglass, D. H., ed. (1972). "Superconductivity in d- and f-Band Metals." Am. Inst. Phys., New York.

Doyle, N. J., Hulm, J. K., Jones, C. K., Miller, R. C., and Taylor, A. (1968). *Phys. Lett. A* **26**, 604.

Ettmayer, P. (1970a). *Monatsh. Chem.* **101**, 127.

Ettmayer, P. (1970b). *Monatsh. Chem.* **101**, 1720.

Fleischer, L. R., and Tobin, J. M. (1969). "High Temperature Technology." IUPAC, Butterworth, London.

Fleischer, L. R., and Tobin, J. M. (1971). *J. Cryst. Growth* **8**, 235.

Fomenko, V. (1966). "Handbook of Thermionic Properties." Plenum, New York.
Gasser, R. P. H., Gowan, P. M., and Newman, D. C. (1968). *Trans. Faraday Soc.* **64**, 2196.
Geils, R. H., and Potter, D. I. (1973). *Met. Trans.* **4**, 1469.
Gingerich, K. A. (1968). *J. Chem. Phys.* **49**, 14.
Guha, J. P., and Kolar, D. (1972). *J. Less-Common Met.* **29**, 33.
Hannay, N. B., ed. (1973). "Treatise on Solid State Chemistry," Vol. 1. Plenum, New York.
Hastie, J. W., Hauge, R. H., and Margrave, J. L. (1970). *Annu. Rev. Phys. Chem.* **21**, 475.
Haughton, J. L., and Prince, A. (1956). "The Constitutional Diagrams of Alloys: A Bibliography," Inst. Metals, London.
Hausner, H. H., and Bowman, M. G., eds. (1968). "Fundamentals of Refractory Compounds." Plenum, New York.
Hiraga, K. (1973). *Philos. Mag.* [8] **27**, 1301.
Hörz, G. (1971). *Z. Mettallkd.* **62**, 208.
Hörz, G., and Lindenmaier, K. (1972a). *Z. Mettalkd.* **63**, 103.
Hörz, G., and Lindenmaier, K. (1972b). *Z. Metallkd.* **63**, 240.
Hulm, J. K., and Blaugher, R. D. (1972). *Proc. Int. Conf. Low Temp. Phys., 13th, 1972*, Vol. 3, p. 3.
Jansson, S. A. (1970). *J. Vac. Sci. Technol.* **7**, 5.
Johansen, H. A. (1970a). U.S. Patent 3,523,044.
Johansen, H. A. (1970b). *Planseeber. Pulvermetall.* **18**, 16.
Johansen, H. A., and Asai, G. (1954). *J. Electrochem. Soc.* **101**, 604.
Johansen, H. A., and Cleary, J. G. (1966). *J. Electrochem. Soc.* **113**, 378.
Johansen, H. A., and Comenetz, G. (1969). *High Temp. Technol., Proc. Int. Symp., 3rd, 1969*, p. 525.
Jun, C. K., and Shaffer, P. T. B. (1971). *J. Less-Common Met.* **24**, 323.
Kelley, K. K. (1963). *In* "Selected Values of Thermodynamic Properties of Metals and Alloys," (R. Hultgren, ed.), Suppl. Wiley, New York.
Kelley, K. K., and King, E. G. (1961). *U.S., Bur. Mines, Bull.* **592**.
Kelly, A., and Rowcliffe, D. J. (1967). *J. Am. Ceram. Soc.* **50**, 253.
Khromov, Yu. F., Yanchur, V. P., and Yeremeyev, V. S. (1972). *Phys. Metals Metallogr.* (*Engl. Transl.*) **33**, 182.
Kieffer, R., and Benesovsky, F. (1963). "Hartstoffe." Springer-Verlag, Vienna.
Korshunov, I. G., Zinovev, V. E., Geld, P. V., Chernyaev, V. S., Borukovich, A. S., and Shveikin, G. V. (1973). *Teplofiz. Vys. Temp.* **11**, 889; *High Temp.* (*Engl. Ed.*) **11**, 803 (1973).
Kossowsky, R., and Harvey, F. J. (1974). *Met. Trans.* **5**, 790.
Kroll, W. J., Hergert, W. F., and Carmody, W. R. (1950). *U.S., Bur. Mines, Rep. Invest.* **4752**.
Levy, R. B., and Boudart, M. (1973). *Science* **181**, 547.
Mathias, B. T. (1962). *IBM J. Res. Dev.* **6**, 250.
Matskevich, T. L. (1968). *Sov. Phys.—Tech. Phys.* **13**, 295.
Mills, T. (1970). *J. Less-Common Met.* **22**, 373.
Mills, T. (1971). *J. Less-Common Met.* **23**, 317.
Mills, T. (1972). *J. Less-Common Met.* **26**, 223.
Muller, J. M., and Gault, F. G. (1970). *Bull. Soc. Chim. Fr.* p. 416.
Nowotny, H. (1972). *Angew. Chem., Int. Ed. Engl.* **11**, 906.
Nowotny, H., and Benesovsky, F. (1968). *Planseeber. Pulvermetall.* **16**, 204.
Parks, R. D., ed. (1969). "Superconductivity," Vols. 1 and 2. Dekker, New York.
Pessall, N., and Hulm, J. K. (1966). *Physics* (*Long Island City, N.Y.*) **2**, 311.
Pessall, N., Jones, C. K., and Johansen, H. A. (1965). *Proc. Int. Symp. Magnet Technol., 1965*.
Pessall, N., Hulm, J. K., and Walker, M. S. (1967). Final Report AF33(615)-2729.

Pessall, N., Gold, R. E., and Johansen, H. A. (1968). *J. Phys. Chem. Solids* **29**, 19.
Phillips, J. C. (1972). *In* "Superconductivity in d- and f-Band Metals" (D. H. Douglass, ed.), p. 339. Am. Inst. Phys., New York.
Phillips, J. C. (1973). *In* "Treatise on Solid State Chemistry," Vol. 1, "The Chemical Structure of Solids" (N. B. Hannay, ed.). Plenum, New York.
Rudy, E. (1969). "Compendium of Phase Diagram Data," AFML-TR-65-2, Part V.
Rudy, E. (1973). *J. Less-Common Met.* **33**, 327.
Rudy, E., and Benesovsky, F. (1960). *Planseeber. Pulvermetall.* **8**, 72.
Samsonov, G. V. (1964). "Handbooks of High-Temperature Materials," "Properties Index." Plenum, New York.
Samsonov, G. V., Podchernyaeva, I. A., Fomenko, V. S., Laurenko, V. A., Okhremchuk, L. N., and Protsenko, T. C. (1972). *Poroshk. Metall*, p. 62; *Sov. Powder Metall. Met. Ceram.* (Engl. *Transl.*) **11**, 389.
Santoro, G. (1963). *Trans. AIME* **227**, 1361.
Santoro, G., and Dolloff, R. (1968). *J. Appl. Phys.* **39**, 2293.
Sarian, S. (1969). *J. Appl. Phys.* **40**, 3515.
Savage, H., and Alstetter, C. (1970). *J. Less-Common Met.* **22**, 399.
Schick, H. L., ed. (1966). "Thermodynamics of Certain Refractory Compounds," Vols. 1 and 2. Academic Press, New York.
Schmidt, F. A., and Carlson, O. N. (1972). *J. Less-Common Met.* **26**, 247.
Schwarzkopf, P., and Kieffer, R. (1953). "Refractory Hard Metals." Macmillan, New York.
Searcy, A. W. (1963). *Surv. Prog. Chem.* **1**, 35.
Schaffer, P. T. B. (1964). "High-Temperature Materials, Index." Plenum, New York.
Shepela, A. (1972). *J. Less-Common Met.* **26**, 33.
Sinfelt, J. H., and Yates, D. J. C. (1971). *Nature (London), Phys. Sci.* **229**, 27.
Storms, E. K. (1967). "The Refractory Carbides." Academic Press, New York.
Storms, E. K. (1968). *In* "Fundamentals of Refractory Materials" (H. H. Hausner and M. G. Bowman, eds.). Plenum, New York.
Stull, D. R., and Prophet, H., proj. directors. (1971). "JANAF Thermochemical Tables," 2nd ed. US Govt. Printing Office, Washington, D.C.
Sully, A. H., and Brandes, E. A. (1967). "Metallurgy of the Rarer Metals," "Chromium," 2nd ed., p. 182. Plenum, New York.
Toth, L. E. (1971). "Transition Metal Carbides and Nitrides." Academic Press, New York.
Toth, L. E., Zackay, V. F., Wells, M., Olson, J., and Parker, E. R. (1965). *Acta Metall.* **13**, 379.
Tsuchiya, S., and Ozaki, A. (1969). *Bull. Chem. Soc. Jpn.* **42**, 344.
Wain, H. L., Henderson, F., and Johnston, S. T. M. (1955). *J. Inst. Met.* **83**, 133.
Warren, R. (1969). *J. Less-Common Met.* **17**, 65.
Weaver, D. E. (1972). *Diss. Abstr. Int. B.* **33**, 1134.
Weger, M. (1964). *Rev. Mod. Phys.* **36**, 175.
Wicks, C. E., and Block, F. E. (1963). *U.S. Bur. Mines, Bull.* **605**.
Wood, F. W. (1974). *Diss. Abstr. Int. B.* **34**, 3347.
Yamamoto, S., Susa, K., and Kawabe, U. (1974). *J. Chem. Phys.* **60**, 4076.
Zainulin, Yu. G., Alyamovski, S. I., Shveikin, G. P., and Geld, P. V. (1972). *Dokl. Chem. (Engl. Transl.)* **207**, 843.

Living Polymers: Their Significance and Applications

MICHAEL SZWARC

Department of Chemistry,
State University of New York,
College of Environmental Science and Forestry,
Syracuse, New York

I. INTRODUCTION

Polymerization is a process converting small molecules, referred to as monomers, into large, covalently bonded aggregates—the polymers. A particularly important polymerization is that yielding a long, linear array of covalently bonded monomeric units—a macromolecule known as a linear polymer. Two routes lead to this outcome: polyaddition and polycondensation. The first one yields a linear polymer by consecutive additions of monomers to a growing center, i.e.,

$$X + M \rightarrow XM \qquad XM + M \rightarrow XM_2 \qquad \ldots \qquad XM_n + M \rightarrow XM_{n+1}$$

where X is the growth initiating species and M is the monomer. The alternative route forms a long polymer by repetitive linkage of small polymers into still larger ones, a process exemplified by the following scheme:

$$M + M \rightarrow M_2 \qquad M_2 + M_2 \rightarrow M_4 \qquad M_4 + M_2 \rightarrow M_6$$

and generally

$$M_n + M_m \rightarrow M_{n+m}$$

It is the linear polyaddition that will interest us here. Three basic steps are involved in this process: (1) initiation, (2) propagation, and (3) termination or chain transfer. Initiation is responsible for the formation of centers to which the monomers may add and start the growth. Propagation is the growth process, the repeating steps, $XM_n + M \rightarrow XM_{n+1}$, that eventually produce a long, linear polymer. The growth ultimately ceases. Termination is the process that deprives a growing polymer of its ability to grow without forming a new growing center, whereas chain transfer terminates the growth of one polymeric molecule but simultaneously produces a new growing center. Initiation may be thought of as the birth of a polymer, propagation as its growth, and termination or chain transfer as its death. Hence, not surprisingly, a terminated polymer has been called a dead polymer.

An intriguing question may be posed: Is it possible to visualize a polyaddition involving initiation and propagation but not termination or chain transfer? Such a process should yield a polymer that retains its ability to grow indefinitely, provided that we deal with an ideal system. Such a polymer, in contradistinction to a dead polymer, will be referred to as a living polymer (Szwarc, 1956).

II. Consequences of Elimination of Termination or Chain Transfer

Does a polymer become infinitely long when its growth continues and termination and chain transfer are eliminated? The answer is negative. In any polymerizing system, initiation produces some amount of growing centers, say I mole/liter. The amount of the available monomer must be finite; let us denote its initial concentration by M_0. Hence, the uninterrupted growth distributes M_0 moles of monomer among I centers and thus the number-average degree of polymerization of the resulting polymers is $\overline{DP}_n = M_0/I$.

The growth ceases eventually not due to termination or chain transfer but because the supply of the available monomer is exhausted. The living polymers formed in the process do not die; they retain their capacity of growth and wait for the next prey. Therefore they may be fed by adding more monomer to the reactor. Indeed, this has been realized (Szwarc, 1956); the growth resumes and the polymers become longer. This procedure allows us to attain any desired degree of polymerization of the ultimate product by a controlled rate of addition of monomer. Since most polymerizations are exothermic, overheating of the system is prevented by this device.

It is implicitly assumed that *all* the monomer present in a polymerizing system is consumed by the propagation if termination or chain transfer are eliminated. This is not rigorously correct. Any elementary reaction is reversible, this being the consequence of the principle of microscopic balancing. The propagation is represented by a general equation:

$$\mathrm{XM}_n + \mathrm{M} \underset{k'_n}{\overset{k_n}{\rightleftharpoons}} \mathrm{XM}_{n-1}$$

and the rate of monomer consumption is given by

$$\frac{-d[\mathrm{M}]}{dt} = [\mathrm{M}]\sum k_n[\mathrm{XM}_n]$$

the summation being carried over all the growing species, i.e., from $n = 0$ to ∞. The rate of depropagation is given by $\Sigma\, k_n'[\mathrm{XM}_n]$ with n varying from 1 to ∞ because no depropagation is possible for $n = 0$. In a system in which initiation ceased to produce new growing centers, i.e., at that stage of the process when the concentration of the growing species remains constant in time, the depropagation balances the propagation at a definite concentration of monomer, $[\mathrm{M}]_e$, and then

$$[\mathrm{M}]_e\sum_0^{\infty} k_n[\mathrm{XM}_n] = \sum_1^{\infty} k_n'[\mathrm{XM}_n]$$

The living polymers are now in equilibrium with their own monomer and no further polymerization takes place, although the supply of the monomer has not been exhausted.

It is reasonable to expect k_n and k_n' to be constant, independent of n, provided that n is sufficiently large. Let us denote these constants by k_p and k_d respectively, and by n_0 the degree of polymerization for which $k_n = k_p$ and $k_n' = k_d$ if $n > n_0$. In systems involving high molecular weight polymers the concentration of growing species with $n < n_0$ is negligible, and then the following approximate equation is valid:

$$k_p[\mathrm{M}]_e \sum_{n_0}^{\infty} [\mathrm{XM}_n] = k_d \sum_{n_0}^{\infty} [\mathrm{XM}_n]$$

leading to the relation

$$K_p = \frac{k_p}{k_d} = \frac{1}{[\mathrm{M}]_e}$$

Therefore, for high molecular weight living polymers, the equilibrium concentration, $[\mathrm{M}]_e$, of the monomer is independent of their degree of polymerization (Bywater and Worsfold, 1962b; Dreyfuss and Dreyfuss, 1965, 1966; Vofsi and Tobolsky, 1965).

The equilibrium constants K_p in high molecular weight systems are uniquely determined for reactions such as

$$\sim\!\sim\mathrm{M}\cdot\mathrm{M}\sim\!\sim + \mathrm{M} \rightarrow \sim\!\sim\mathrm{MMM}\sim\!\sim$$

irrespective of the mode of monomer's insertion. Of course, addition of monomer to the end of a polymer chain is equivalent to its insertion, provided that the nature of the end groups remains unaltered. This condition is fulfilled in the propagation of living polymers, i.e., in the reaction

$$\sim\!\sim\mathrm{M}\cdot\mathrm{M}^* + \mathrm{M} \rightarrow \sim\!\sim\mathrm{MMM}^*$$

where M* denotes a reactive terminal unit. Thus, although $[\mathrm{M}]_e$ is determined for a system involving a particular type of living polymers propagating by a specific mechanism, its significance is applicable to any system, living or dead, involving polymers of the same structure as those of the investigated living polymers. This generalization makes studies of monomer–living polymer equilibria valuable because it allows us to investigate the thermodynamics of polymerizing systems by a simple, nonlaborious technique. In a sufficiently dilute solution, when the activity coefficients are unity, K_p and $[\mathrm{M}]_e$ become independent of polymer concentration, i.e., such systems resemble saturated solutions in which the concentration of solute is independent of the amount of the undissolved material remaining in contact with the solution.

Determination of $[\mathrm{M}]_e$ provides a convenient method of studying the dependence of activities of polymer segments on polymer concentration and the nature of solvent. It suffices to add some living polymers to a solution of dead polymers in a required solvent at a required concentration and determine the concentration of the monomer after establishing the equilibrium. The analytical techniques now available make such an analysis feasible, even if $[\mathrm{M}]_e \approx 10^{-6}$ *M*, or slightly lower. Examples of such studies have been reported in the literature (see Szwarc, 1968, for the pertinent references).

III. Molecular Weight Distribution of Living Polymers

Conventional polymerization processes produce polymers of nonuniform molecular weights. Heterogeneity of the product results from the statistical character of chemical reactions. For example, in most polymerizing systems the growing chain has only two alternatives: to add monomer and increase its length or to be terminated and converted into a dead polymer. If the probabilities of these two events remain constant during the process, say p for the former and $1 - p$ for the latter, then the probability P_j of j-mer formation is

$$P_j = p^{j-1}(1 - p)$$

The mole fraction of j-mers in the final product is therefore equal to P_j, and their weight fraction equal to $jp^{j-1}(1 - p)$. The resulting molecular weight distribution is known as the most probable one and the ratio of weight-average molecular weight to its number average is then $1 + p$ (being ≈ 2 for high molecular weight polymers) because then p approaches unity.

Utilization of living polymers permits us to produce polymers of a much more uniform molecular weight distribution. Low molecular weight living polymers, or living oligomers, can be readily prepared by initiating polymerization in a nonterminating system with a relatively small amount of monomer. These are grown to a required length by adding monomer to their well-stirred solution. The molecular weight distribution of the final product approaches Poisson distribution provided that the final length of the polymers prepared in this way is substantially greater than the size of the oligomers used at the start of the preparation. This distribution arises because the probability of growth is the same for each polymer molecule while termination is eliminated. For a high molecular weight polymer the Poisson distribution negligibly differs from the uniformity of molecular sizes, the ratio of the weight-average molecular weight to the number average being insignificantly greater than unity.

Although polymers of highly uniform molecular weight were prepared by this technique (Altares *et al.*, 1964; Calderon and Scott, 1965; Yen, 1967; Fujimoto and Nagasawa, 1975), some difficulties and pitfalls have to be mentioned. Impurities present in the monomer, insufficiently vigorous mixing of the reagents, temperature gradients in the reactor, etc., broaden the molecular weight distribution of the product. A comprehensive treatment of these problems is reported elsewhere (Szwarc, 1968).

Monomer–living polymer equilibrium causes interesting changes in the molecular weight distribution of living polymers. Poisson distribution is obtained when the rate of propagation is much greater than the rate of depropagation, i.e., when $[M] \gg [M]_e$, and therefore the latter step virtually does not affect the course of polymerization at this stage of the process. However, when the propagation is completed and $[M] \approx [M]_e$ an unconventional reaction still prevails. The dynamic equilibrium

$$\underset{(n+1)\text{-mer}}{\sim\sim M \cdot M \cdot M^*} \rightleftarrows \underset{n\text{-mer}}{\sim\sim MM^*} + M$$

cuts off a monomer from one polymer and adds it to another. This process slowly changes the Poisson molecular weight distribution into the most probable molecular weight distribution, and thus the ratio of weight-average molecular weight to number-average molecular weight increases from 1 to 2.

In some respects this process is unique; the concentration of polymer and, to a lesser extent, of the monomer remain virtually constant in its course. For example, spectrophotometric analysis of the polymer concentration, or any measurements based on colligative properties of the system, will not reveal the occurrence of any process. However, measurement of the viscosity of the solution, or of any of its properties determined by higher moments of molecular weight distribution, will show changes in time. Approximate mathematical treatment of this reaction was reported by Brown and Szwarc (1958), and the more rigorous derivation revealing the transients was outlined by Miyake and Stockmayer (1965).

IV. Block Polymers and Their Properties

Conventional linear polymers are built by a sequence of identical monomeric segments, say A, strung together into long, covalently bonded molecules. If two kinds of units, say A and B, form a polymeric molecule, their distribution along the chain may be alternative, ∿∿ ABABAB ∿∿ or random, ∿∿ AABBBABAAB ∿∿, or the A's may be linked together forming a long sequence followed by a long sequence of B's, ∿∿ AAAAA · BBBB ∿∿. The first kind of polymer is referred to as an alternating copolymer, the second type is known as a random copolymer, while the last ones form block polymers—a block of A's followed by a block of B's.

The simplest two-block polymer is composed of a chain of A's followed by a chain of B's, as depicted in the preceding paragraph. However, more complex block structures may be visualized, e.g., a terblock,

$$AA \cdots A \cdot BB \cdots B \cdot AA \cdots A$$

or

$$AA \cdots A \cdot BB \cdots B \cdot CC \cdots C$$

Generalization of these structures is obvious and thus we may deal with a tetrablock polymer, pentablock polymer, etc., or even with a polyblock polymer.

Properties of conventional homopolymers are determined by the nature of monomeric units and the mode of their linkage, while the nature of the —AB— units determines the properties of alternating copolymers. In fact, we may treat an alternating copolymer as a homopolymer built from the —AB— monomeric units. Random copolymers have properties that often substantially differ from those of the homopolymers formed from the respective monomeric units, and these need not be given by some average of the properties of the latter. Indeed, some new properties, not encountered in the respective homopolymers, may be exhibited by the copolymers.

In many respects the properties of block polymers resemble the properties of a blend of the respective homopolymers. Nevertheless, some new and often unexpected properties shown by blockpolymers make them interesting to study and valuable in technological applications. Growth in sophistication of modern technology imposes ever greater and more intricate demands on the properties of materials used in industry and everyday life. Judiciously designed block polymers may meet such a challenge and provide the necessary material.

Although the concept of block polymers was recognized a long time ago and various techniques for their synthesis were developed and described in the literature (Smets and Hart, 1960; Woodward and Smets, 1955; Smets, 1965), the real impetus to study this type of macromolecules was provided by the discovery of living polymers (Szwarc, 1956; Szwarc *et al.*, 1956). Previous synthetic techniques often resulted in mixtures of homopolymers and blockpolymers. The sizes of the individual blocks were ill defined and the distribution of monomers along the chains could not be controlled. Consequently, the investigators were frequently at a loss to decide what kind of materials they would study and how to relate their properties to their constitution. Not surprisingly, the academic and industrial progress in this field was slow and painful.

Application of living-polymers technique changed this situation radically. Living polymers composed of monomer A may be prepared at desired length and, as explained in the previous section, they may have a nearly uniform molecular weight distribution. Their ends remain active and hence they may initiate polymerization of another monomer, say B, and thus continue their growth. Therefore, the following procedure suggests itself immediately.

To a solution of living poly-A, a required amount of monomer B is added under vigorous stirring. In this way the living poly-A is "fed" by B and is eventually converted into a two-block living polymer AA···A·B··· B. Of course, the obvious condition has to be fulfilled to make such a synthesis feasible, namely, the active living terminal units A* should add monomer B and form living terminal units ∿∿ AB*. This condition not always is encountered. Which pair of monomers is suitable for such a synthesis depends on the mechanism of growth of the living polymers. For example, anionically growing living polystyrene may continue its growth with isoprene or methylmethacrylate (Szwarc *et al.*, 1956; Korotkov *et al.*, 1960), but anionically growing living methylmethacrylate cannot add styrene (Graham *et al.*, 1960).

The technique just outlined offers two important advantages to an experimenter: (1) the prepared block polymers are free of homopolymers, and (2) the preparation method allows one to control the sizes of both blocks and therefore to produce a material not only of the desired composition (a required A/B ratio) but also uniform in molecular weight of both blocks Systematic studies of the properties of block-polymers as functions of their composition and length became possible.

A further advantage of the living-polymer technique arises in those systems where living poly-A can continue its growth with B and living poly-B may grow further by adding A. Under these conditions, terblock, and more generally, polyblock polymers of well-defined architecture may be easily synthesized. The following experiment illustrates the potential of this method.

Let two reservoirs containing, say, one mole of monomer A and one mole of monomer B, respectively, be attached to a vessel containing a solution of a suitable rapidly reacting initiator of living polymerization. The polymerization may be performed by slowly adding all the A and thereafter all the B. This yields a two-block polymer of 50:50 composition and of a degree of polymerization uniquely determined by the amount of initiator present at the onset of the reaction, i.e.,

AA ··· A · BB ··· B

Let us now repeat this preparation with the following modification of the procedure: We add first only $\frac{1}{2}$ of the available A, then $\frac{1}{2}$ of the B, thereafter we add the remaining A and finally the remaining B. The resulting macromolecules should have the same composition (50:50) and the same length as the previously described polymers, but the distribution of monomers along the chains would be different, namely,

A ··· A · BB ··· B · AA ··· A · BB ··· B.

Generalization of this procedure is obvious and thus one may systematically vary, in a great number of ways, the distribution of the monomers along the polymeric chain and study its effect on the properties of the resulting materials (Schlick and Levys, 1960).

Still another modification of this technique becomes possible if the living poly-B may grow by addition of monomer C. Then a controlled terblock polymer, AA ··· A · BB ··· BBC ··· C, may be formed.

There are techniques [e.g., initiation by electron transfer (Szwarc, 1956)] that lead to the formation of living polymers endowed with two active ends, *viz.*,

$$*AA \cdots A*$$

Feeding of such polymers with a suitable monomer B yields directly terblock polymers.

All the novel and intriguing properties of block polymers stem from their semi-microscopic heterogeneity. As a rule, polymers with differently built chains are incompatible and in concentrated solution or in melt they separate into distinct phases (Dobry and Boyer-Kavenoki, 1974; for theoretical justification, see Scott, 1949). Since each block of a block polymer retains virtually all the characteristics of the respective homopolymer, the identically built blocks tend to aggregate with each other and exclude from their domains the blocks of a different nature. Such a segregation was observed in solutions of block polymers and studied by investigating their viscosity and light-scattering characteristics. However, while a concentrated solution of two incompatible homopolymers eventually splits into two separated phases, a block-polymer system is segregated by the formation of mesomorphic phases, i.e., it leads to the development of an organized spacial structure of the swollen or melted mass of the block polymers.

The bonds which hold the two incompatible blocks together are responsible for this phenomenon. As the poly-A blocks aggregate they form domains having their extension limited at least in one dimension. Otherwise, had the size of those domains been too large, the poly-B blocks, that are incompatible with poly-A, would be dragged into the poly-A domains from which they should be excluded. The sizes of the domains are therefore determined by the length of the blocks, and their structure depends on the composition of the macromolecules and the nature and concentration of the solvent, if any, that swells the polymeric mass.

What kind of structures are formed and observed? Basically, three types are possible: lamellae, cylinders of poly-A protruding through a continuous phase of poly-B, and spheres of poly-A imbedded in the continuous phase of poly-B. The lamellar structure is formed in a mass containing approximately equal volume fractions of poly-A and poly-B blocks, while a large excess of one component over the other favors the cylindrical or spherical structures.

Formation of lamellar structures in a colorless block polymer may lead to a spectacular result. Since the refractive indexes are different for poly-A and poly-B domains, light is reflected from their interphases. If the thicknesses of each kind of layer are constant, the system behaves like a diffraction grating and interference colors appear provided that the molecular weight of the blocks is sufficiently high. For example, films developed from polystyrene–polybutadiene block polymers may be red, green, or blue depending on molecular weight, although either homopolymer forms a colorless, transparent film.

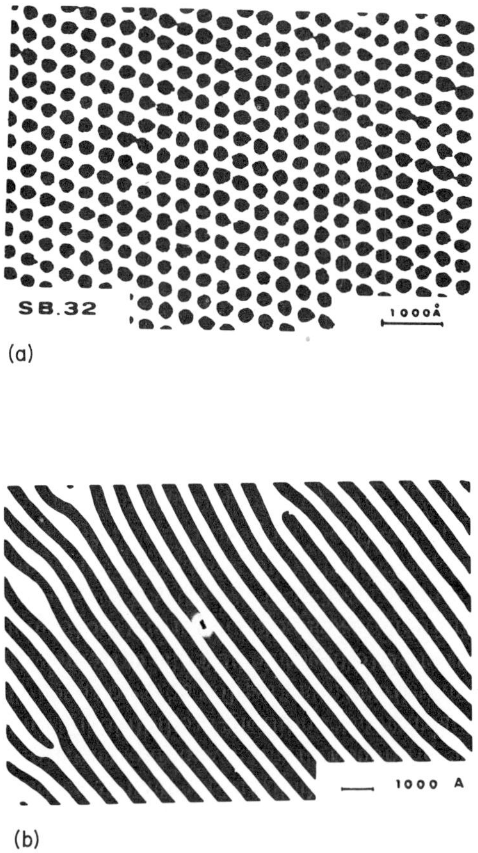

FIG. 1. Electron micrographs of thin films of block polymers shaded by OsO_4. The black areas result from the polybutadiene domains; the white represent the polystyrene domains. Slicing parallel (a, c) and perpendicular (b, d) to the direction of "wires."

Another interesting effect may be observed when filsm are cast from solutions of a polystyrene–polybutadiene block polymer. Using a good solvent for the glassy polystyrene and bad for the rubbery polybutadiene, one expands the domains of the former and collapses those of the latter. In a film cast from such a solution, polybutadiene forms beads imbedded in the continuous phase of glassy polystyrene and this makes the resulting material brittle. However, the opposite structure is developed when a good solvent for the polybutadiene but bad for polystyrene is used in casting the film. Such a film exhibits the rubbery properties of the polybutadiene that now forms the continuous phase. Hence, two different materials may be obtained from the same block polymer. This phenomenon was first reported by Merrett (1957; see also Breslow *et al.*, 1962).

Most unusual materials were described by Sadron and Gallot (see Gallot *et al.*, 1966). Using anionically produced two-block polystyrene–polybutadiene polymers of uniform molecular weight and swelled with

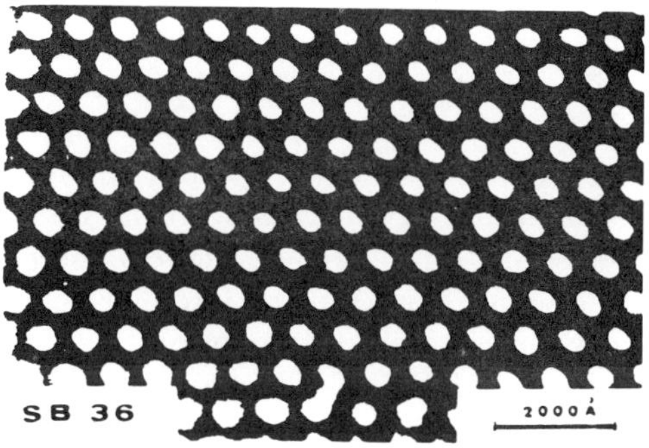

(c)

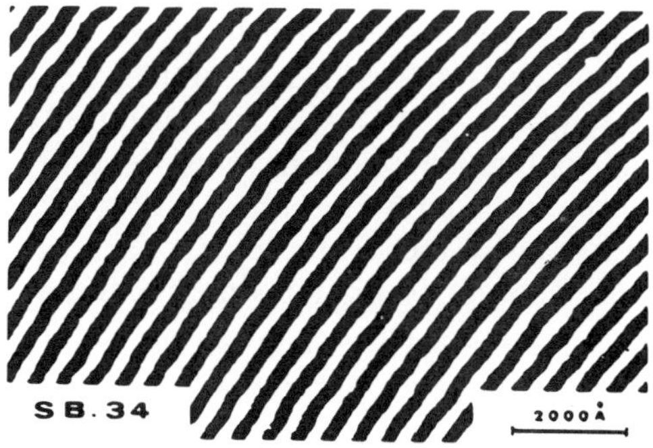

(d)

appropriate solvents, they developed highly regular mesomorphic phases having long wirelike cylinders composed of one kind of block regularly placed parallel to each other in the continuum of the other blocks. After evaporating the swelling agent, a hard mass is obtained in which the mesomorphic phases are frozen. Thin slices, cut with the aid of a microtome, display electron microscope patterns such as shown in the lower pictures of Fig. 1 provided that the slicing was performed in the direction parallel to the "wires." However, if by chance the direction of slicing was perpendicular to the direction of the wires, the upper pictures of Fig. 1 are obtained. These two kinds of pictures resulting from the same material unequivocally prove its structure.

The black domains result from the osmium tetroxide shaded polybutadiene while the polystyrene domains are white in the pictures. The two pictures on the right resulted from polymers having relatively long blocks of polybutadiene and short polystyrene, the former therefore composed the continuous phase. Material of reverse composition led to the pictures shown on the left, there polystyrene was the continuous phase. The regularity of these structures is remarkable.

Finally, one more example illustrates the unexpected properties of blockpolymers and emphasizes the importance of appropriate distribution of monomers along polymeric chains. Terblock polymers, polystyrene–polybutadiene (or isoprene)—polystyrene, form interesting and novel thermoplastic rubber, a material which is rubbery and shows considerable tensile strength at ambient temperature (Holden and Milkovich, 1966). It behaves, therefore, like a vulcanized rubber, i.e., a cross-linked one, although no chemical cross-linking has been introduced. At high temperatures its toughness disappears and it flows freely. Neither the terblocks polybutadiene—polystyrene—polybutadiene nor the diblocks polystyrene—polybutadiene show these remarkable properties, even if the molecular weight and composition of these polymers are identical to those of thermoplastic rubber (Holden and Milkovich, 1966). Obviously, the proper distribution of monomers is essential for developing the desired property.

Morphological studies of thermoplastic rubber reveal the cause of its behavior. Glassy polystyrene blocks form discrete domains acting as giant cross-linking centers and holding *both* ends of the rubbery polybutadienes that form the continuous phase. It is essential, therefore, to provide two anchors (polystyrene blocks) for each polybutadiene chain. At high temperature the glassy domains melt and the material flows like any non cross-linked rubber. The cross-linking centers develop again as the temperature of the mass is lowered. The system is reversible—the cross-linking is physical and not chemical in its nature.

Several reviews concerned with the behavior of block and graft polymers were presented in a recent symposium and were published by the Syracuse University Press (Burke and Weiss, 1973).

V. Polymers with Functional End Groups

Although termination and chain transfer are eliminated in living polymerizing systems, the resulting living polymers are not immortal. Addition of an appropriate reagent to their solutions may convert the active end groups, capable of resuming the growth, into groups that are inactive in propagation but still functional. This "killing" process may be performed at a time determined by the experimenter with a reagent of his choice.

To illustrate the procedure let us consider a living, anionically propagating polystyrene, i.e.,

$$\cdots CH_2CH(Ph)^-, Cat^+$$

These carbanions may add a variety of monomers and increase their length. Addition of water or any other proton donor converts them into inactive, dead, hydrocarbon end groups,

$$\cdots CH_2CH(Ph)^-, Cat^+ + H_2O \rightarrow \cdots CH_2CH_2(Ph) + Cat^+, OH^-$$

which are nonfunctional, while the addition of CO_2 results in the formation of terminal, functional carboxylic group (Richards and Szwarc, 1959), i.e.,

$$\cdots CH_2CH(Ph)^-, Cat^+ + CO_2 \rightarrow \cdots CH_2CH(Ph)COO^-, Cat^+ \xrightarrow{H_2O}$$

$$\cdots CH_2CH(Ph)COOH + Cat^+, OH^-$$

Similarly, a hydroxylic functional end group is formed on addition of ethylene oxide (Richards and Szwarc, 1959), *viz.*,

$$\cdots CH_2CH(Ph)^-, Cat^+ + \underset{\diagdown O \diagup}{CH_2-CH_2} \rightarrow \cdots CH_2CH(Ph)CH_2CH_2O^-, Cat^+ \xrightarrow{H_2O}$$

$$\cdots CH_2CH(Ph)CH_2CH_2OH + Cat^+, OH^-$$

A variety of other end groups can be attached to the polymer chains by letting living polymers react with judiciously chosen reagents (Rempp and Loucheux, 1958; Uraneck *et al.*, 1964; Zelinski *et al.*, 1963).

Introduction of functional end groups becomes especially valuable when it leads to bifunctional polymers. This can be accomplished if the reactions mentioned above are performed with living polymers having both ends active. Such polymers are formed by electron transfer initiation (Szwarc, 1956) but alternative methods for their production are also available.

VI. Conditions Required for Elimination of Termination or Chain Transfer

In the previous sections the characteristic features of living polymers and their utilization in synthetic polymer chemistry were discussed. It was shown how their use allows us to produce polymers of uniform molecular weight, to synthesize block polymers of preconceived architecture and to introduce functional end groups of a desired nature. Other synthetic applications have been reviewed elsewhere (Szwarc, 1968, p. 90).

We must now inquire as to what mechanisms and under what conditions living polymers may be formed. Four basic mechanisms operate in addition polymerization: radical, anionic, cationic, and coordination. In radical polymerization termination results from combination or disproportionation of two polymeric radicals which takes place during their encounters. In solution or melt the polymeric radicals are mobile and their encounters cannot be avoided, although their frequency may be greatly reduced by increasing the viscosity of the polymerizing medium. Hence, radical polymerization cannot yield living polymers under these conditions. However, if the growing polymers precipitate, the radical ends could be trapped, and hence immobilized, in a rigid matrix and then their encounters would be prevented. Consequently, their lifetime could be long enough to permit us to treat such systems as "living radicals."* However, even these systems do not offer all the advantages of the truly living systems. The initiation usually proceeds for a relatively long time; it arises from a thermal, chemical, or photochemical decomposition of a suitable radical generator such as peroxide, azo-compound, etc., which often continues throughout the course of the polymerization. The primary radicals of low molecular weight formed in such a reaction are mobile; they could therefore diffuse into the matrix and terminate the immobilized polymeric "living radicals." Furthermore, chain transfer is still feasible in the matrix, although by proper choice of the system it may be virtually eliminated.

There are other ways to avoid termination in radical polymerization. For example, emulsion polymerization initiated by a pulse of extremely short duration may be useful in the preparation of block polymers.

Ionic polymerization provides the most convenient method for the preparation of living polymers. In fact, the first living polymers were obtained by anionic polymerization, and in most studies of living polymers the anionic systems were utilized. In contrast to radical polymerization, encounters of two ionically growing polymers do not result in their termina-

*Such radicals are usually referred to as buried radicals.

tion. Their growth may be terminated by chain transfer or by isomerization into inactive species.

Cations and anions have to be present in ionically propagating systems. In anionic propagation the end groups of the growing chains are negatively charged, their charge being neutralized by the respective cations. Association of cations, such as alkali or alkali-earth ions, with the growing anions does not cause their permanent deactivation. For example, carbanions remain still active when coordinated with Li^+, Na^+, K^+, or Cs^+ although in many systems their reactivity could be substantially lower than that of free ions. Moreover, the associates, whenever formed, remain in dynamic equilibrium with the free carbanions and therefore still have the chance to grow rapidly during some periods of their life.

The high reactivity of carbanions allows them to interact with various agents and these reactions might convert them into inactive species. Such "killing" reactions were discussed in the preceding section and protonation provides the most common cause of the destruction of carbanions. Hence, living anionic polymers, especially those having carbanions on their end groups, have to be produced in an environment free of moisture, oxygen, CO_2, etc. If these stringent conditions are fulfilled, living polymers formed from many monomers remain active for hours or even days.

It might be instructive to describe here the extremely simple experiments that for the first time demonstrated conclusively the formation of living poly-

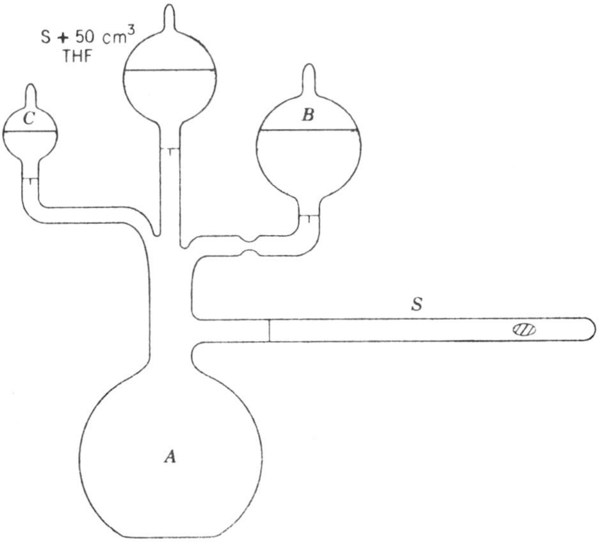

FIG. 2. The apparatus designed to demonstrate the existence of living polymers.

mers (Szwarc, 1956; Szwarc *et al.*, 1956). Polymerization of rigorously purified styrene was performed in the apparatus shown in Fig. 2. The monomer (10 ml) was kept in evacuated bulb C, while the initiator, sodium naphthalenide in 50 ml of tetrahydrofuran, was sealed in bulb B. By crushing the breakseal of bulb B the initiator and the solvent were let into the previously evacuated and flamed reactor A equipped with a magnetic stirrer. Bulb B was sealed off at the constriction while the contents of reactor A were frozen with liquid nitrogen. Thereafter, the initiating solution was brought to ambient temperature, the stirrer switched on, and the monomer introduced by crushing the breakseal on bulb C. Polymerization ensued and the green color of the solution, due to sodium naphthalenide, turned cherry red as the polystyryl carbanions were formed. The reaction was over within seconds; nevertheless, the red color persisted, implying that the resulting carbanions were not destroyed. By tilting the reactor the solution was poured into side tube S containing a piece of iron enclosed in a glass envelope. Magnetically lifting the iron to a marked place and measuring its time of fall provided a way to determine the solution's viscosity. Thereafter, the solution was returned into the reactor and 50 ml of fresh solvent and 10 ml of monomer were added under stirring by crushing the breakseal of bulb D. The added monomer polymerized again within a few seconds and by repeating the previously described procedure the viscosity of the resulting solution was remeasured. A substantially longer time of falling was noted (about 50 sec) while previously it was 5 sec only. Since the concentration of the polymer remained unchanged (10 in 50 ml compared with 20 in 100 ml) the increase in the viscosity had to be attributed to the increase of the molecular weight of the previously formed polymers, i.e., the polymers formed in the first stage of the experiment remained living and continued their growth when the additional monomer was supplied in the second stage of the experiment.

Two more examples may illustrate the variety of forms of living polymers that propagate by anionic mechanism. Salts of polyglycol,

$$\sim\!\sim\!\sim\text{OCH}_2\text{CH}_2\text{OCH}_2\text{CH}_2\text{O}^-,\ \text{Cat}^+$$

are capable of resuming their growth when epoxide is added. This system has to be protected from moisture, alcohols, etc.; however, it is inert to O_2 or CO_2. Polymerization of chloral initiated by organo–tin compounds yields

$$\sim\!\sim\!\sim\underset{\displaystyle \text{CCl}_3}{\underset{|}{\text{CHO}}}-\underset{\displaystyle \text{CCl}_3}{\underset{|}{\text{CH}}}-\text{OSn(CH}_3)_3$$

a living polymer that resumes its growth after addition of fresh chloral (Bloodworth and Davies, 1963). The last example is a borderline case;

the growth may be considered as resulting from the operation of an anionic or a coordination mechanism causing monomer insertion.

Living polymers have not yet been observed in cationic polymerization of vinyl, vinylidene, or diene monomers. Most of these reactions were initiated by Friedel–Crafts reagents which require for their action addition of a suitable co-catalyst. Polymerization of isobutene initiated by BF_3 and a trace of water illustrates the reason for the failure to obtain living polymers. The BF_3,OH_2 complex is the actual initiator; it donates a proton to the monomer, converting it to carbenium ion

$$BF_3,OH_2 + CH_2{=}C(CH_3)_2 \rightarrow CH_3{-}C^+(CH_3)_2, (BF_3OH)^-$$

Propagation leads to a macro-cation,

$$CH_3-C(CH_3)_2-CH_2C(CH_3)_2- \cdots -CH_2C^+(CH_3)_2, (BF_3OH)^-,$$

which eventually loses a proton, forming a dead polymer and regenerating the initiator

$$CH_3-C(CH_3)_2-CH_2C(CH_3)_2- \cdots -CH_2\underset{\displaystyle CH_3}{\underset{|}{C}}{=}CH_2 + BF_3(OH_2).$$

Other proton donating acids may act in a similar way. However, if the proton affinity of the anion (its basicity) is sufficiently low this type of termination may be avoided.

Other modes of termination of cationic polymerization are also possible. Carbenium ions may combine with their counteranions and form esters. An example of such a reaction is provided by cationic polymerization of styrene initiated by trifluoroacetic acid (Throssell *et al.*, 1956),

$$\cdots CH_2\overset{+}{C}H(Ph) + CF_3COO^- \rightarrow \cdots CH_2CH(Ph)O\cdot CO\cdot CF_3$$

A similar reaction, but more complex, terminates the polymerization of styrene initiated by perchloric acid. At low temperatures, $\sim -90°C$, the recombination

$$\cdots CH_2\cdot\overset{+}{C}H(Ph) + ClO_4^- \rightarrow \cdots CH_2CH(Ph)OClO_3$$

yields an inactive, covalently bonded ester and results in an irreversible termination (Pepper, 1974), while at ambient temperature the ester is capable of propagating the growth, although only slowly. The phenomenon of reversible termination will be thoroughly discussed later.

Complex anions such as $AlCl_4^-$, BF_4^-, AsF_6^-, etc., neither abstract protons from the growing carbenium ions nor recombine with them into esters. However, they might terminate the polymerization by transfering negative halogen to the positive centers. Such a termination was observed (Bawn *et al.*, 1971) in cationic polymerization of vinyl ethers initiated by trityl$^+$, $SbCl_6^-$, i.e.,

$$\cdots CH_2\underset{\displaystyle OR}{\underset{|}{C}H}\ {}^{+}, SbCl_6^- \rightarrow \cdots CH_2\underset{\displaystyle OR}{\underset{|}{C}H}Cl + SbCl_5$$

The halogen ion transfer is most facile with $AlCl_4^-$, $AlBr_4^-$, or $SbCl_6^-$; it can be virtually avoided, at least in some systems, with AsF_6^- or SbF_6^-.

Although no living polymers were observed in cationic polymerization of vinyl, vinylidene or diene monomers, numerous examples of cationically growing living polymers have been reported for hetero-ring opening polymerization. The first and the most thoroughly studied living, cationically propagating polymer was reported simultaneously by two groups of investigators (Dreyfuss and Dreyfuss, 1965; Vofsi and Tobolsky, 1965). Other examples will be discussed later.

There are reports (Bier, 1961) that termination may be avoided in some coordination polymerizations; the growing polymers remain nonterminated for many minutes and preparation of block polymers becomes feasible under these conditions.

VII. Kinetics of Propagation of Living Polymers

Since solutions of living polymers can be prepared and then fed by monomer, the propagation can be studied without the interference of initiation or termination. The concentration of living polymers remains constant throughout the whole reaction; it can be determined before, during, and after completion of the polymerization by spectrophotometric methods. Hence, the propagation is a simple first-order reaction, i.e., $-d[M]/dt = \text{const.}\,[M]$ and plots of ln $[M]$ versus time are linear.

A simple example of such a reaction was reported by Allen, Gee, and Stretch (1960; Stretch and Allen, 1961) who investigated anionic polymerization of styrene initiated in dioxane by sodium naphthalenide. The apparent unimolecular rate constant, k_u, was found to be proportional to the concentration of living polymers and therefore the calculated bimolecular rate constant, k_p, was a genuine propagation constant. Extension of this work by others (Bhattacharyya *et al.*, 1965b; Dainton *et al.*, 1965, 1969) has demonstrated that k_p depends on the nature of cations which implies that ion pairs are the propagating species in this system.

The propagation is more complex in tetrahydrofuran (Bhattacharyya *et al.*, 1964a, 1965a). In this solvent the apparent first-order constant, k_u, is not proportional to the concentration of living ends. In fact, the calculated

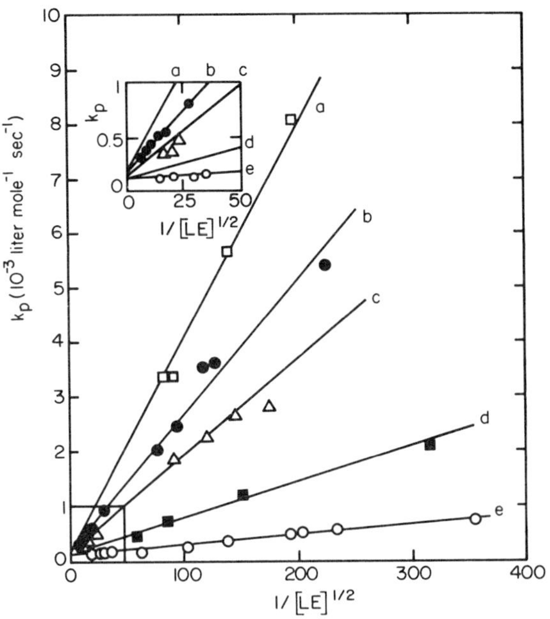

FIG. 3. Plot of k_p (apparent bimolecular rate constants of propagation of living polystyrene salts in THF) versus the reciprocal of the square root of the living-ends (LE) concentration: (a) Li^+, (b) Na^+, (c) K^+, (d) Rb^+, and (e) Cs^+.

apparent bimolecular constant, $k_p = k_u/[\text{living polymers}]$, increases with dilution of living polymers, LP, and in the concentration range, $[LP] = 10^{-3}$–10^{-5} *M*, it is linear with the reciprocal of $[LP]^{1/2}$. This is shown in Fig. 3.

It was realized that two species propagate anionic polymerization of styrene in tetrahydrofuran; the free ~~~$CH_2\bar{C}H(Ph)$ carbanions and their ion pairs ~~~$CH_2\bar{C}H(Ph)$, Cat^+. Denoting by f the fraction of living polymers present in the form of free ions, by $1 - f$ the fraction of ion pairs, and by k_- and $k_\pm$ their respective propagation constants, one finds,

$$k_p = fk_- + (1 - f)k_\pm$$

Since the following equilibrium is established in tetrahydrofuran,

$$\sim\sim\sim CH_2\bar{C}H(Ph), Cat^+ \rightleftarrows \sim\sim\sim CH_2\bar{C}H(Ph) + Cat^+ \qquad (K_D)$$

the mass law leads to the equation

$$\frac{f^2[LP]}{(1 - f)} = K_D$$

which simplifies to $f = (K_D/[LP])^{1/2}$ whenever $f \ll 1$. Thus, at low degree of dissociation of ion pairs

$$k_p = k_{\pm} + \frac{(k_{-} - k_{\pm})K_D^{1/2}}{[\mathrm{LP}]^{1/2}}$$

in agreement with the observation. The slope of the line k_p versus $1/[\mathrm{LP}]^{1/2}$ gives therefore $(k_{-} - k_{\pm})K_D^{1/2}$ and the intercept is equal to $k_{\pm}$.

Within the scale of these measurements $k_{\pm}$ is small and therefore with difficulty distinguished from 0. This drawback can be remedied by performing the experiments in the presence of an easily dissociated salt sharing with the living polymer a common counterion. The addition of salt represses the dissociation of the living polymer, retards its propagation, and thus makes the value of the intercept comparable to the measured k_p. Such a retardation effect is shown in Fig. 4. By studying the conductance of the salt its dissociation constant could be determined and thereafter one may calculate the concentration of the counterions in the investigated mixture of living polymers and the salt, provided that the former contributes negligibly to the dissociation. Having determined $[\mathrm{Cat}^+]$, we can plot the observed k_p versus $1/[\mathrm{Cat}^+]$ as shown in Fig. 4. Such plots are linear because $f = K_D/[\mathrm{Cat}^+]$ and hence $k_p = k_{\pm} + (k - k_{\pm})K_D/[\mathrm{Cat}^+]$. Their slopes are given by $(k_{-} - k_{\pm})K_D$ and the intercepts provide experimentally reliable values of $k_{\pm}$.

The results of both experiments permit us to determine K_D, k_{-}, and $k_{\pm}$, namely, $K_D^{1/2}$ is given by the ratio of the slopes of the lines shown in Figs. 3 and 4, respectively. Alternatively, the dissociation constant K_D may be deter-

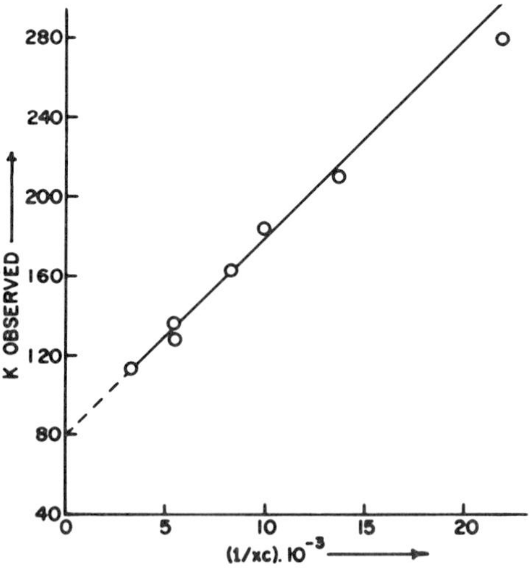

FIG. 4. Plot of the apparent bimolecular propagation constant of sodium polystyrene in THF versus the reciprocal concentration of the added Na^+ ions.

mined directly by studying the conductance of living polymers. However, it is necessary to find the value of Λ_0 (the limiting conductance) in order to complete the calculation. The method providing this information has been described elsewhere (Bhattacharyya *et al.*, 1964a), and it is gratifying to note the agreement between the values of K_D obtained by both methods.

The observations discussed above fully confirm the existence of two distinct species, namely, free polystyryl carbanions and their associates with the appropriate counterions, i.e., ion pairs. Both participate in the propagation, and in THF the free polystyryl anions are substantially more reactive than the ion pairs, i.e., $k_- \gg k_\pm$. The former propagation constant is independent of cation's nature and only slightly affected by the nature of ethereal solvent in which the polymerization takes place. Thus, k_- does not vary from 60,000 $M^{-1}\text{sec}^{-1}$ by more than a factor of two when the propagation proceeds in tetrahydrofuran, tetrahydropyran, dimethoxyethane, 2-methyltetrahydrofuran, mixtures of tetrahydrofuran with dioxane, etc.

On the other hand, the values of $k_\pm$ depend on the nature of the cation and are strongly affected by the nature of the solvent. This is clearly shown by the data collected in Table I. For example, $k_\pm$ of sodium polystyryl increases from as little as 3–6 $M^{-1}\text{sec}^{-1}$ when the reaction proceeds in dioxane to as

TABLE I

PROPAGATION BY LIVING POLYSTYRENE SALTS IN VARIOUS SOLVENTS AT 25°C

	$k_\pm$(liter mole^{-1} sec^{-1})				
Solvent	Li^+	Na^+	K^+	Rb^+	Cs^+
		Living Polystyrene Salts			
Dioxane	0.9	3.4	20	21.5	24.6
		6.5	28	34	15
Benzene (nonassociated)			47	24	18
Tetrahydropyrane	<10[a]	14,12[a]	73	83	53
		10			
2-Methyltetrahydrofuran	57	11	7.5		22
Tetrahydrofuran	~160	80	~100(?)	~50	25
Dimethoxyethane		3600			
					~150
		Living Poly-α-Methylstyrene Salts			
Dioxane		0.02	0.1	0.06	
Tetrahydropyran	2.6	0.047	0.25	0.35	0.26
Tetrahydrofuran		0.04			

[a] Aging the solution inhibits the reaction, making $k_\pm$ smaller.

much as $\sim 4000\ M^{-1}sec^{-1}$ for the polymerization in dimethoxyethane. These variations of $k_{\pm}$ with the solvent and the changes in the gradation of $k_{\pm}$ with the cation's nature as observed in different solvents suggest that the ion pairs may have different structures in different solvents.

VIII. Different Kinds of Ion Pairs

Winstein *et al.* (1954) and Sadek and Fuoss (1954) suggested, that two kinds of ion pairs may coexist in solutions: the tight contact pairs and the loose solvent-separated pairs. Since then voluminous theoretical and experimental material has been amassed which shows that ion pairs may possess a variety of forms (see, e.g., Szwarc, 1972, 1974). It seems that the association of polystyrylcarbanions with alkali cations in dioxane results in the formation of tight ion pairs, but both the tight and loose pairs coexist in equilibrium when the association takes place in tetrahydrofuran or dimethoxyethane.

The existence of two kinds of ion pairs in the above solvents was suggested by the investigation of the temperature dependence of $k_{\pm}$ (Fig. 5). In tetrahydrofuran its value *increases* as the temperature decreases from 25°C down to −70°C (Shimomura *et al.*, 1967) whereas in dimethoxyethane $k_{\pm}$ increases in the same temperature range to a maximum value and then decreases. This behavior of polymerization was accounted for by assuming that loose and

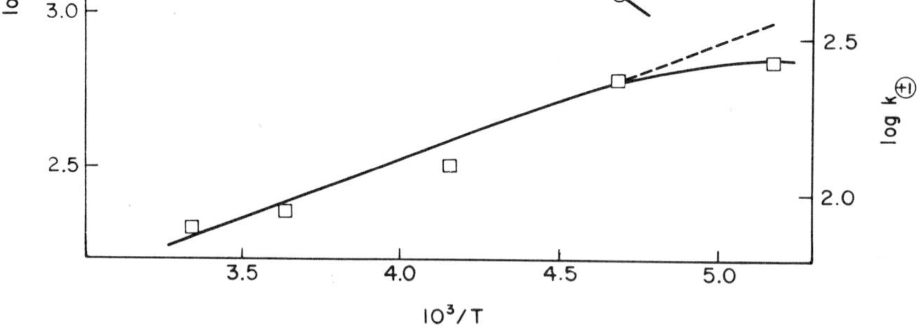

FIG. 5. Temperature dependence of $k_{\pm}$ of sodium polystyrene propagation in DME and in THF.

tight ion pairs propagate at different rates, namely k_l and k_t, respectively, the former being much greater than the latter. If the activation energy E_l is smaller than ΔH, the exothermicity of the conversion, tight pairs $\rightleftharpoons$ loose pairs, and if the proportion of the loose pairs is very small, then the lowering of the temperature speeds up the reaction because the increase in the fraction of the more reactive loose pairs more than counterbalances the decrease of k_l resulting from the genuine activation energy. Utilizing all of the available experimental data it was concluded that $k_l \approx$ 20,000–30,000 M^{-1} sec^{-1} at 25°C, while k_t is probably smaller than 10 M^{-1}sec^{-1}. The observed $k_{\pm}$ is therefore given by $\gamma k_l + (1 - \gamma)k_t$, where γ is the fraction of the reactive loose ion pairs present in the polymerizing system. These ideas have been further explored by various investigators (see, e.g., Schulz *et al.*, 1972).

The nature of ion pairs may be modified by adding powerful complexing agents to the investigated solutions. For example, sodium polystyryl exists in tetrahydropyran in the form of slowly propagating tight pairs. Addition of tetraglyme enhances the rate of polymerization (Shinohara *et al.*, 1968, 1969), as shown in Fig. 6. In fact, two effects contribute to this result: (1) the

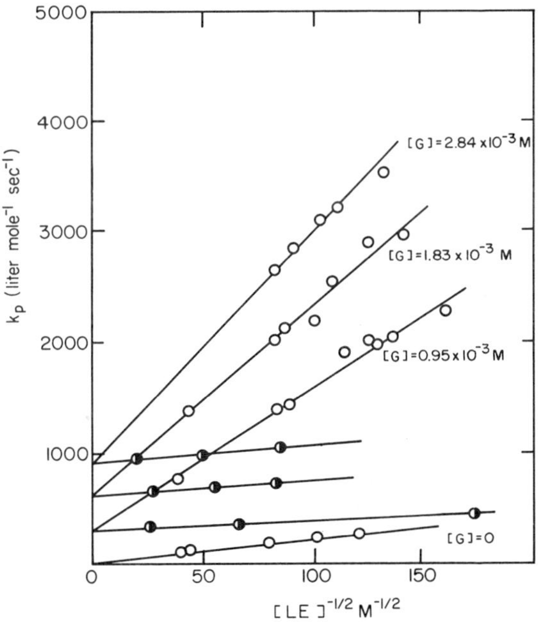

FIG. 6. Effect of added tetraglyme (G) on the apparent bimolecular rate of propagation of sodium polystyrene in THF. Note the increase in the intercepts, indicating higher reactivity of "glymated" sodium polystyrene pairs, and the increase in the slopes arising from their higher degree of dissociation.

unreactive tight pairs are converted into much more reactive tetraglyme-solvated pairs, and (2) the proportion of the highly reactive free polystyryl anions increases because the ionic dissociation of the tetraglyme-solvated pairs is greater than of the tight pairs. The first effect is reflected by the increase of the intercept of the plot of k_p versus $1/[\mathrm{LP}]^{1/2}$; indeed, its value increases 100-fold as the concentration of tetraglyme slightly exceeds 10^{-3} *M*. The second effect is shown by the increase in the steepness of the respective lines. Analysis of the data shows that the propagation constant of tight pairs in THP is about 10 $M^{-1}\mathrm{sec}^{-1}$, whereas that of the tetraglyme-solvated pairs is about 4000 $M^{-1}\mathrm{sec}^{-1}$ at about 25°C.

Similar phenomena have been reported for other systems. For example, small amounts of hexamethylphosphorictriamide enhance the polymerization of potassium polystyryl in tetrahydrofuran (Levresse *et al.*, 1971), and the addition of triglyme accelerates anionic polymerization of α-methylstyrene (Ginn and Ivin, 1970).

IX. How Structure of Ionic Species Affects Reactivity

Early studies of anionic polymerization created the impression that free carbanions have to be more reactive than ion pairs. This conclusion has been enhanced by studies of cationic polymerization of vinyl monomers. The propagation constant of some free carbenium ions was found to exceed 10^6 M^{-1} sec^{-1}—a value much higher than that expected for ion pairs. However, it was pointed out by the present writer that this conclusion needs not be universally valid (Szwarc, 1968). If the polarization of the oncoming monomer is the main driving force of the reaction, then the free ion should be more reactive than an ion pair because the field of an ion is more powerful than that generated by a dipole–ion pair. On the other hand, it is conceivable that the push–pull mechanism arising from the simultaneous operation of anion and cation may provide a more facile route for the reaction, and in such a case the reactivity of ion pairs may be higher than that of free ions. Indeed, it seems that ion pairs propagate anionic polymerization of some polar monomers more rapidly than the respective free ions.

An interesting example of such a situation was reported by Hemery *et al.* (1975), who studied anionic polymerization of propylene sulfide with sodium ions complexed with [222] cryptate in THF at −30°C. In this system $k_{\pm} = 11.3$ $M^{-1}\mathrm{sec}^{-1}$ while $k_{-} = 3.8$ $M^{-1}\mathrm{sec}^{-1}$, i.e., on dilution the apparent bimolecular constant of propagation, k_p, *decreases*, in contrast with most of the systems showing an *increase* of k_p with dilution of growing polymers. However, the noncomplexed ion pairs, $\cdots \mathrm{CH_2 \cdot CH(CH_3)S^-, Na^+}$ are less reactive than the free thiolate anions.

These results indicate that complexation increases the reactivity of ion pairs. This need not always be true. Although an appropriate example concerned with polymerization is not yet available, such a situation was observed in protonation of anthracenide salts by ethanol (Minnich *et al.*, 1973). In that reaction, protonation of potassium salt complexed with a crown is, by powers of ten, slower than of the noncomplexed salt.

The penultimate unit of the growing polymer, or even a segment placed further away, may act as the complexing agent. Such a situation is encountered in kinetic studies of anionic polymerization of 2-vinyl pyridine (Tardi *et al.*, 1967; Fisher and Szwarc, 1970). In spite of the very low dissociation constant of the respective ion pairs, which could imply that these living polymers form tight, contact pairs, their reactivities are very high, and, unexpectedly, higher in a "poor" solvent like dioxane than in a "good" solvent like tetrahydrofuran. These observations can be rationalized by assuming intramolecular complexation of the cation with the segments of the polymer, e.g.,

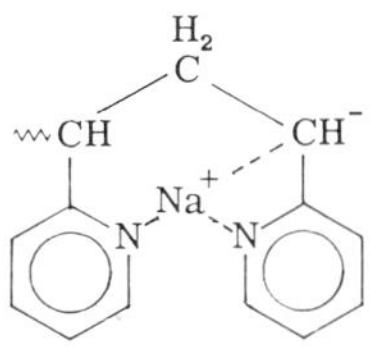

The complexation increases the reactivity of the pair, but on its "dissociation" the cation remains bound to the segments of the chain and therefore does not contribute to conductance of the solution.

X. Higher Aggregation of Living Polymers

A strange phenomenon was observed in the course of kinetic studies of polymerization of living polystyrene associated with Cs^+ cations (Bhattacharyya *et al.*, 1964b). The cesium salt of living polystyrene endowed with one growing end showed conventional behavior. Its dissociation constant could be determined either from conductance studies or by comparing the effect of dilution and the effect of added Cs^+, BPh_4^- on rates of its propagation (see p. 101). Both approaches yield the same results. However, the salt formed from living polymers endowed with two active ends seems to be less reactive than expected, its apparent dissociation constant determined by the previously mentioned kinetic technique was too low when compared with that of the salt formed from polymers possessing one active end only. On the other hand, direct conductance studies led to an apparent dissociation constant which was too high.

These apparently contradictory findings were rationalized by invoking the formation of triple ions. These species were identified by Fuoss and Kraus (1933) and their formation results from the association of ions with ion pairs,

$$A^- + A^-, B^+ \rightleftarrows A^-, B^+, A^-$$

or

$$B^+ + A^-, B^+ \rightleftarrows B^+, A^-, B^+$$

In the cesium salt of living polystyrene endowed with two growing ends the ionic dissociation produces,

$$Cs^+, \bar{C}H(Ph) \cdot CH_2 \cdots CH_2\bar{C}H(Ph) + Cs^+$$

The polystyryl carbanion remains in the vicinity of an ion pair. Consequently, an *intra*molecular association takes place yielding

$$\widehat{CH_2\bar{C}H(Ph)Cs^+, \bar{C}H(Ph)CH_2}$$

a cyclic triple ion. Thus, the free polystyryl carbanions produced by the original dissociation are removed from the system. Since the equilibrium

$$Cs^+, \bar{C}H(Ph)CH_2 \cdots CH_2\bar{C}H(Ph), Cs^+ \rightleftarrows$$

$$Cs^+, \bar{C}H(Ph)CH_2 \cdots CH_2\bar{C}H(Ph) + Cs^+$$

has to be maintained, more polystyryl carbanions and Cs^+ cations are formed. The former are partially converted into the cyclic triple ions, while the latter remain in the solution. Thus, $[Cs^+]$ becomes larger than $[\ldots\bar{C}H(Ph)]$, and the concentration of the reactive free polystyryl carbanions becomes smaller than would be expected had the cyclization been prevented. This decrease in $[\ldots\bar{C}H(Ph)]$ is reflected in the low rate of propagation and conventionally would be interpreted as an indication of a low degree of dissociation. On the other hand, the total concentration of the conducting ions, namely Cs^+, free polystyryl carbanions, and the negative triple ions is higher than would be expected in the absence of the cyclization. Hence, the results of the conductance studies create the impression of a too high degree of dissociation.

Mathematical treatment of this problem led to determination of the propagation constant of the cyclic triple ions and the equilibrium constant of the cyclization. The latter depends on the length of the chain separating the active end groups: the longer the chain, the lower the degree of cyclization. This conclusion was verified by experiment.

Formation of triple ions may be visualized by an alternative process to the ion–ion pair association, namely by a transfer of one ion of an ion pair to another ion pair, i.e.,

$$2A^-, B^+ \rightleftarrows A^- + B^+A^-B^+$$

The position of this equilibrium is concentration independent because two particles are formed by two particles reacting. Since the degree of dissociation

$$A^-, B^+ \rightleftarrows A^- + B^+$$

decreases with increasing concentrations of ion pairs, the alternative route of the triple ions formation become dominant at higher concentrations of salts.

Polymerization of barium salt of living polystyrene illustrates the effects arising from the preceding reaction. Barium salt of living polystyrene endowed with one active end was prepared and the kinetics of its propagation were studied (De Groof *et al.*, 1975). In every run the reaction, measured by the rate of styrene consumption, obeyed the first order law, i.e.,

$$-d[CH_2{:}CHPh]/dt = k_u[CH_2{:}CHPh]$$

However, k_u was found to be a constant, independent of the concentration of living polymers, in contrast to the behavior of alkali salts, where k_u increased with increasing [LP]. The dissociation

$$Ba^{2+}(\bar{C}HPh\cdot CH_2\cdots X)_2 \rightleftarrows Ba^{2+}(\bar{C}HPh\cdot CH_2\cdots X) + \bar{C}HPh\cdot CH_2\cdots X \qquad (K_{diss})$$

is extremely low, a not unexpected finding because the cation is doubly charged. However, the equilibrium

$$2Ba^{2+}(\bar{C}HPh\cdot CH_2\cdots X)_2 \rightleftarrows Ba^{2+}(\bar{C}HPh\cdot CH_2\cdots X) + Ba^{2+}(\bar{C}HPh\cdot CH_2\cdots X)_3 \qquad (K_{triple})$$

forms a measurable concentration of ions yielding a positively charged $Ba^{2+}(\bar{C}HPh\cdot CH_2\cdots X)$ and a negatively charged $Ba^{2+}(\bar{C}HPh\cdot CH_2\cdots X)_3$. Their concentration is given by $K_{triple}[Ba^{2+}(\bar{C}HPhCH_2\ldots X)_2]$.

The free polystyrl anions turned out to be the only species that contributes to the observed propagation; the growth of $Ba^{2+}(\bar{C}HPh\cdot CH_2\cdots X)$, $Ba^{2+}(\bar{C}HPh\cdot CH_2\cdots X)_2$, and $Ba^{2+}(\bar{C}HPh\cdot CH_2\cdots X)_3$ is vanishingly slow. The ionic dissociation, very low to begin with, becomes further suppressed by the formation of the positive $Ba^{2+}(\bar{C}HPh\cdot CH_2\cdots X)$, and hence the concentration of the free polystyryl carbanions is given by $K_{diss}/K_{triple}^{1/2}$, i.e., it is independent of the concentration of living polymers. Thus, the observed k_u is equal to $k_-\ K_{diss}/K_{triple}^{1/2}$. Conductance studies confirmed the proposed mechanism.

It would be interesting to study the kinetics of propagation of a barium salt of living polystyrene endowed with two active ends. Such a salt is expected to be in a cyclic, virtually inactive form most of the time

$$\mathrm{CH_2\cdot\bar{C}H(Ph)Ba^{2+},\bar{C}H(Ph)\cdot CH_2}$$

The cyclic form should coexist in equilibrium with the open, highly reactive form, $\mathrm{\bar{C}H(Ph)\cdot CH_2\cdots CH_2\cdot\bar{C}H(Ph)}$, Ba^{2+}, responsible for the polymerization, if any. The equilibrium between the cyclic and open forms should shift in favor of the latter as the length of the chain increases, and therefore any propagation, if observed, should be autoaccelerated. Moreover, those chains that started their growth earlier will grow faster than those that remained relatively short, and hence the molecular weight distribution of the product should change with time.

An attempt to produce this type of polymer was reported by François (Mathis *et al.*, 1975); however, in spite of his claim, the resulting polymer apparently is composed of polymers endowed with one growing end group only.

A higher degree of association is encountered in living polymers involving Li^+ counterions when dissolved in hydrocarbon solvents. For example, viscometric and light-scattering studies, as well as studies involving colligative properties of solution, have demonstrated that the lithium salt of living polystyrene is dimeric in benzene solution (see, e.g., the review by Bywater, 1965). However, a minute fraction of the nonassociated polymers, coexisting in equilibrium with the associates, is responsible for the observed polymerization. Hence, the rate of the reaction is proportional to the square root of living polymers concentration. This relation was reported first by Worsfold and Bywater (1960) and later confirmed by other investigators. The addition of ethers, e.g., tetrahydrofuran, dissociates the aggregates, and therefore the rate of polymerization dramatically increases when THF is added to a benzene solution of lithium polystyrene (Bywater and Worsfold, 1962a). Interestingly, the highest rate is observed at low concentration of THF; its increase reduces the rate, suggesting that some lower etherates are more reactive than living polymers solvated by a greater number of THF molecules.

Lithium salts of living polyisoprene in hydrocarbon medium show the remarkable ability to polymerize this monomer in 1–4, cis fashion. This propensity of lithium polyisoprene is lost on addition of ethers; the nature of the ion pair determines the stereochemistry of polymerization, which is drastically changed on modification of the ion pair structure. We may remark that the ion pairs involving other alkali metals as counterions do not induce the 1–4, cis polymerization of isoprene even in hydrocarbon solution.

The effect of ion pair structure on the stereochemistry of polymerization has also been observed in other systems (Yohsino *et al.*, 1964, 1965; Schuerch *et al.*, 1964, 1967).

XI. Ion Pairs and Covalent Bonds

Does an ion pair differ from a polar covalent bond? Of course it is easy to conceive the difference between a loose, solvent-separated pair and a polar, covalently bonded species which might be treated as if it were a tight, contact ion pair. For example, a gaseous molecule of NaCl may be described as a tight, contact ion pair or as two atoms bonded by a polar covalent bond. Certainly its geometry, reflected in the Na–Cl distance, and the electronic structure governing the average distribution of the negative charge of the surrounding electrons, is uniquely determined for its ground state, and there is no way to visualize two distinct species: NaCl and Na^+, Cl^-. Under the influence of a solvent the gaseous species may be transformed into a solvent-separated pair and eventually dissociate into free ions.

However, the situation is more complex when we deal with a polyatomic species. Consider, for example, tritylchloride. The covalent form has a tetrahedral configuration around the central carbon atom and the C–Cl bond may be described as resulting from utilization of the sp^3 hybridized orbital of the carbon. On the other hand, the Ph_3C^+, Cl^- *tight*, contact ion pair should have a trigonal configuration around the carbon and a different electronic charge distribution than that established in Ph_3C-Cl. In this example the distinction is clear and two forms could be differentiated by their optical spectra, although in both cases, NaCl and tritychloride, it would be possible to discuss a ground-state molecule referred to as a covalently bonded species and its excited state described as a tight ion pair. In this respect the problem encountered here is similar to that leading to a concept of a ground state and excited state of a charge-transfer complex.

In ionic and particularly in cationic polymerization, we find examples of even more distinct species: one treated as a covalently bonded molecule while the other is referred to as an ion pair.

Let us consider cationic polymerization of cyclic monomers like tetrahydrofuran yielding, under proper conditions, a living polymer. The growing species has the structure of an oxonium ion pair (Matyjaszewski and Penczek, 1974a).

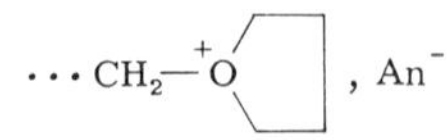

which grows by a nucleophilic attack of the monomer on the carbon atom adjacent to the O^+ center. Thus,

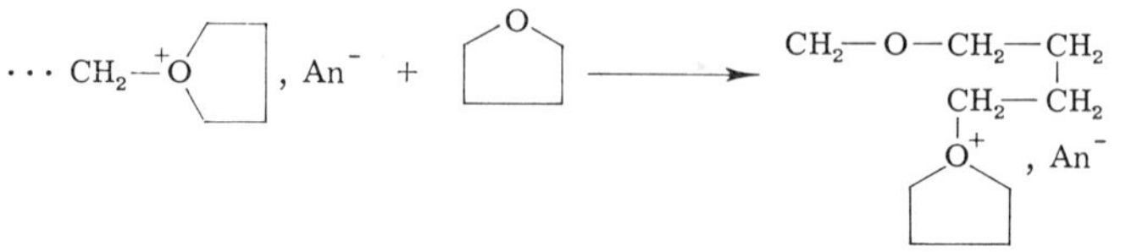

the O–CH_2 bond being ruptured and a new oxonium ion being formed. However, some anions may react with the growing polymer in a fashion similar to the monomer addition and form a macroester, e.g., an oxonium ion associated with trifluoromethane sulfonic acid residue yields the ester, *viz.*,

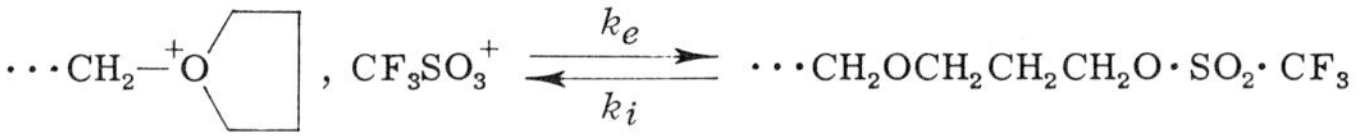

This equilibrium is greatly affected by the nature of solvent, for example, in CCl_4 solution the macroester is the dominant species whereas the ion pairs are the most abundant in nitromethane (Matyjaszewski and Penczek, 1974a). The two forms can be distinguished by their different nmr spectra (Kobayashi *et al.*, 1974; Matyjaszewski and Penczek, 1974b; Win and Pruckmayr, 1975; Pruckmayr and Win, 1976).

In methylene chloride the concentrations of both forms are comparable. Their contributions to the propagation have been determined, namely $\sim 10^{-4}\ M^{-1}\text{sec}^{-1}$ for the macroester and $4 \times 10^{-2}\ M^{-1}\text{sec}^{-1}$ for the macro-ion pairs at 25°C when $[\text{THF}] \sim 8\ M$.

Two distinct mechanisms may lead to the conversion of a macroester into a macro-ion pair; the unimolecular process described above, and a bimolecular reaction resembling the propagation step of the ester-terminated polymers or the depropagation step of the macro-ion pairs. Thus,

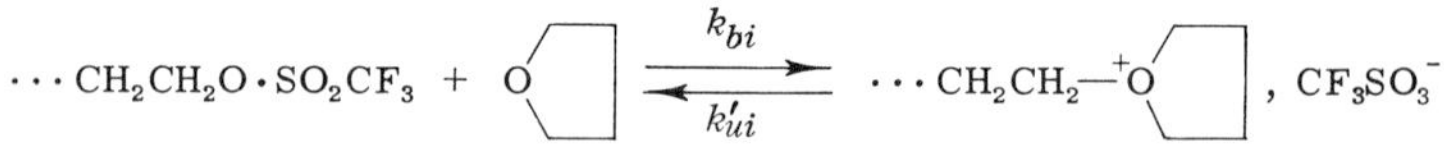

The rate constants k_{ui} and k_{bi} were determined for the reaction performed in methylene chloride solution, namely, the former was found to be 2×10^{-2} sec^{-1}, whereas the latter is about $10^{-4}\ M^{-1}\text{sec}^{-1}$, i.e., comparable to the propagation rate constant of the macroester. At a typical concentration of THF ($\sim 8\ M$), the conversion of the ester into an ion pair under the above conditions takes place vitrually exclusively by intermolecular mechanism.

The above discussion seems to imply that the equilibrium between covalent and ionic form is affected by the presence of monomer. However, the principle of microscopic reversibility is not violated. The system as a whole is in equilibrium only when the concentration of the monomer is equal to its equilibrium concentration with respect to living polymers, $[\mathrm{THF}]_e$. We have to conclude, therefore, that

$$\frac{k_p}{k_d} = \frac{k'_p}{k'_d} = 1/[\mathrm{THF}]_e$$

where k_p and k'_p denote the propagation constants of the macroester and macro-ions, respectively, whereas k_d and k'_d are the corresponding rate constants of the depropagation. In addition, the ratio of the bimolecular rate constant of conversion of macroester to macro-ion pair, to the unimolecular rate constant of the process yielding the macroester and the monomer (i.e., k_{bi}/k'_u) has to be equal to $K/[\mathrm{THF}]_e$, K being the equilibrium constant of the unimolecular conversion of the ester into the ion pair (i.e., $K = k_i/k_e$). When $[\mathrm{THF}] > [\mathrm{THF}]_e$, the excess of THF leads to the conversion of the ester into the ion pair, making the concentration of the latter higher than its equilibrium concentration.

XII. Living and Dormant Polymers

Coexistence of several species of vastly different reactivities introduces interesting complications in the kinetics of propagation. In an extreme case living polymers may exist in two forms, those propagating, i.e., living and the other inert and hence dormant. The conversion of one form into another allows all the polymers to grow provided that this transformation is not too slow. The first example that led to the concept of dormant polymers was discussed by the present author in 1962 when the system living polystyrene–anthracene was investigated (Khanna *et al.*, 1962). Addition of anthracene, A, to living polystyrene led to the formation of an adduct,

$$\cdots \mathrm{CH_2\overline{C}H(Ph), Cat^+} + \mathrm{A} \rightleftarrows \cdots \mathrm{CH_2CH(Ph)\cdot A^-, Cat^+}$$

in which the polystyryl carbanion is linked by a covalent bond with the ninth carbon atom of anthracene, while the negative charge resides on the tenth carbon atom. This addition is reversible as was demonstrated by the exchange studies involving labeled radioactive anthracene (Asami *et al.*, 1962). The addition of styrene to the living polystyrene-anthracene complex results in a slow polymerization in which *all* the polymers contribute to the growth because the molecular weight of all of them increases. However, the rate of propagation was found to be inversely proportional to the concentration of the excess of anthracene present in the system (Khanna *et al.*, 1962). All these observations demonstrate that the complexed polymers, $\cdots CH_2CH(Ph)A^-, Cat^+$, do not grow; they are dormant, and the propagation is carried on by the nonassociated living species which, however, exchange with the associated ones.

These ideas can be generalized. For example, in a system containing ion pairs of vanishingly low reactivity and highly reactive free ions, the latter may be treated as the truly living species while the former are the dormant polymers. For a very high rate of association–dissociation and a relatively low rate of propagation, the molecular weight distribution of the resulting polymers should be virtually of Poisson type. In the reverse case, only the living polymers would grow, forming polymers of very high molecular weight, thus leaving the dormant polymers as low molecular weight material, i.e., the resulting product should have a bimodal distribution of molecular weights.

The exchange taking place between the living and dormant polymers leads, in every case, to some broadening of molecular weight distribution from that predicted by the Poisson law. This distortion of the distribution was investigated theoretically (Szwarc and Hermans, 1964; Figini, 1967) and utilized by Schulz and his co-workers (Löhr and Schulz, 1964; Figini *et al.*, 1965) in their attempt to determine the rate of association and dissociation in the ion–ion-pair system. Their results, although interesting, might be questioned because the rate of ion association derived from their data is lower by two or three powers of ten than expected on the basis of the Debye equation. Moreover, an alternative mechanism of exchange, not taken into account in their treatment, is also possible (Szwarc, 1967), namely,

$$\sim\!\sim\!M^-, Cat^+ + \sim\!\sim\!M^- \rightleftarrows \sim\!\sim\!M^- + \sim\!\sim\!M^-, Cat^+$$

In the same vein the nonassociated lithium ion pairs of living polystyrene or polydienes may be treated as truly living, whereas their dimeric aggregates would represent the dormant species (Szwarc, 1967). Again the dissociation–association or metathetic exchange should broaden the molecular weight distribution of the resulting polymers.

XIII. Degradation of Living Polymers Formed by Ring-Opening Polymerization

In ring-opening polymerizations, the release of a strain of monomers contributes to the driving force of the polymerization. This is reflected in the thermodynamics of the reaction, i.e., the equilibrium concentration of a strained cyclic monomer in equilibrium with its living polymer is lower than that expected for a similar but strainless monomer. For example, thermodynamic data show that cyclopropane can be polymerized nearly quantitatively and form polymethylene

$$X + n\,C_3H_6 \rightleftarrows X(CH_2)_{3n}$$

whereas cyclohexane could form only a low molecular weight oligomer. Unfortunately, this undoubtedly correct statement cannot be verified experimentally because a suitable catalyst X needed for this reaction is unknown. However, there are examples of such reactions involving more complex monomers. For example, the trimeric cyclic siloxane $(SiMe_2O)_3$ readily polymerizes by anionic mechanism, whereas polymerization of larger rings is slow and thermodynamically unfavored. Hence, the trimer can be readily converted into a living polymer, which on standing degrades into a cyclic tetramer, i.e.,

$$4n\,(SiMe_2O)_3 \rightarrow -(SiMe_2O)_{12n}- \rightarrow 3n\,(SiMe_2O)_4$$

This, in fact, is the simplest mechanism by which four moles of the cyclic trimer could be converted into three moles of the cyclic tetramer.

The relatively rapid conversion of strained monomers into living polymers that slowly degrade into larger rings has recently been observed in a variety of ring-opening polymerizations proceeding by a cationic mechanism (Goethals, 1976). These involve monomers derived from rings such as

Some ring compounds, e.g., crown ethers, act as powerful complexing agents for some cations. The presence of such cations shifts the equilibrium between the cyclic monomers and living polymers in favor of this ring which is the most powerful complexing agent. Hence, the following situation may be envisioned:

$$\text{Strained ring monomer} \rightarrow \text{living polymer} \rightleftarrows \text{variety of ring compounds}$$

Let us denote the latter by C_4, C_5, etc., and assume that C_5 is the most powerful complexing agent for a cation Cat^+. Then the reaction

$$C_5 + Cat^+ \rightarrow Cat^+, C_5 \text{ complex}$$

removes C_5 from the system and the equilibrium proceeding via living polymers converts the other ring compounds into C_5.

XIV. Copolymerization in Living Polymer Systems

Copolymerization of two monomers involves four reactions, *viz.*,

$$\cdots A + A \rightarrow \cdots AA \qquad (k_{11})$$
$$\cdots A + B \rightarrow \cdots AB \qquad (k_{12})$$
$$\cdots B + A \rightarrow \cdots BA \qquad (k_{21})$$
$$\cdots B + B \rightarrow \cdots BB \qquad (k_{22})$$

Classic studies of radical copolymerization in stationary state provide the ratios $r_1 = k_{11}/k_{12}$ and $r_2 = k_{22}/k_{21}$. Some nonstationary methods are needed to determine the absolute values of k_{11} and k_{22} and then having r_1 and r_2, one calculates k_{12} and k_{21}, the cross-propagation constants.

In previous sections of this review it was shown how the homopropagation constants are determined in living polymer systems. The method takes advantage of the feasibility of direct determination of the concentration of growing polymers, which under proper conditions remains constant during the whole polymerization process. Addition of monomer B to living polymers $\cdots A$ permits a direct determination of the cross-propagation constant. The simplest approach is based on the change in the absorption spectrum of the active end groups $\cdots A$ when they are converted into active groups $\cdots A \cdot B$ (or $\cdots A \cdot B \cdots B$) (Morton and Ells, 1962). Thus the rate of the reaction

$$\cdots A + B \rightarrow \cdots AB$$

is measured by monitoring the absorbence of $\cdots A$ which decays or of $\cdots B$ which increases since the monitoring is not affected by the subsequent homopropagation of B. Provided that the concentration of $\cdots A$ and of the monomer B is known, the rate $-d[\cdots A]/dt$ permits one to calculate directly the value of k_{12}.

Alternatively, one may measure the rate of disappearance of B, and having the initial concentration of $\cdots A$, the k_{12} could be calculated by extrapolation of the results to zero time (Bhattacharyya *et al.*, 1963). This method allows also the investigation of the effect of penultimate unit on the value of k_{12} (Lee *et al.*, 1963).

Let us consider one example illustrating the complexity of some copolymerization in the living polymer system. Copolymerization of styrene and butadiene proceeding in benzene by anionic mechanism with Li^+ counterion was first investigated by Rakova and Korotkov (1958) who noted that the reaction starts slowly and consumes butadiene. After exhaustion of this monomer the reaction speeds up and styrene becomes polymerized. The correct explanation of this phenomenon was provided by O'Driscoll and Kuntz (1962) who pointed out that the addition of styrene to living polybutadiene is slower than the homopropagation of the latter, while the addition of butadiene to living polystyrene is extremely fast, much faster than the homopropagation of living polystyrenes which is faster than homopropagation of living polybutadiene. Thus, the growing ends are mainly terminated by slowly propagated butadiene and are converted into styrene terminated ends after butadiene was consumed. The homopropagation of living polystyrene, being relatively fast, results in acceleration of the reaction. The latter is prevented in the presence of butadiene by conversion of $\cdot\cdot\cdot CH_2\overline{C}HPh, Li^+$ into $\cdot\cdot\cdot CH_2CH{:}CH \cdot CH_2^-, Li^+$.

Let us close this review by discussing an interesting feature of the copolymerization of living polyvinylnaphthalene with styrene. The terminal carbanion of living polyvinylnaphthalene, $\cdots CH_2\overline{C}H(C_{10}H_7)$ absorbs at λ_{max} of 558 nm. On addition of about two mole equivalents of styrene this peak disappears and a new peak appears at λ_{max} of 440 nm instead of the one expected at $\lambda_{max} = 340$ nm, characteristic of the $-CH_2-CH(Ph)$ anion. On standing overnight, the absorbence of the solution changes, the 440-nm peak vanishes, and the 558-nm peak reappears.

Careful investigation of this system shows that the addition of the first styrene unit leads to a complexed anion, namely

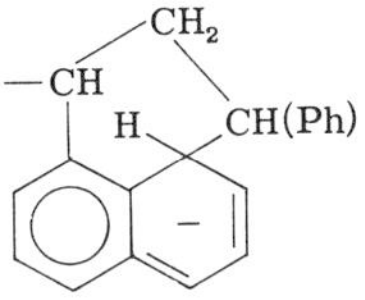

The benzylic $-\overline{C}H(Ph)$ carbanion finds itself in the vicinity of a naphthyl moiety and strongly interacts with it. This leads to two results: (1) the 340-nm absorbence of $-CH_2{}^-\overline{C}H(Ph)$ is replaced by the 440-nm absorbence of the complex, and (2) the reactivity of the $-CH_2\overline{C}H(Ph)$ group is greatly reduced. Thus, the rate constant of addition of the next styrene unit is reduced from ~ 1000 $M^{-1}sec^{-1}$ to 2 $M^{-1}sec^{-1}$. However, those few complexed ends that managed to add another styrene unit continue to grow rapidly because steric causes prevent the complexation of the second, third, etc., unit with the naphthyl moiety. Consequently, most of the end groups form the complexes, and very few produce long chains of polystyrene.

The equilibrium

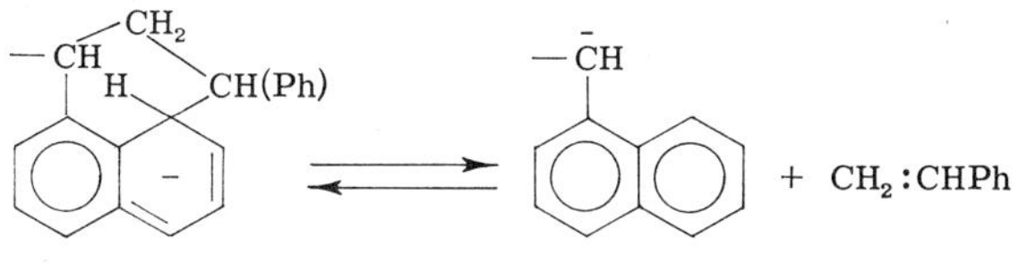

produces higher concentrations of styrene than the equilibrium of this monomer with its living polymer. Hence, on standing, the complexes slowly dissociate and the long chain of polystyrene becomes still longer. The saying, "The rich are getting richer and the poor are getting poorer," applies to this system, and thus the original polyvinylnaphthalene end groups reappear. The kinetic studies led to the determination of all three rate constants: first styrene addition ($\sim$27 M^{-1}sec^{-1}), second addition ($\sim$2 M^{-1}sec^{-1}), and the subsequent propagation ($\sim$1000 M^{-1}sec^{-1}).

It is hoped that this review has shown the utility of living polymers, their unusual features, and the ways in which their studies enriched our knowledge of some organic reactions and especially ionic reactions of organic compounds.

References

Allen, G., Gee, G., and Stretch, C. (1960). *J. Polym. Sci.* **48**, 189.

Altares, T., Wyman, D. P., and Allen, V. R. (1964). *J. Polym. Sci. A* **2**, 4533.

Asami, R., Khanna, S. N., Levy, M., and Szwarc; M. (1962). *Trans. Faraday Soc.* **58**, 1821.

Bawn, C. E. H., *et al.* (1971). *Polymer* **12**, 119.

Benoit, H. (1964). *J. Polym. Sci., Part C* **4**, 1509.

Bhattacharyya, D. N., Lee, C. L., Smid, J., and Szwarc, M. (1963). *J. Am. Chem. Soc.* **85**, 533.

Bhattacharyya, D. N., Lee, C. L., Smid, J., and Szwarc, M. (1964a). *Polymer* **5**, 54.

Bhattacharyya, D. N., Smid, J., and Szwarc, M. (1964b). *J. Am. Chem. Soc.* **86**, 5024.

Bhattacharyya, D. N., Lee, C. L., Smid, J., and Szwarc, M. (1965a). *J. Phys. Chem.* **69**, 612.

Bhattacharyya, D. N., Smid, J., and Szwarc, M. (1965b). *J. Phys. Chem.* **69**, 624.

Bier, G. (1961). *Angew. Chem.* **73**, 186.

Bloodworth, A. J., and Davies, A. G. (1963). *Proc. Chem. Soc., London, Ser. A* p. 315.

Breslow, S. Ye., Pyrkov, L. M., Frenkel, S. A., Lains, L. A., and Klenin, S. I. (1962). *Vysokomol. Soedin.* **4**, 250.

Brown, W. B., and Szwarc, M. (1958). *Trans. Faraday Soc.* **54**, 416.

Burke, J. J., and Weiss, V., eds. "Block and Graft Copolymers." (1973). Syracuse Univ. Press, Syracuse, New York.

Bywater, S. (1965). *Adv. Polym. Sci.* **4**, 66.

Bywater, S., and Worsfold, D. J. (1962a). *Can. J. Chem.* **40**, 1564.

Bywater, S., and Worsfold, D. J. (1962b). *J. Polym. Sci.* **58**, 571.

Calderon, N., and Scott, K. W. (1965). *J. Polym. Sci., Part A* **3**, 551.

Dainton, F. S. *et al.* (1965). *Makromol. Chem.* **89**, 257.

Dainton, F. S., Hui, K. M., and Ivin, K. J. (1969). *Eur. Polym. J.* **5**, 382.

De Groof, B., Van Beylen, M., and Szwarc, M. (1975). *Macromolecules* **8**, 396.

Dobry, A., and Boyer-Kavenoki, F. (1947). *J. Polym. Sci.* **2**, 90.

Dreyfuss, M. P., and Dreyfuss, P. (1965). *Polymer* **6**, 93.

Dreyfuss, M. P., and Dreyfuss, P. (1966). *J. Polym. Sci.* **4**, 2179.

Figini, R. V. (1967). *Makromol. Chem.* **107**, 170.

Figini, R. V. *et al.* (1965). *Z. Phys. Chem.* (*Frankfurt am Main*) **45**, 269.

Fisher, M., and Szwarc, M. (1970). *Macromolecules* **3**, 23.

Fujimoto, T., and Nagasawa, M. (1975). *Polym. J.* **7**, 397.
Fuoss, R. M., and Kraus, C. A. (1933). *J. Am. Chem. Soc.* **55**, 2387.
Gallot, B., Mayer, R., and Sadron, C. (1966). *C. R. Hebd. Seances Acad. Sci.* **263**, 42.
Ginn, J. M., and Ivin, K. J. (1970); *Makromol. Chem.* **139**, 57.
Goethals, E. (1976). Akron Meeting on Cationic Polymerization, June 1976.
Graham, R. K., Dunkelberger, D. L., and Goode, W. E. (1960). *J. Am. Chem. Soc.* **82**, 400.
Hemery, P., Boileau, S., and Sigwalt, P. (1975). *Polym. Lett.* **13**, 49.
Holden, G., and Milkovich, R. (1966). U.S. Patent 3,265,765.
Hostalka, H., and Schulz, G. V. (1965). *Z. Phys. Chem.* (*Frankfurt am Main*) **45**, 286.
Khanna, S. N., Levy, M., and Szwarc, M. (1962). *Trans. Faraday Soc.* **58**, 747.
Kobayashi, S., Danda, H., and Saegusa, T. (1974). *Macromolecules* **7**, 415.
Korotkov, A. A. *et al.* (1960). *Vysokomol. Soedin.* **1**, 157.
Lee, C. L., Smid, J., and Szwarc, M. (1963). *Trans. Faraday Soc.* **59**, 1192.
Levresse, B., Franta, E., and Rempp, P. (1971). *Makromol. Chem.* **142**, 111.
Löhr, G., and Schulz, G. V. (1964). *Makramol. Chem.* **77**, 240.
Mathis, C., Lamaude-Christmann, L., and François, B. (1975). *Makromol. Chem.* **176**, 931.
Matyjaszewski, K., and Penczek, S. (1974a). *J. Polym. Sci.* **12**, 1905.
Matyjaszewski, K., and Penczek, S. (1974b). *Macromolecules* **7**, 136.
Merrett, F. M. (1957). *J. Polym. Sci.* **24**, 467.
Minnich, E. R., Long, L. D., Ceraso, J. M., and Dye, J. L. (1973). *J. Am. Chem. Soc.* **95**, 1061.
Miyake, A., and Stockmayer, W. H. (1965). *Makromol. Chem.* **88**, 90.
Morton, M., and Ells, F. R. (1962). *J. Polym. Sci.* **61**, 25.
O'Driscoll, K. F., and Kuntz, I. (1962). *J. Polym. Sci.* **61**, 19.
Pepper, D. C. (1974). *Int. Symp. Macromol. 23rd, 19* p.
Pruckmayr, G., and Win, T. K. (1976). *Macromolecules* **9**,
Rakova, G., and Korotkov, A. A. (1958). *Dokl. Akad. Nauk SSSR* **119**, 982.
Rempp, P., and Loucheux, M. H. (1958). *Bull. Soc. Chim. Fr.* p. 1497.
Richards, D. H., and Szwarc, M. (1959). *Trans. Faraday Soc.* **55**, 1644.
Sadek, H., and Fuoss, R. M. (1954). *J. Am. Chem. Soc.* **76**, 5905 and 5897.
Schlick, S., and Levy, M. (1960). *J. Phys. Chem.* **64**, 883.
Schuerch, C. *et al.* (1964). *J. Am. Chem. Soc.* **86**, 4481. Schuerch, C. *et al.* (1967). *J. Am. Chem. Soc.* **89**, 1396.
Schulz, G. V. *et al.* (1972). *Adv. Polym. Sci.* **9**, 1.
Scott, R. L. (1949). *J. Chem. Phys.* **17**, 279.
Shimomura, T., Toelle, K. J., Smid, J., and Szwarc, M. (1967). *J. Am. Chem. Soc.* **89**, 976.
Shinohara, M., Smid, J., and Szwarc, M. (1968). *J. Am. Chem. Soc.* **90**, 2175.
Shinohara, M., Smid, J., and Szwarc, M. (1969). *Chem. Commun.* p. 1232.
Smets, G. (1965). Bunsen-Discussionstagung, Ludwigshaften.
Smets, G., and Hart, R. (1960). *Fortschr. Hochpolym. Forsch.* **2**, 173.
Stretch, C., and Allen, G. (1961). *Polymer* **2**, 151.
Szwarc, M. (1956). *Nature* (*London*) **178**, 1168.
Szwarc, M. (1967). *Prepr. Polym. Sect., Am. Chem. Soc. Meet., 1967.*
Szwarc, M. (1968). "Carbanions, Living Polymers, and Electron-Transfer Processes," pp. 44–69. Wiley (Interscience), New York.
Szwarc, M. ed. (1972). "Ions and Ion Pairs in Organic Reactions," Vol. I. Wiley, New York.
Szwarc, M., ed. (1974). "Ions and Ion Pairs in Organic Reactions," Vol. II, Wiley, New York.
Szwarc, M., and Hermans, J. J. (1964). *Polym. Lett.* **2**, 815.
Szwarc, M., Levy, M., and Milkovich, R. (1956). *J. Am. Chem. Soc.* **78**, 2656.

Tardi, M., Rougé, D., and Sigwalt, P. (1967). *Eur. Polym. J.* **3**, 85.
Throssell, J. J., Sood, S. P., Szwarc, M., and Stannett, V. (1956). *J. Am. Chem. Soc.* **78**, 1122.
Uraneck, C. A., Short, J. N., and Zelinski, R. P. (1964). U.S. Patent 3,135,716.
Vofsi, D., and Tobolsky, A. V. (1965). *J. Polym. Sci. A* **3**, 3261.
Win, T. K., and Pruckmayr, G. (1975). *Macromolecules* **8**, 77.
Winstein, S., Clippinger, E., Fainberg, A. H., and Robinson, G. C. (1954). *J. Am. Chem. Soc.* **76**, 2597.
Woodward, A. E., and Smets, G. (1955). *J. Polym. Sci.* **17**, 51.
Worsfold, D. J., and Bywater, S. (1960). *Can. J. Chem.* **38**, 1891.
Yen, S. P. S. (1967). *Makromol. Chem.* **81**, 152.
Yoshino, T., Komiyama, J., and Shinomiya, N. (1964). *J. Am. Chem. Soc.* **86**, 4482.
Yoshino, T., Komiyama, J., and Shinomiya, N. (1965). *J. Am. Chem. Soc.* **87**, 387.
Zelinski, R. P., Hsieh, H. L., and Strobel, C. W. (1963). U.S. Patent 3,109,871.

Microbial Degradation of Organic Compounds in the Biosphere

Stanley Dagley

Department of Biochemistry,
College of Biological Sciences,
University of Minnesota,
St. Paul, Minnesota

I. Introduction

In a preface to the second edition of the very successful textbook, "Dynamic Aspects of Biochemistry," the late Professor Ernest Baldwin reminded his readers of the debt that biochemists owe to Frederick Gowland Hopkins who did so much to foster and develop a dynamic habit of thought about biochemical systems. For one of the most striking features of a living cell is indeed the dynamic state of its constituent molecules. The soundness of this concept was emphasized by results of isotopic studies in the late 1930s, and its interpretation in terms of chemical reactions became possible with the discovery of the cycles and sequences of modern metabolism. Thus, a metabolic wall chart displays an extensive network of reactions in which one reaction sequence is repeatedly linked to another through those chemical intermediates that are common to each one. A dozen sequences, for example, are linked through the common metabolite, pyruvate; some of them are degradative reactions that give rise to pyruvate while in other cases this same compound initiates a pathway of biosynthesis. It is evident that the reaction network of the metabolic map provides a means by which pyruvate and many other compounds can be rapidly synthesized and degraded by enzymes. However, at any moment of time the concentrations of these metabolites within the cell remain, in general, nearly constant not because the molecules are chemically stable but because a steady state is achieved with a balance maintained between the rates at which particular molecules are broken down and reformed. To maintain this steady state, the cell requires a continuous supply of energy and materials. As Sir Hans Krebs pointed out, this feature provides the chemical background of excitability, one of the fundamental properties of living matter. A change in the environment of a cell may be described in the last resort in chemical terms and the very life of the cell may depend upon its response. This response is not likely to need some entirely new activity. It may draw upon one or more of the multitude of reactions already existing in the resting condition. The vast interrelated network of reactions maintained in the steady state confers upon the cell a chemical resilience upon which its survival in a changing environment may depend.

When we consider the global chemical processes that operate to sustain life on this planet, we perceive certain overall similarities, or at least analogies, with the systems mentioned. The whole biosphere, like any individual organism that lives inside it, exists in a chemically dynamic state. The biosphere is bathed in a flow of energy that is received as solar radiation in the daytime and is largely radiated as heat to outer space at night, with a fraction of the input harnessed by photosynthetic reactions to serve the purposes of life. A great number of organic compounds are synthesized, transformed, and decomposed continually; together, these segments of the steady state chemistry of our planet constitute the carbon cycle. Other cyclic chemical systems, e.g., those for nitrogen and sulfur, have been identified. It would appear that the basic conditions for the existence of life include those that permit the establishment of such cycles, operating within a stream of solar energy. Some philosophical systems find difficulty in adjusting to this concept; for example, the fact that in one year the spiders of Great Britain consume, on a conservative estimate, a weight of insects greater than that of its human population may not be welcomed. Nevertheless, cycles of synthesis and degradation, of life and death, are basic to our existence and it is wise to accept the fact. For, as I shall attempt to show, the environmental consequences of man's technological activities cannot be adequately assessed without an understanding of the biochemistry of the carbon cycle.

I have argued elsewhere that despite its importance in nature, the carbon cycle receives inadequate treatment in textbooks and biological curricula, an observation which is particularly relevant to that section of the cycle concerned with biological degradation (Dagley, 1975a, 1976). Such enzyme-catalyzed reactions continuously release vast quantities of carbon, sequestered in organisms that have died, to be used again by those that are alive. To the biochemist concerned exclusively with mammalian metabolism some of the compounds of these catabolic pathways possess unusual, not to say bizarre, structures. Their relevance for pollution problems will be indicated, but the reader is referred to a previous article (Dagley, 1975b) for more detailed biochemical discussion. When insults to the environment are perpetrated, protests have their proper place; but educators should also be concerned about presenting a conceptual framework in order to enlarge our understanding of the biosphere, and to approach environmental problems in a rational manner.

II. Biological Degradations

A. General Features of Catabolism

The release of carbon for recycling throughout the biosphere is mainly accomplished by microorganisms. Commensurate with the part they have to play, the total mass of microbes on this planet greatly exceeds the total mass of animals. This fact is rarely appreciated by science undergraduates and least of all by that section of the general public that acquires its information about microorganisms from television advertisements for disinfectants. I shall first discuss those special biochemical aptitudes that enable microorganisms to degrade and utilize for growth the vast range of natural products present in dead organic matter. In this discussion I shall confine myself to aerobic microbial catabolism, partly because more is known about aerobes than anaerobes, but mainly because we shall be concerned with the complete oxidation of these structures to release carbon dioxide which can be used again by other organisms in photosynthesis. However, it is appropriate at this point to recognize that the chemical structures of many organic compounds, natural and man made, can be extensively modified by anaerobes; moreover, in the presence of nitrate or sulfate ions that may serve as electron acceptors, such compounds may be degraded completely to carbon dioxide under anaerobic conditions. Like mammals, microbes of all types initiate attack upon proteins, polysaccharides, nucleic acids and fats by hydrolyzing them to their respective monomers; but some microbial hydrolytic enzymes display unique features, indispensable for the operation of the carbon cycle. Thus, the most abundant of all biopolymers is cellulose, and its hydrolysis is catalyzed exclusively by microbes. For the degradation (and synthesis) of smaller molecules, aerobic microorganisms make use of most of the biochemical sequences and cycles that occupy a central position in metabolic charts and are generally found in other living forms. In microbes and mammals alike, biochemicals are seldom degraded by reversing metabolic pathways of synthesis, despite the fact that the individual enzymic steps in either direction do not involve large changes in free energy. Thus, palmitic acid is synthesized in the cytosol and degraded in the mitochondria by separate and distinct reaction sequences that involve totally different metabolites and are catalyzed by unrelated enzymes. In this case, such a separation of the opposing processes of synthesis and degradation confers advantages for regulating reaction rates and consequently the concentrations of products. In other instances, as in the biosynthesis of proteins, the assembly of monomers in their correct linear sequence calls for elaborate biochemical equipment which would not be required for the simple, reverse process of hydrolysis. Accordingly we find, for example, that the microbial degradation of aromatic compounds, which accounts for a significant segment of the carbon cycle, is accomplished by reactions that are completely different from those used to synthesize the benzene nucleus. Indeed, the microbial metabolic pathways by which plant products are degraded may involve reactions seldom encountered in other organisms.

B. Aspects of Microbial Catabolism

To what biochemical attributes do microbes owe their exceptional degradative versatility? All living things catalyze a range of hydrolytic reactions; and lack of ability to hydrolyze cellulose, in the way that various microbial species are capable of doing, can be attributed to restrictions on substrate specificity rather than to failure to catalyze a particular type of reaction. However, there is one *type* of reaction that microbes catalyze much more extensively than do any other forms of life, namely the direct incorporation of atoms from molecular oxygen into organic molecules. The plant kingdom accumulates great quantities of polymers, such as lignin, and innumerable compounds, like alkaloids, terpenes, and flavonoids, that are biochemically inert, and the release of their carbon for recycling depends almost entirely upon the action of microbial degradative enzymes. By direct incorporation of oxygen, microbes are able to bring these metabolically unreactive compounds into chemical sequences that lead to the central pathways of metabolism, such as the Krebs cycle or the fatty acid "spiral." Thus, an alkane remains biochemically inert until a terminal carbon has been oxidized. The fatty acid so formed may then enter a main channel of metabolism, namely β oxidation. A saturated ring structure is also inert until a hydroxyl group is inserted, and, if this occurs at an appropriate point in the molecule, the possibility then arises that further oxidation may break the ring, providing fragments that can be degraded. The unsubstituted benzene nucleus is an inert resonance structure, but when suitably substituted with two hydroxyl groups, enzymic ring fission may then take place. Enzymes that catalyze the fission of molecular oxygen and incorporate the atoms into chemical structures are designated oxygenases. Their importance can be appreciated when viewed in the perspective obtained when we consider the role of oxygen in evolution.

III. The Biochemistry of Oxygen

A. Oxygen in Evolution and in the Carbon Cycle

The accumulation of oxygen gas in our atmosphere was a development that made possible the full evolution of living forms in all their present diversity. Any organism that remained anaerobic was handicapped by its inability to take full advantage of the energy gradient that came to be established between biochemicals and carbon dioxide. With O_2 abundant in our atmosphere, the stablest compound of carbon is CO_2. Under these conditions all terrestial compounds are thermodynamically unstable to varying degrees. All organic compounds found on earth have been synthesized by a living organism, whether the organism was a plant or an organic chemist. The energy released by the ancient, anaerobic process of glycolysis is only a fraction of what is made available when glucose is burned to carbon dioxide and water.

However, hazards to life as well as thermodynamic advantages and opportunities are presented by an atmosphere of O_2. Since living forms are made of biochemicals, indiscriminate combustion of these materials must be avoided. Accordingly, if we inspect a map that summarizes metabolic sequences of importance in mammalian biochemistry, we find that oxygen is effectively excluded until such time as the cytochrome chain is ultimately reached and protons and electrons finally combine directly with oxygen to give water.

Before this point is reached, biochemicals have been oxidized not by using O_2 but by removing H atoms and transferring them to NAD or NADP. Anaerobic processes analogous to those that operated when life emerged have been preserved in aerobic organisms. However, this apparent exclusion of O_2 from general metabolism simply arises from a fundamental property of the molecule without which the present range of life on earth could not be sustained. This property is the sluggishness shown by oxygen towards organic compounds at physiological temperatures, despite the fact that a great deal of energy is released once the reaction is started and complete combustion is able to occur. Oxygen exists in a triplet ground state: it has two unpaired electron spins, and the molecule is paramagnetic. Since carbon in organic compounds is in a singlet state, concerted reactions with oxygen are "spin forbidden;" and in order to react, oxygen must be "activated" by radiation energy to attain a singlet state, for example, or by specifically combining with an enzyme. Haphazard activation would be catastrophic.

The possibility that highly reactive oxygen might gain random access to cellular constituents poses a threat to any form of life; and the fact that superoxide dismutase is apparently ubiquitous in aerobic organisms (McCord *et al.*, 1971) suggests that this enzyme constitutes an essential part of life's defense against the toxicity of oxygen (Fridovich, 1974). Strict anaerobes do not possess superoxide dismutase, and they die on contact with oxygen. As for higher organisms, one toxic effect of active oxygen may be ascribed to the peroxidation of unsaturated lipids which are important for the structural and functional integrity of biological membranes. Superoxide dismutase and catalase inhibit this peroxidation; scavengers of singlet oxygen also inhibit, whereas scavengers of hydroxyl radical do not. Although this radical is amongst the most vigorous and indiscriminate oxidizers known, it now seems that the more discriminating singlet oxygen may be effective in causing damage at critical sites in living cells (Pederson and Aust, 1975; Kellogg and Fridovich, 1975). It appears that hydrogen peroxide and superoxide anion, which is generated in small amounts during the normal operation of some flavoproteins in aerobic organisms, may react to give singlet oxygen (O_2*) as follows:

$$O_2^- + H_2O_2 \rightarrow OH^- + OH\cdot + O_2*$$

The selenoenzyme glutathione peroxidase (EC 1.11.1.9) catalyzes reduction of lipid peroxides as well as of H_2O_2 and it has been suggested that it breaks the autocatalytic chain reaction of lipid peroxidation, so protecting the cell membranes from oxidative damage (Flohé, 1971; Christophersen, 1969). Many of the pathological conditions of selenium deficiency seem to be related to peroxidation of unsaturated lipids in membranes (Underwood,

1962) and dietary selenium (the metal present in glutathione peroxidase) has been shown to exert a protective effect against the oxidative damage to rat erythrocyte membranes (Rotruck *et al.*, 1971).

It is clear, therefore, that the formation of highly reactive forms of oxygen constitutes an extreme hazard for any aerobic organism. Nevertheless, it became necessary early in evolution to catalyze this process in a controlled manner in order to initiate degradation of innumerable natural products. This special function is performed by oxygenases which are mainly, though by no means exclusively, elaborated by aerobic microorganisms. Without this activation, and the subsequent reactions in degradative sequences that are also catalyzed, the carbon cycle could not function continuously; neither would this planet be cleansed from the comtaminating activities of man. However, although the amount of oxygen incorporated into organic compounds by microorganisms greatly exceeds that fixed by all other forms of life, the qualitative significance of mammalian oxygenases deserves mention; for without hydroxylases the biochemistry of man would, no doubt, have been very different from what it is. Prostaglandins are synthesized from unsaturated long-chain fatty acids by reaction sequences that involve hydroxylations, and the same is true for the formation of testosterone and estrone from cholesterol; indeed reactions involving O_2 directly are crucial for the synthesis of the last-named compound from squalene. Moreover, as will be mentioned later, mammals would not have survived the poisons that abound in nature without the ability to activate oxygen using cytochrome P_{450}.

B. Benzene Ring-Fission Dioxygenases

One of the most widely distributed structural units in nature is the benzene nucleus. Thus, the three-dimensional polymer, lignin, which appears to be derived from the enzymic coupling of three aromatic alcohols (*p*-coumaryl, coniferyl, and synapyl alcohols) constitutes 18–35% of woody tissues. It is estimated that about 1.5×10^{10} tons of carbon in the form of carbon dioxide is annually transformed into wood (Bolin, 1973). Microbes are responsible for degrading lignin, and it appears that aromatic compounds of lower molecular weight are often released first from the polymer by fungal action. Decomposition is then completed by bacteria which also utilize as nutrients great quantities of other benzenoid compounds synthesized in the plant kingdom. Since this stable chemical structure is so common, bacterial attack upon the benzene ring will be chosen to illustrate the properties of dioxygenases. These enzymes incorporate both atoms of O_2 into the substrate molecule, and at the same time they cleave the benzene nucleus. In the case of catechol (I, Fig. 1) ortho fission of the ring may occur [reaction (a)], or

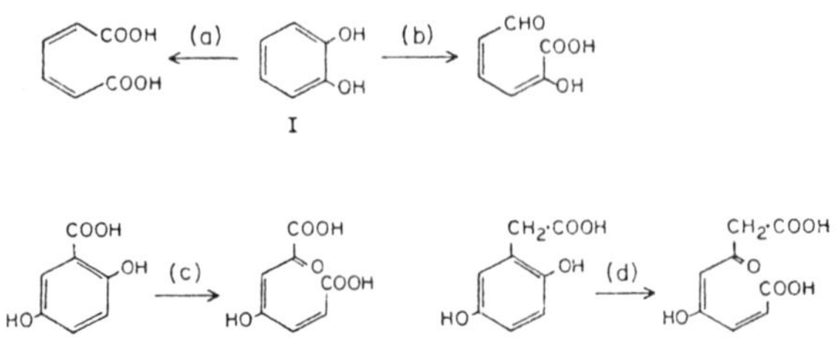

FIG. 1. Fission of the benzene nucleus by dioxygenases. Separate enzymes catalyze ring cleavage of: I, catechol, by *ortho* fission (a) or by *meta* fission (b); II, gentisic acid; III homogentisic acid.

meta fission [reaction (b)]; the former is catalyzed by catechol 1,2-dioxygenase and the latter by a different enzyme, catechol 2,3-dioxygenase, Substrates for dioxygenases show a common structural feature: the benzene ring must carry at least two hydroxyl groups placed either ortho or para to each other. Thus, compounds II and III, namely gentisic and homogentisic acids (Fig. 1), carry hydroxyl groups in the para position and the nucleus is cleaved by gentisate 1,2-dioxygenase [reaction (c)] or homogentisate 1,2-dioxygenase [reaction (d)] which again are separate and distinct enzymes. The requirement for two hydroxyl groups, positioned as stated, is no doubt determined by the shifts of electrons that take place during the ring-fission reaction. A feasible mechanism has been proposed by Hayaishi (1964). As mentioned earlier, since carbon in organic compounds is in a singlet state, direct combination with O_2 cannot occur while the gas remains in a triplet state. However, unlike carbon, transition elements have one or more unpaired electrons in *d* orbitals so that direct reaction between these metals and O_2 is not spin forbidden; thus those dioxygenases that have been examined in detail are iron–sulfur proteins. We can envisage O_2 as combining with the iron present at the catalytic site of the enzyme to give $Fe(II){-}O_2$, and then the electronic state $Fe(III){-}O_2\cdot^{-}$ and $Fe(IV){-}O_2^{2-}$ may readily be assumed, thereby making reactive forms of O_2 available. However, recent studies with crystalline protocatechuate 3,4-dioxygenase show that catalytic events follow a less direct course than this (J. M. Wood, personal communication). It appears that the hydroxyl group at C-4 chelates iron at the active site, the C-3 hydroxyl group loses a proton, and the attack by oxygen upon the bound substrate anion is then permitted. Properties of the ortho-fission enzymes catechol 1,2-dioxygenase and protocatechuate 3,4-dioxygenase, and also of the meta-fission enzymes catechol 2,3-dioxygenase and 3,4-dihydroxyphenylacetate (homoprotocatechuate) 2,3-dioxygenase have been reviewed by Hayaishi (1966) in whose laboratory their purification and crystallization was accomplished.

The ortho-fission enzymes are red (beautiful photographs are given in the above review) and the meta-fission enzymes are colorless; these properties appear to reflect the valence state of the bound iron. Despite the similarities of the various reactions catalyzed, one oxygenase seldom attacks the natural substrate of another at a significant rate, that is, fast enough to supply the requirements of growth. This is not to say that each is absolutely specific for one substrate: thus catechol 1,2-dioxygenases from various organisms will tolerate the presence of methyl and other alkyl substituents at C-3 or C-4, the rate of attack depending upon the source of the enzyme (Patel *et al.*, 1976; Nakazawa *et al.*, 1967). More surprising is the recent discovery that catechol 1,2-dioxygenase from *Pseudomonas* catalyzes a detectable meta-fission reaction (also referred to as *extradiol cleavage*) for both 3-methylcatechol and 3-methoxycatechol. The rates of attack were small but significant, the ratios of rates being respectively 17 and 5 for intra-diol fission relative to extradiol fission (Fujiwara *et al.*, 1975).

The substrate specificities of various dioxygenases may be illustrated with reference to Fig. 2. Each of these eight dioxygenases shows little or no activity against the substrates of the other seven. The main exception to this generalization is the ability of enzyme (2), protocatechuate 4,5-dioxygenase, to oxidize rapidly both gallate and 3-0-methylgallate (Tack *et al.*, 1972): these are the substrates for enzymes (3) and (4) respectively.

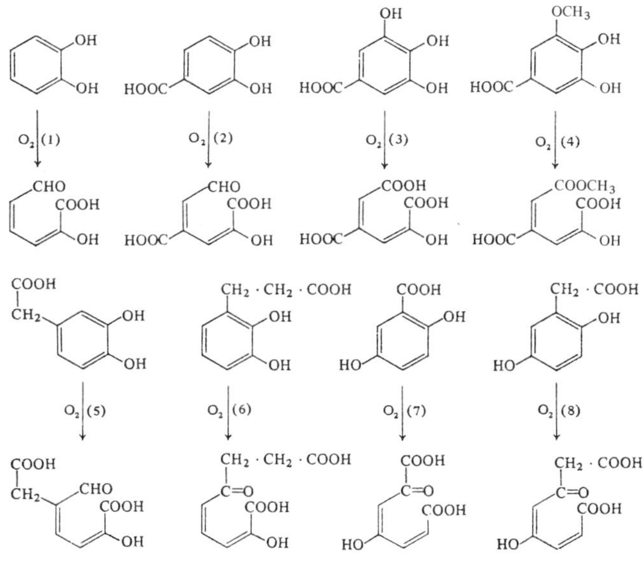

FIG. 2. Some substrate-specific dioxygenases. With the few exceptions mentioned in the text, any one of these enzymes attacks only the substrate shown, and has little or no action upon the remainder.

However, one gallate dioxygenase was found to catalyze reaction (3) but not reaction (4), whereas another enzyme from the same species of *Pseudomonas* catalyzed both of these reactions. Of these two enzymes, the one selected for induction was determined by the carbon source used by the organism for growth (Sparnins and Dagley, 1975). In fact, each enzyme of Fig. 2 is inducible, that is, the amount present in the cells is usually increased at least tenfold on exposure to its substrate or to a precursor in the metabolic pathway. The dioxygenases of Fig. 2 have been studied in various species of microorganisms, and no single strain is likely to possess all of them. No data are available concerning amounts of each substrate encountered in the free condition in nature, but those for enzymes (2)–(5) might arise from the biodegradation of lignin and numerous other abundant natural materials. Thus, as a component of tannins, gallic acid is widespread. The substrate for enzyme (1) (catechol) is known to be formed by microbial oxidation of benzoic and mandelic acids, that for enzyme (6) is formed from melilotic, cinnamic and hydrocinnamic acids; and the substrate for enzyme (8) (homogentisic acid) is formed from tyrosine. All these substrate precursors are found in nature. Moreover, when the cells of higher plants are disrupted, with the object of releasing enzymes for isolation, it is usually found that numerous phenolic compounds are present in extracts at concentrations sufficient to inhibit activity of the isolated enzyme. Progress in plant enzymology was greatly assisted when Loomis and Battaile (1966) introduced polyvinylpyrrolidone as an absorbent for these inhibitory phenolic compounds frequently present in plant enzyme preparations.

Functional specialization, which emerges from this survey as the most striking feature of these enzymes, prompts the question as to how this came about when, apparently, one or two dioxygenases of broad substrate specificities might have served just as well. Moreover, each dioxygenase is found to be induced separately and independently by its growth substrate or by a related metabolite. In attempting an explanation, we may speculate that the first event in the evolution of a new pathway of aromatic catabolism was the acquisition of the appropriate primitive dioxygenase by a microorganism. For this, a protein that could bind the potential substrate must have been available in the first place; but only the simplest chemical reagent was needed to activate oxygen, namely ionic iron: this, too would have to be bound by the protein. On this view we might expect to find a separate dioxygenase for each of the common dihydric or trihydric phenols that an organism became capable of utilizing for growth. We might also predict differences in properties for dioxygenases catalyzing the same reaction in various microbial species.

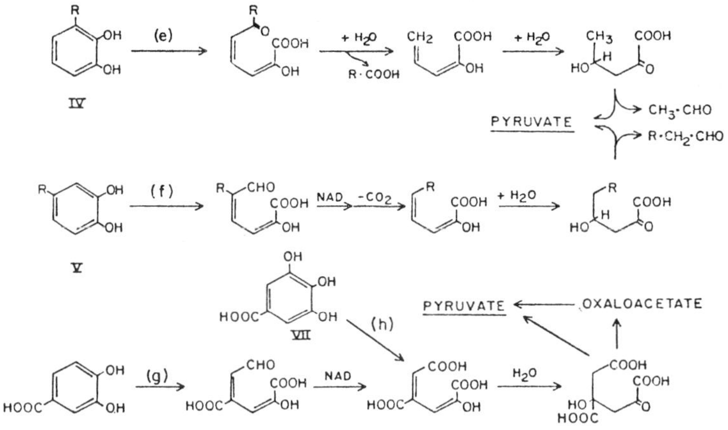

FIG. 3. Reaction sequences that give rise to pyruvate and other metabolites after the benzene nucleus has been opened by dioxygenases (e)–(h).

C. Degradation of Ring-Fission Products

Enzymes that convert the products of ring-fission into metabolites of the Krebs cycle are also usually substrate specific. These reactions have been reviewed (Dagley, 1971) and only references to more recent work will be given here; general schemes are shown in Fig. 3. If R = H for compounds IV and V two pathways for the same compound, catechol, are then designated; and in some species of *Pseudomonas* alternative routes are, in fact, available for this substrate. Enzymes (e) and (f) now represent one and the same dioxygenase, namely catechol 2,3-dioxygenase [enzyme (1) of Fig. 2, or enzyme (b) of Fig. 1]. When R = CH_3 in the general structures IV and V, we obtain respectively, 3-methylcatechol and 4-methylcatechol. These compounds are again attacked by the catechol 2,3-dioxygenase, but the first follows the hydrolytic route for IV and the second, the alternative route of oxidative decarboxylation shown for V. However, when R = $CH_2 \cdot CH_2 \cdot$ COOH in IV, a different dioxygenase is involved: this is enzyme (6) of Fig. 2. Catabolites and enzymes become identical in this pathway following the hydrolytic step, because the release of R · COOH removes the distinction between initial dihydric phenols of structure IV. But for pathway V when R = CH_2COOH, every enzyme involved in the sequence is different from those indicated when R = CH_3 (Sparnins *et al.*, 1974). Despite similarities in structure, no metabolite of one pathway is acted upon by the analogous enzyme from the other. When R = COOH in V enzymes are again implicated that differ from those for R=CH_2COOH. Indeed, there is also a slight variation in the general sequences and accordingly the degradative pathway for protocatechuate VI is shown separately. It is a striking feature that the introduction of a methylene group into the substituent of the benzene nucleus should necessitate the elaboration of a separate suite of enzymes, from the initial dioxygenase all the way to the Krebs cycle. This feature is again demonstrated for gentisic and homogentisic acids: not only are enzymes (c) and (d) of Fig. 1 separate and specific catalysts, but so are those that follow these initial reactions, giving rise to pyruvate and acetoacetate respectively (fumarate is formed from both ring-fission substrates). It may be noted that gallic acid VII enters sequence VI by reaction (h) [dioxygenase (3) of Fig. 2].

D. Function and Properties of Flavoprotein Hydroxylases

From the foregoing account we may list the following reaction sequences that channel aromatic substrates by separate and distinct pathways to the Krebs cycle. These central metabolic channels are the meta-fission pathways for catechol (also used for 2,3-dihydroxyphenylpropionic acid after ring fission and hydrolysis), for 3,4-dihydroxybenzoic acid (protocatechuic acid), and for 3,4-dihydroxyphenylacetic acid (homoprotocatechuic acid); the degradative pathways for gentisic and homogentisic acid; and those reactions that follow ortho fission of catechol and protocatechuate (Fig. 4). (Further research will undoubtedly add to this list.) Now, if an organism grows at the expense of a benzoid compound that carries only one hydroxyl group, then a second must be introduced. Since the positioning of this group will commit the organism to using one, and only one, pathway into the Krebs cycle, it is not surprising that each bacterial aromatic hydroxylase is usually found to exhibit stringent substrate specificity. Several of these enzymes have been purified to homogeneity, and in some instances crystallized, and most of them have been shown to be flavoproteins. They belong to a category of enzymes termed monooxygenases, or mixed function oxygenases, since the reactions they catalyze require NADH or NADPH and result in the incorporation on one atom of oxygen into the carbon substrate (Mason *et al.,* 1955). Thus, melilotate hydroxylase catalyzes the reaction.

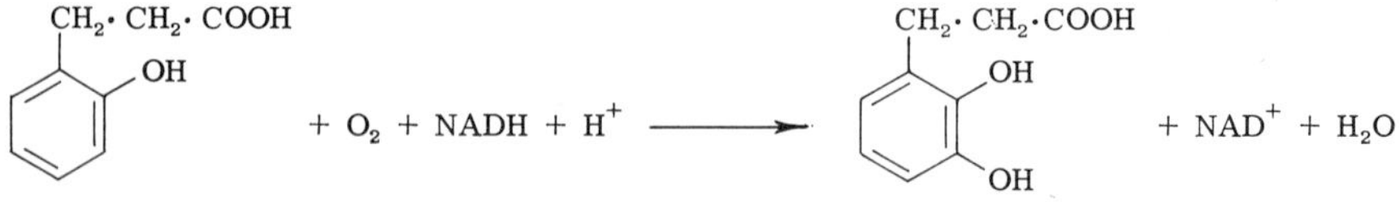

Other purified flavoprotein hydroxylases include 4-hydroxybenzoate-3-hydroxylase (Howell *et al.*, 1972) and 3-hydroxybenzoate-4-hydroxylase

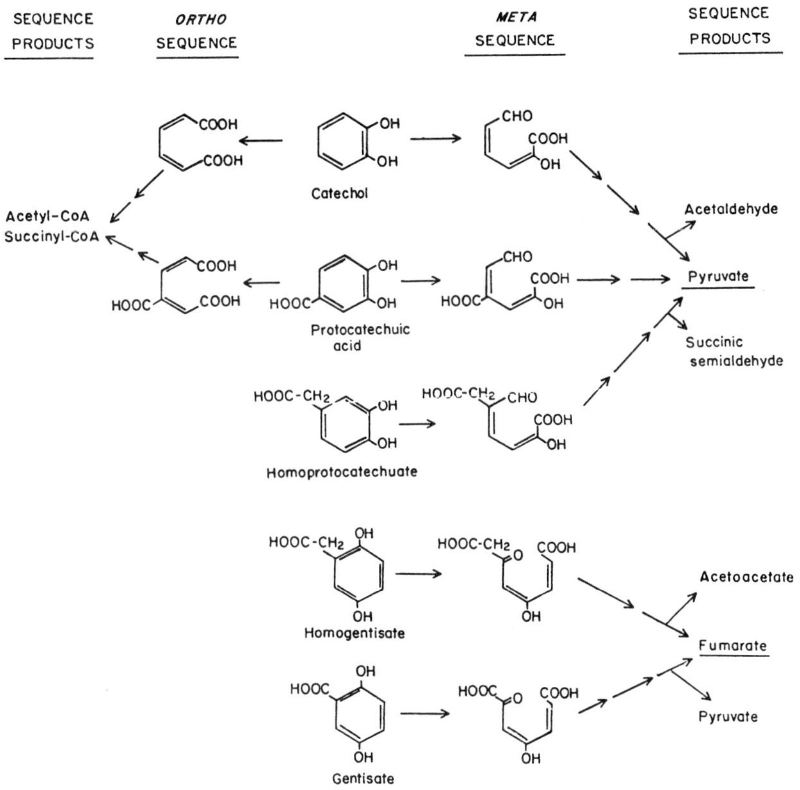

FIG. 4. Summary of central reaction sequences in bacterial aromatic catabolism. The products are pyruvate and metabolites of the Krebs cycle: acetaldehyde and acetoacetate give rise to acetyl CoA; succinic semialdehyde is oxidized to succinate.

(Michalover and Ribbons, 1973); these are different enzymes that give the same product, protocatechuate, from their respective substrates. Salicylate hydroxylase, which produces catechol by a simultaneous oxygenation and decarboxylation of its substrate, has been intensively investigated (Takemori *et al.*, 1969; White-Stevens and Kamin, 1972). Orcinol hydroxylase was crystallized by Ohta and Ribbons (1970) and shown to introduce a third hydroxyl group into its substrate, giving 2,3,5-trihydroxytoluene. This compound serves as substrate for a dioxygenase whereas orcinol cannot since its two hydroxyl groups are situated meta. Chapman and Ribbons (1976) have shown that ring fission of 2,3,5-trihydroxytoluene initiates reactions of a new catabolic pathway that involves formation of acetate and pyruvate from acetopyruvate.

The studies of Strickland and Massey (1973) have furnished an outline of the mode of action of one of these flavoprotein hydroxylases, melilotate hydroxylase (melilotate 3-monooxygenase; EC 1.14.13.4). Their work provides a fine example of how kinetic studies that give Lineweaver-Burk plots for different concentrations of substrates (O_2, NADH, melilotate) can provide data for steady state analysis, enabling a mechanism of enzyme action to be suggested. The mechanism they proposed, which was confirmed by separate measurements made with a stopped-flow spectrophotometer, show the following (six) steps during hydroxylation: the FAD-enzyme combines with meliotate and then with NADH to given an enzyme complex with substrate, $FADH_2$ and NAD^+, from which the latter is released as the next reaction in the sequence. The reduced flavoprotein, which still binds melilotate, now combines with oxygen to give an oxygenated species which can be detected in the spectrophotometer. This intermediate loses water, concomitant with the formation of 2,3-dihydroxyphenylpropionate; this is released from the enzyme in the final step, leaving the oxidized flavoprotein to initiate another six-step cycle. Similar sequences, in which a substrate combines with the flavoprotein which is then reduced before taking up O_2, have been proposed for salicylate hydroxylase (White-Stevens *et al.*, 1972) and 4-hydroxybenzoate 3-hydroxylase (Nakamura *et al.*, 1970; Spector and Massey, 1972). Strickland and Massey (1973) consider that the oxygenated species, which furnishes the "active" O_2 for hydroxylation, might be reduced flavoprotein substituted by ·OOH at either C-10a or C-4a of the isoalloxazine ring system. Hamilton (1971) favors the latter, and has proposed that the middle ring of this system opens between N and C-4a to give a carbonyl oxide which would then become the oxygenating species. Resonance structures for this proposed intermediate can be written for which the terminal oxygen is positively charged and hence is electrophilic in nature. This is a necessary feature of any feasible proposal about the nature of enzymic "active" O_2, which always attacks the benzene nucleus ortho or para to the existing hydroxyl group, namely where electrons are most readily available. This condition would also be satisfied by an oxaziridine which Orf and Dolphin (1974) have proposed as the active oxygenating agent of flavin monooxygenases.

An interesting consequence of the electrophilic nature of "active" oxygen generated in enzymic hydroxylation is seen when the flavoprotein 4-hydroxyphenylacetic acid 1-hydroxylase is considered (Hareland *et al.*, 1975). The hydroxylation at C-1 is accompanied by a shift of the side chain from C-1 to C-2 [Fig. 5(a)] and this can be explained by the shifts of electrons depicted in Fig. 5(b). An absolute requirement for enzymic activity is that the substrate molecule shall contain a hydroxyl group at C-4; and it is assumed that the indispensible function of this group is to ionize, thereby supplying the electrons required for hydroxylation at C-1 when attacked by electrophilic oxygen, shown symbolically in Fig. 5(b) as [OH^+]. A rearrangement of electrons then leaves C-2 with a positive charge and promotes a shift of the

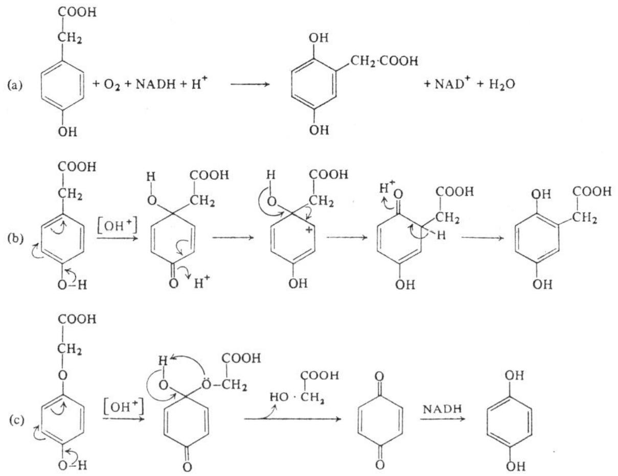

FIG. 5. The reaction catalyzed by the enzyme 4-hydroxyphenylacetic acid 1-hydroxylase, (a). Mechanism proposed for this reaction, (b); and for conversion of 4-hydroxyphenoxyacetic acid into hydroquinone by the same enzyme, (c). The hydroxylating, or electrophilic, species of oxygen is shown symbolically as $[OH^+]$.

carboxymethyl side chain to this position. Molecular rearrangement by 1,2 shifts of alkyl groups or hydrogen atoms are not uncommon when carbonium ions are generated by electrophilic reagents (Hendrickson *et al.*, 1970), and such a mechanism has been suggested to account for the migration of groups that is observed during enzymic conversion of squalene to cholesterol (Barker, 1971). Two other features of this enzyme may be noted:

(i) 4-hydroxyphenoxyacetate, which is identical to 4-hydroxyphenylacetate except that an oxygen atom has been inserted between the benzene nucleus and the side chain is attacked by this enzyme to give benzoquinone with release of glycollate, that is, the side chain of this substrate does not migrate. This behavior has been explained on the basis of the mechanism of Fig. 5(c).

(ii) Although the side chain is said to migrate, there is good evidence that this group remains bound to the enzyme surface throughout the reaction: the benzene ring itself appears to move so as to bring C-2 into proximity with the bound side shain, probably at the same time as the enzyme undergoes a conformational change (Hareland *et al.*, 1975).

E. DIHYDROXYLATION OF HYDROCARBONS

Benzene, toluene, ethylbenzene, and naphthalene are important products of industries based upon coal and petroleum; and for this reason, as well as for the academic interest involved, the utilization of these compounds

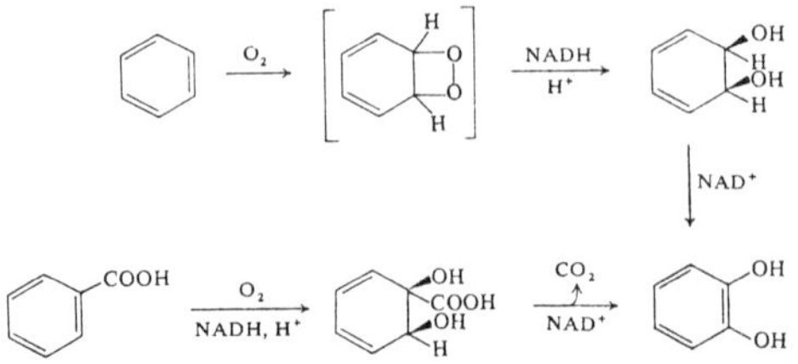

FIG. 6. Conversion of benzene and benzoic acid into catechol. The peroxide intermediate is hypothetical.

by microorganisms has received much attention during the last decade. It has emerged that the degradation of aromatic hydrocarbons is initiated by a mechanism that differs from that of any monohydroxylase discussed in previous sections, insofar as both atoms of molecular oxygen are incorporated, so that two hydroxyl groups are inserted into the substrate simultaneously (Gibson, 1972). It appears that one molecule of oxygen may be taken up by benzene to form a peroxide, and that the NADH required in the dihydroxylation is utilized for reducing the peroxide to a dihydrodiol. Gibson *et al.* (1970b) selected a mutant of *P. putida* that accumulated *cis*-benzene glycol from benzene (Fig. 6) and also incorporated two atoms of $^{18}O_2$ into this compound. Cell-free extracts were also found to contain a NAD-dependent dehydrogenase that converted *cis*-benzene glycol into catechol. Further, the mutant oxidized toluene to its corresponding *cis*-glycol, namely (+)-*cis*-2,3-dihydroxy-1-methylcyclohexa-1,4-diene (Gibson *et al.*, 1970a). The presence of a methyl group entails molecular asymmetry, and the absolute stereochemistry of this diol has now been determined (Kobal *et al.*, 1973; Ziffer *et al.*, 1973).

Reiner and Hegeman (1971) have also studied the enzymic oxidative decarboxylation of benzoate to give catechol, a reaction frequently encountered in pseudomonads but obscured for many years. They have established a reaction sequence involving the *cis*-dihydrodiol of benzoate as the precursor of catechol (Fig. 6). The formation of *cis*-diols by microbial systems, probably by the reduction of peroxides, stands in contrast to hydroxylations catalyzed by rat liver microsomes. These preparations oxidize naphthalene, for example, to give an epoxide that is then hydrolyzed to a *trans*-, rather than a *cis*-, dihydrodiol (Jerina *et al.*, 1968) whereas *Pseudomonas* forms *cis*-1,2,-dihydro-1,2-dihydroxynaphthalene from the hydrocarbon (Jerina *et al.*, 1971). Other compounds that are oxidized to *cis*-diols, and then to substituted catechols, prior to benzene ring fission by microorganisms include ethylbenzene, 4-fluorotoluene, 4-bromotoluene, chlorobenzene (Gibson *et al.*, 1973a), and biphenyl (Gibson *et al.*, 1973b). The metabolism of the last-named compound is of particular interest since polychlorinated diphenyls are worldwide environmental pollutants. Most of this work was done by bacteria, but there is mounting evidence that the aryl hydroxylating systems of fungi may be different: they appear to resemble those of liver microsomes. The migration and retention of deuterium (NIH Shift) that occurred during aromatic hydroxylation by several fungal species could best be accounted for in terms of epoxide intermediates (Auret *et al.*, 1971; Ferris *et al.*, 1973). Moreover, some organisms converted *o*-, *m*- and *p*-xylenes to hydroxylated products that were the same as those formed by mammalian systems (Smith *et al.*, 1976). Perhaps it is too early to speculate that a general difference between procaryotic and eucaryotic organisms is to be found in their different enzymic mechanisms of oxygenation.

F. Choices of Alternative Metabolic Pathways

We have seen that there are several main channels of metabolism available for degrading dihydric and trihydric phenols and that when bacteria prepare for growth at the expense of an aromatic compound they appear to "choose" one, and only one of these sequences. This commitment is assured by the induction of one of the substrate-specific hydroxylases described in Section III,E. Indeed, when one views the rich diversity of metabolic pathways available, and the precision with which one or another is chosen, one gains the superficial impression that microbes are bent upon displaying a rather pointless chemical expertise—that of metabolizing stable chemical structures, containing six to nine carbon atoms, in every manner conceivable. This impression is heightened when one considers instances when

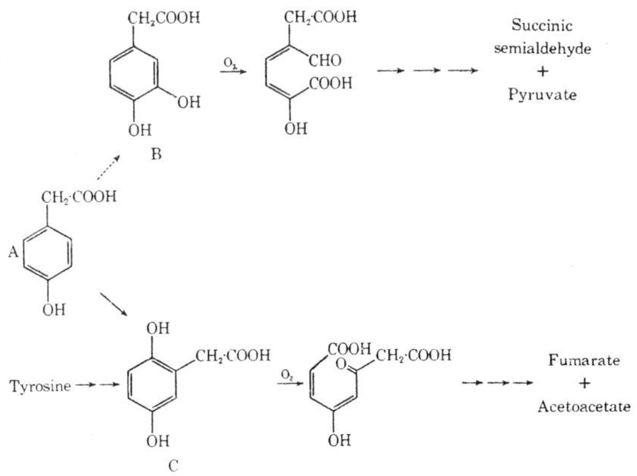

FIG. 7. Choice of degradative pathway by *P. acidovorans*. The cells are able to elaborate enzymes to degrade compound B to succinic semialdehyde and pyruvate, but when grown with compound A they convert it into C and produce fumarate and acetoacetate.

an organism is able to catalyze reactions of a sequence that we should consider to be perfectly suitable for degrading its growth substrate, and yet we find that it prefers another metabolic route. An example is provided by the observations of Sparnins and Dagley (1975) using *P. acidovorans*. When this organism is grown with succinate as carbon source and the cells are then exposed to 3,4-dihydroxyphenylacetic acid (compound B of Fig. 7), they elaborate all the enzymes required for rapid transformation of this substrate into Krebs-cycle metabolites. This is the reaction sequence shown for compound V, Fig. 3, where R = $CH_2 \cdot COOH$. It would be reasonable to anticipate that when the organism grows with 4-hydroxyphenylacetic acid as sole carbon source (compound A of Fig. 7), a hydroxylase for C-3 would be induced, as shown by the dashed line, so that the above sequence of reactions would become available for degradation. This hydroxylase is elaborated by some species, but not by this one. Instead, compound A is hydroxylated at C-1 to give homogentisic acid (III of Fig. 1), which then undergoes ring-cleavage, as shown in Fig. 1, and is degraded to Krebs-cycle metabolites by an entirely different pathway that involves fumaric and acetoacetic acids.

The experiments just described can be interpreted if we assume, first, that *P. acidovorans* has long possessed, since early times in its evolution, the suite of enzymes needed to degrade the natural product tyrosine, which can provide all the carbon and energy required for growth. Homogentisate is a catabolite of tyrosine in mammals and in many microbes. Second, we may assume that 4-hydroxyphenylacetic acid 1-hydroxylase (see Fig. 5) was evolved later in specific response to the presence of 4-hydroxyphenylacetate in some natural habitat, in order that this compound could be degraded by the existing enzymes of the homogentisate pathway and thereby satisfy growth requirements. Apparently, this organism never evolved a 3-hydroxylase for 4-hydroxyphenylacetate. By contrast, a group of Gram positive bacteria have recently been shown to possess this enzyme, and they metabolize tyrosine by way of 4-hydroxyphenylacetate and homoprotocatechuate (Sparnins and Chapman, 1976). These organisms stand in contrast to *P. acidovorans* because they do not use the homogentisate pathway.

Like tyrosine, certain dihydric and trihydric phenols have served as nutrients for soil bacteria during long periods of evolution. It is probable that those degradative products of lignin that fall into this category have provided an abundant and continual supply of initial substrates needed for establishing the main channels of bacterial degradation (Fig. 4). Large numbers of other aromatic natural products that did not already possess hydroxyl groups suitably positioned for attack by dioxygenases would also be encountered, and it would be necessary to prepare them for entry into functioning catabolic pathways by the action of specific hydroxylases. Such compounds as these are synthesized by higher plants or may be derived from natural products by fungal action; for example, a notable feature of the metabolism of aryloxyacetic acids by fungi is the predominance of apparent nonspecific nuclear hydroxylation without ring fission (for references, see Faulkner and Woodcock, 1964).

Every biochemical system of living matter must, of course, accord with chemical logic, but the range of activities possessed by an organism is a legacy of its evolution. Thus, as life evolved in an atmosphere of increasing oxygen concentration, many toxic materials made their appearance and threatened survival. It is a mistake to imagine that all compounds synthesized in nature are harmless whereas chemicals introduced by man invariably involve hazards. Of those people who know that there are deadly poisons in certain fungi and toxic alkaloids in many plants, many may not appreciate that our daily food contains real or potential toxicants which are of natural occurrence. Crosby (1969) has observed that "eating represents the greatest direct exposure to exotic compounds that most of us ever received." Indeed, the American Chemical Society's symposium on natural food toxicants considered many natural compounds which belong to categories that are associated exclusively in the public mind with manufactured products. These included a class of herbicides and fish poisons (furanocoumarins, present in certain plants), a plant growth regulator (indoleacetic acid), insecticide-like compounds found in carrot and potato, and a variety of fungicides, bacteriocides, and antibiotics. To microbes, during the course of evolution, such compounds came to serve as sources of energy, to be oxidized completely to carbon dioxide. Thus, atropine, a natural product probably more toxic than most pesticides, serves as the sole source of carbon for certain microorganisms (Niemer *et al.*, 1959; Niemer and Bucherer, 1961). To other living forms these compounds were xenobiotics, evoking mechanisms of detoxication and elimination. Accordingly, as we have seen, the cytochrome P_{450} systems of liver hydroxylate a wide range of compounds, and epoxides and *trans*-diols are intermediates in elimination processes. The flavoprotein hydroxylases of bacteria are by contrast stringently substrate specific and they direct metabolism to equally specific degradative sequences. Far from being eliminated by bacteria these substrates, by serving as sources of carbon and energy, have assured the organisms of certain evolutionary advantages.

IV. Microbial Degradation of Alicyclic and Heterocyclic Compounds

Although degradation of benzenoid compounds is quantitatively more important, alicyclic and heterocyclic compounds are widely distributed in nature and are utilized by microorganisms as sources of carbon for growth.

Natural alicyclic compounds range from the structurally simple cyclohexane, found in small quantities in crude oil, through simple terpenes such as camphor and menthol, to the complex sesquiterpenes and sterols. Natural pesticides such as the pyrethrins contain both cyclopentene and cyclopropane rings and the cyclohexane ring is a component of such synthetic pesticides as tricyclohexyltin hydroxide, cycloate (S-ethyl-N-ethyl-N-cyclohexylthiolcarbamate) and lenacil (3-cyclohexyl-5,6-trimethyleneuracil). Other compounds of natural occurrence include aflatoxins, griseofulvin, streptomycin, and morphine which are heteropolycyclic, and contain alicyclic, aromatic, and furan ring elements; the furan ring is a component of menthofuran, benzofurans, furanocoumarins, and furanochromones (Brown *et al.*, 1970; Harrison *et al.*, 1971). The pyridine ring occurs in nature mainly in the form of pyridine coenzymes (nicotinamide and pyridoxal derivatives) and plant alkaloids, while other pyridine compounds find their way into the environment as industrial effluents and herbicides. Two modes of fission of the pyridine nucleus have been described which show resemblances to the gentisate and catechol meta-fission pathways. Thus, Berhman and Stanier (1957a) showed that 2,5-dihydroxypyridine, in which the hydroxyl groups are positioned across the nucleus in an analogous fashion to those in gentisic acid, was metabolized to maleic acid by *P. fluorescens*. In this case, an isomerase converted maleate to fumarate. Watson *et al.* (1974b) described another pathway in which 3,4-dihydroxypyridine, formed from 4-hydroxypyridine by hydroxylation, is cleaved between C-2 and C-3 and is then metabolized to formate and pyruvate by reactions analogous to those employed in the meta-fission pathway for catechol. These authors have purified this hydroxylase which is a typical NADH-dependent flavoprotein (Watson *et al.*, 1974a). Another specific hydroxylase in this category converts 2,6-dihydroxypyridine to 2,3,6-trihydroxypyridine during the bacterial oxidation of nicotine (Holmes and Rittenberg, 1972). In some instances, therefore, degradative sequences are initiated by reactions similar to those I have described for benzenoid compounds; accordingly, attention will now be focussed upon reaction mechanisms that appear to apply specifically to microbial attack upon alicyclic and heterocyclic compounds. These include hydroxylation by addition of water, and the formation of lactones or other esters by the action of monooxygenases.

When a microbial growth substrate is attacked by a flavin hydroxylase there is an initial investment of energy, through the NAD(P)H required, which is more than balanced by energy released by later catabolic reactions. Reoxidation of NADH through the operation of the electron transport system of an aerobic organism generates three molecules of adenosine triphosphate (ATP) by oxidative phosphorylation of ADP. However, quite

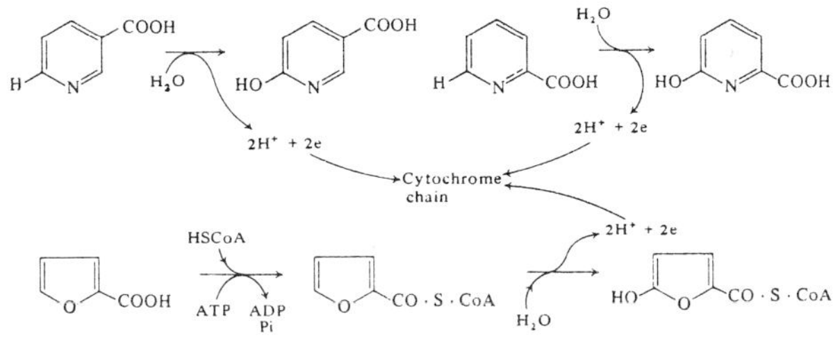

FIG. 8. Hydroxylations that use water. Substrates: nicotinic acid, picolinic acid, and 2-furoic acid.

stable ring systems are susceptible to attack without the participation of oxygen and without the expenditure of reducing power. The sequences of Fig. 8 show that the oxygen atom of water is incorporated enzymically, as hydroxyl, into the pyridine rings of nicotinic acid (Hunt *et al.*, 1958) and picolinic acid (Dagley and Johnson, 1963). Similarly a species of *Pseudomonas* uses water to hydroxylate the furan ring of furoic acid, converting the substrate first into its coenzyme A ester (Trudgill, 1969). This initial activation involves an investment of energy, but the actual hydroxylation reaction does not. In each of these systems, in contrast to those mentioned previously, electrons are not supplied to the enzyme but are removed through the cytochrome chain. They might be regarded as being "pulled" by this removal, whereas other hydroxylases receive a thermodynamic "push" from electron donors. It may be noted that the conversion of nicotinic acid into 6-hydroxynicotinic acid has been demonstrated for strictly anaerobic systems. Thus, an FAD-containing (and none-heme iron) enzyme from a *Clostridium* species inserts oxygen derived from water into nicotinic acid, and, in this case, NADP serves as the necessary acceptor of electrons (Holcenberg and Stadtman, 1969). 6-Hydroxynicotinate is therefore the initial product formed from nicotinate by aerobic and anaerobic bacteria alike. A probable mechanism for these pyridine-ring hydroxylases would involve nucleophilic attack, by OH^- derived from water, at the C-6 position of an electron-deficient carboxylated pyridine ring. Thus 6-hydroxydihydronicotinate, for example, would be formed by hydration and then dehydrogenated to give 6-hydroxynicotinate. The subsequent metabolism of this compound presents unusual features. Behrman and Stanier (1957a) showed that it is oxidatively decarboxylated to give 2,5-dihydroxypyridine. Gauthier and Rittenberg (1971) purified to homogeneity, and crystallized, a dioxygenase that opens the nucleus of 2,5-dihydroxypyridine between C-5 and C-6, so that N-formylmaleamate would be the ring-fission produced, expected. However, this compound could not be detected, neither was it a substrate for the enzyme: the products were formate and maleamate. It appears that oxygenation and hydrolysis may occur as concerted reactions, so that N-formylmaleamate neither associates with the enzyme, nor is dissociated from it.

In contrast to benzene and pyridine rings, the furan ring is not opened by dioxygenases. Trudgill (1969) has proposed the following reaction sequence for the conversion of 5-hydroxy-2-furoyl-CoA of Fig. 8 into a Krebs cycle intermediate, 2-ketoglutaryl-CoA:

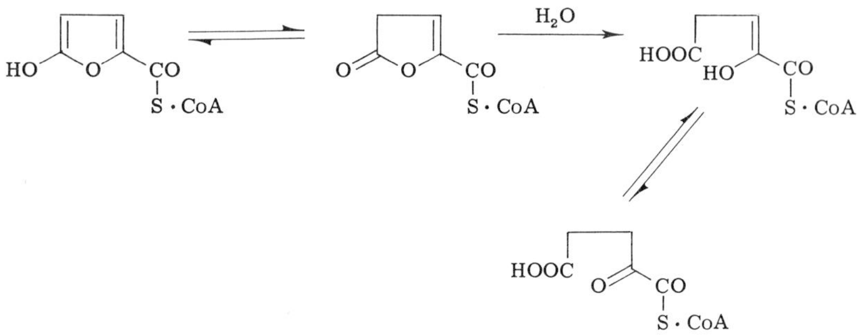

Although the detailed enzymology of this sequence has not yet been established in this case, there are several precedents for the insertion of an oxygen atom between two carbons by monoxygenases prior to hydrolytic ring fission. This reaction is analogous to the Baeyer-Villiger reaction in which peracids are used to convert R.CO.R′ into R.COOR′, and examples are given in Fig. 9. Reaction (a), conversion of 4-androstene-3,17-dione into testolactone by soluble extracts of *Penicillium lilacinum*, was demonstrated by Prairie and Talalay (1963). In reaction (b), oxygen is inserted into one of the rings of 2,5-bornanedione (or 2,5-diketocamphane) to form a lactone, and this incorporation is catalyzed by an enzyme system consisting of three protein fractions. One of these fractions transfers reducing power from NADH to FAD, and a second—the lactonase—contains ferrous iron and uses $FADH_2$ to reduce one of the oxygen atoms to water and to insert the other (Trudgill *et al.*, 1966). The second fraction was shown later to consist of two proteins (Yu and Gunsalus, 1969). The flavoprotein cyclopentanone, NADPH: oxygen oxidoreductase (1,2-lactonizing) which catalyzes reaction (c), giving 1-oxa-2-oxocyclohexane (5-valerolactone), was purified by Griffin and Trudgill (1976) from a species of *Pseudomonas*. Cripps (1975) has shown that reaction (d) initiates the elimination of the side chain of acetophenone in *Arthrobacter*, and Britton *et al.* (1974) showed that 2-tridecanone is converted to undecyl acetate by reaction (e) in a species of *Nocardia* and in *P. cepacia*.

After lactone formation [reactions (b) and (c)], or ester formation [reactions (d) and (e)], enzymatic hydrolysis leads to ring opening or to

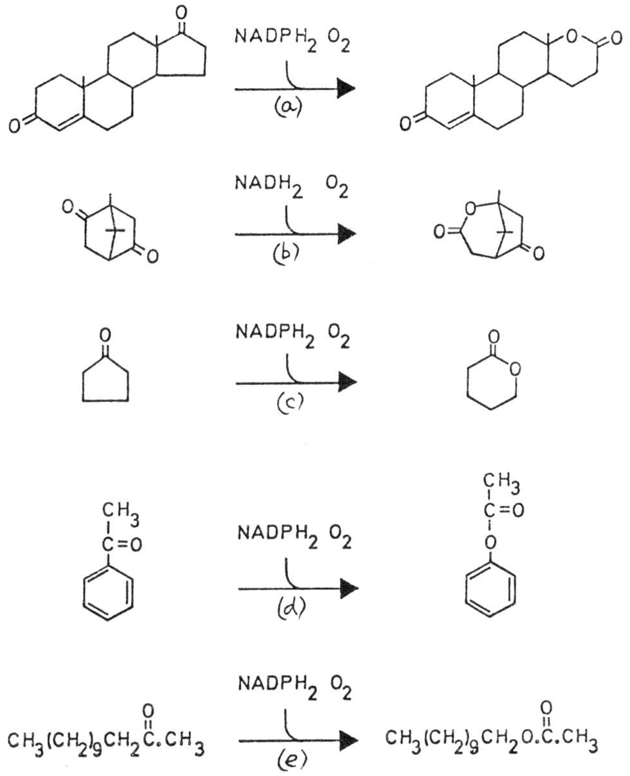

FIG. 9. Enzymatic insertion of an oxygen atom between two carbon atoms.

release of acetate. Thus, 5-valerolactone formed in reaction (c) is hydrolyzed to give 5-hydroxyvalerate; the hydroxymethyl group of this compound is then enzymatically oxidized, giving glutarate, which is then further metabolized to two molecules of acetyl-CoA, plus carbon dioxide (Griffin and Trudgill, 1972). Other oxygenases that catalyze insertion of oxygen into the cyclohexanone ring have been purified from *Acinetobacter* and *Nocardia globerula* (Donoghue *et al.*, 1976). Like cyclopentanone oxygenase they resemble the enzymes that hydroxylate the benzene nucleus and do not appear to require transition-metal ions for activity. The enzymes from *N. globerula* and *Acinetobacter* have molecular weights of 53,000 and 59,000 respectively, and each is a single polypeptide chain with one molecule of bound FAD that does not dissociate during purification. The enzyme from *N. globerula* exists as two electrophoretically distinct forms (Norris and Trudgill, 1976).

The metabolic function of lactone-forming monooxygenases may be

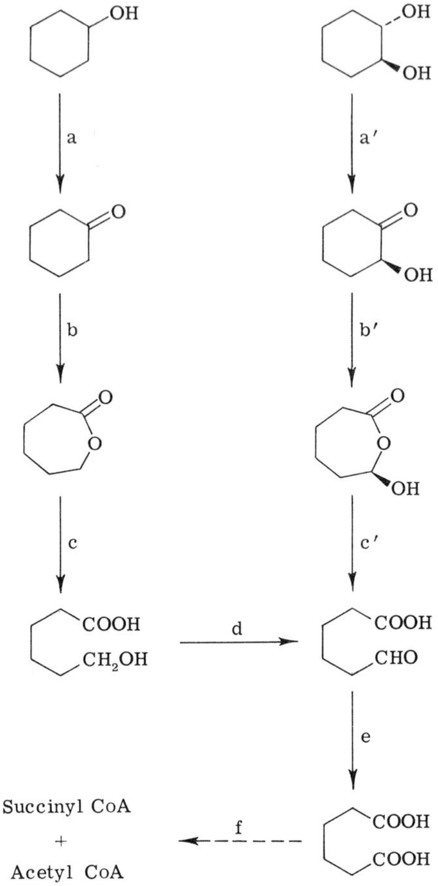

FIG. 10. Bacterial degradation of cycluhexanol and *trans*-cyclohexane 1,2-diol.

illustrated by the catabolic sequences used by *Acinetobacter* to degrade cyclohexanol and *trans*-cyclohexane-1,2-diol (Fig. 10). Dehydrogenases a,a′ form cyclic ketones, into which oxygen atoms are then inserted by monooxygenases b,b′. On the cyclohexanol pathway, a lactone hydrolase c forms 6-hydroxyhexanoate, whereas reaction c′ on the cyclohexanediol route occurs spontaneously to give 6-ketohexanoate (adipate semialdehyde). A dehydrogenase, d, connects the two pathways and adipic acid then is formed from its semialdehyde by another dehydrogenase, e (Donoghue and Trudgill, 1975; Trudgill, 1976). Adipate is metabolized by bacteria to succinyl-CoA and acetyl-CoA (Chapman and Duggleby, 1967).

Up to the time of writing, microbial hydroxylases that contain cytochrome P_{450} have rarely been encountered. A notable, and thoroughly investigated, exception is the enzyme system that hydroxylates camphor at C-5 to give the metabolic precursor of 2,5-diketocamphane, the substrate for monooxygenase (b) of Fig. 9. Hydroxylation of camphor is catalyzed, not by a flavoprotein but by cytochrome P_{450}. Altogether, three proteins are involved in the reaction. First, an iron-sulfur protein, putidaredoxin, is reduced by the electron donor for the system, NADH. The second component, redoxin, is then reduced and a sequence of reactions analogous to those described for the flavoprotein hydroxylases is set in motion: cytochrome P_{450}, in combination with substrate, receives electrons from reduced redoxin and takes up oxygen. The substrate is then hydroxylated and the oxidized cytochrome is regenerated as the oxidized substrate is released (Gunsalus, 1972).

In the case of cyclohexanecarboxylic acid degradation, some microorganisms employ reactions used for the metabolism of benzenoid compounds. Thus, Guyer and Hegeman (1969) showed that the photosynthetic microorganism *P. palustris* could grow aerobically and nonphotosynthetically with cyclohexanecarboxylate [Fig. 11,(I)] as source of carbon, forming compound (II) as the first reaction in a sequence that give rise to pimelic acid (III), which could then furnish metabolites of the Krebs cycle. These reactions are similar to those employed by the same organism growing photosynthetically (and anaerobically) with benzoic acid as source of carbon. For the photosynthetic pathway there is evidence that benzoic acid is first converted into benzoyl-CoA which is then dehydrogenated to give the coenzymeA ester of compound (II); subsequent reactions may then be as shown in Fig. 11, except that the thio esters are intermediates rather than the free acids (Dutton and Evans, 1969; Evans, 1976). In other microorganisms an alternative pathway exists which is initiated by hydroxylation of cyclohexanecarboxylate at C-4. Dehydrogenation yields 4-carboxycyclohexanone (IV) which, by tautomerization provides one double bond toward the process of aromatization. The remaining two double bonds are introduced by an oxidase that uses the electron-transport system of the organism, for which artificial electron acceptors can be substituted (Blakley, 1974; Kaneda, 1974). 4-Hydroxybenzoate results from these dehydrogenations; hydroxylation then furnishes protocatechuate (V) which can undergo either ortho cleavage (Blakley, 1974) or meta cleavage (Smith and Callely, 1975).

Although the thiophen ring is of relatively restricted natural distribution, 2-substituted thiophens are components of tars and oils (Challenger, 1959). The microbial metabolism of thiophen-2-carboxylate by organisms capable of growth with the compound as sole source of carbon has been studied by Amphlett and Callely (1969) and Cripps (1973) and closely parallels the pathway for cleavage of furan-2-carboxylate. Hydroxylation, utilizing the oxygen from water, yields the 5-hydroxythiophen-2-carboxyl-CoA and hydrolytic cleavage of the internal thiol ester with elimination of the sulfur

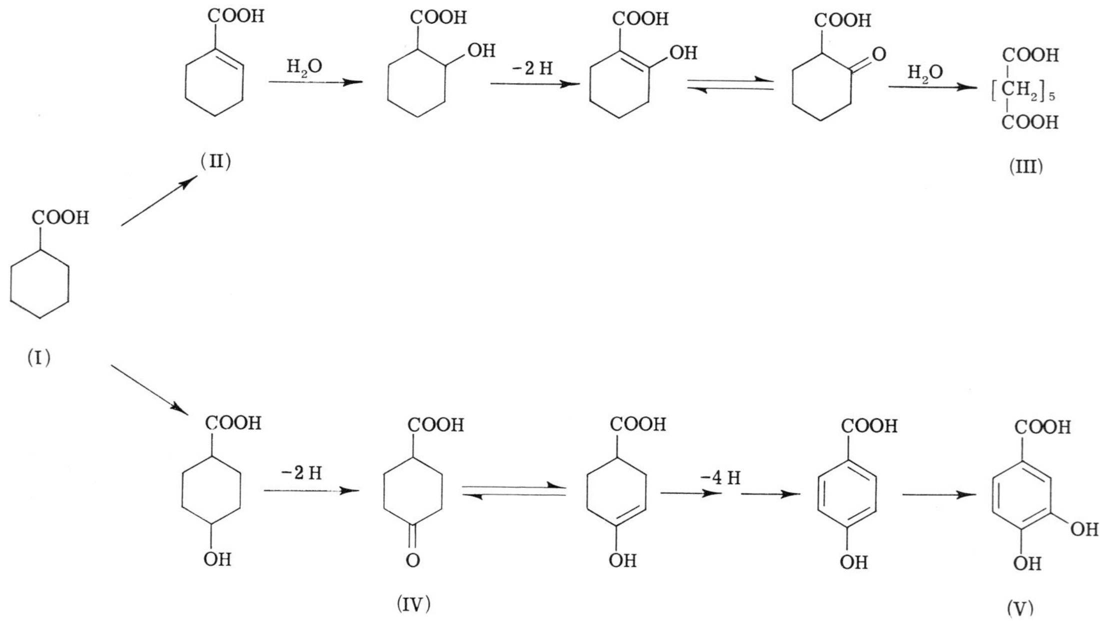

FIG. 11. Alternative pathways for the bacterial metabolism of cyclohexanecarboxylic acid (I), forming pimelic acid (III) or protocatechuic acid (V).

as S^{2-} or SO_4^{2-} results in the formation of 2-oxoglutarate. The close relationship between thiopen and furan metabolism is illustrated by the ability of thiophen-2-carboxylate-grown organisms to metabolize the furan analogue.

V. Microbial Degradation of Man-Made Compounds

A. Introduction

A man-made compound will be biodegradable if it is susceptible to attack by the enzymic apparatus acquired by microbes during the course of evolution. This, in turn, depends upon two factors: first, the ability of microbial enzymes to accept as substrates compounds having chemical structures similar to, but not identical with, those found in nature; and second, the ability of these novel substrates, when in the presence of microbes, to induce or derepress the synthesis of the necessary degradative enzymes. Biodegradability is less likely to be achieved by incorporating into a pesticide molecule structural features that are never encountered in natural products. Conversely, when, for example, the para chlorine atoms of DDT are replaced by methoxyl groups—very common substituents of the benzene nucleus in natural products—the resulting pesticide, methoxychlor, is much more readily metabolized by microbes than was the original DDT. A fruitful application of this principle by Dr. Michael Elliott and his co-workers at Rothamstead Experimental Station, England, has been reported by Aldrich Chemical Company (1976). They have prepared a range of very active synthetic analogues of naturally occurring insecticides, the pyrethrins, which promise to be as effective as other compounds without posing ecological problems. Another example of the advantages gained by preserving structural features found in natural products is provided by the development of synthetic detergents of the alkylbenzenesulfonate class. These compounds formerly created a nuisance in sewage treatment because they possessed highly branched side chains and were very persistent. Biodegradability was achieved by replacing them with linear alkylbenzenesulfonates, thus permitting side chains to be metabolized by β oxidation (Swisher, 1963, 1964). A complete understanding of factors that affect the biodegradability of industrial and agricultural chemicals will only emerge from studies of the induction and the modes of action of these enzymes employed for degrading natural products. In any laboratory study of man-made molecules, there will come a point at which experimental findings must be interpreted within the framework of knowledge of microbial metabolism in general.

B. Industrial Products Readily Degraded by Microorganisms

A large number of naphthoquinones are found in nature as vitamin K and related compounds, and as yellow-red plant pigments; the largest group of natural quinones is made up of anthraquinones (Robinson, 1967). However, most of our current detailed knowledge of the degradative pathways of the naphthalene and anthracene ring structures is due to the investigations of Evans and his colleagues (Davies and Evans, 1964; Evans *et al.*, 1965) working with the unsubstituted hydrocarbons. Polynuclear aromatic hydrocarbons and the organisms that metabolize them are present in soil around oilfields, and naphthalene and anthracene are important products of coal-based industry. On the other hand it is, perhaps, surprising to realize that mankind may have been exposed to small amounts of these potentially carcinogenic compounds throughout the ages. Blumer (1961) has shown that benzpyrenes, phenanthrene and chrysene are present in rural soils distant from highways and industry, so that they are unlikely to have arisen entirely as fall-out from polluted air. Their origin is a matter of speculation and may be due to several natural processes; but since they are formed when wood and other organic matter is burned, they could have arisen from the regular sequence of forest fires that have always occurred even in uninhabited areas. Blumer (1976) has recently presented an extensive survey of their occurrence and a critical account of theories as to their natural origins.

Microbial attack on these hydrocarbons is initiated by dioxygenases that yield *cis*-dihydrodiols from which *ortho*-dihydric phenols are then formed, prior to ring fission by another dioxygenase. Once the substituted ring of, say, 1,2-dihydroxynaphthalene is opened reactions occur similar to those described for catechols (Fig. 3), namely hydration and aldol fission releasing pyruvate and leaving salicylate. This compound is then oxidized to catechol and degraded by meta-fission reactions. Evans *et al.* (1965) have described a fascinating cycle of degradations for phenanthrene. One of the three aromatic rings is dihydroxylated and then degraded to leave 1,2-dihydroxynaphthalene, and this in turn is degraded one ring at a time as described. From the 14 carbon atoms of the stable resonance structure of phenanthrene, the microorganisms obtain three molecules of pyruvate and one of acetaldehyde at a cost of losing only three carbon atoms as carbon dioxide. It is not certain how many enzymes of these various degradative cycles are essentially the same as those studied in other meta-fission pathways; but studies of their substrate specificities might be particularly rewarding.

In addition to the hydrocarbons naphthalene and anthracene, industrial chemicals such as phenol, cresols (methylphenols) and xylenols (dimethylphenols) are furnished by the high temperature carbonization of coal. At least in the United States, which possesses about half of the world's estimated reserves of coal, these processes are likely to increase in importance as alternative supplies of fossil fuels progressively diminish. Bacteria capable of degrading cresols and xylenols are widespread in nature, probably because enzyme sequences used for metabolizing more abundant natural products are also able to function with these phenolics. Their removal from industrial effluents by these microbes call for vigorous aeration in order to provide the oxygen required by hydroxylases and ring-cleaving dioxygenases. Without such aeration, phenols tend to persist and to be converted into objectionable chlorophenols when water for domestic use is treated with chlorine (Davis, 1967; Higgins, 1968).

The ability to degrade a cresol or a xylenol depends mainly upon the choice of catabolic routes available to the organism. Thus, when *P. testeroni* was grown with *p*-cresol (4-methylphenol) as carbon course (Dagley and Patel, 1957) the available pathway consisted in oxidizing the methyl group to carboxyl, giving 4-hydroxybenzoate; this compound was then converted into protocatechuate and degraded by dioxygenase (2) of Fig. 2, followed by reaction sequence VI of Fig. 3. However, these organisms were not able to degrade xylenols because the insertion of a second methyl group entails the formation of reaction intermediates that are not accepted as substrates by later enzymes of this particular catabolic route. By contrast, strains of *P. putida* oxidized cresols by different enzymic reactions that could tolerate the presence of an additional methyl group when suitably placed. In this case the nucleus of the cresol was attacked initially and the methyl group was left intact. Thus, p-cresol was hydroxylated to give 4-methylcatechol, and o- and m-cresols gave 3-methylcatechol (Bayly *et al.*, 1966); the methylcatechols were then degraded by dioxygenase (1), Fig. 2, followed by reaction sequences IV or V of Fig. 3. In these reactions the methyl group appears either in acetic acid, which is formed by hydrolysis of the ring-fission product in sequence IV or else it appears in propionaldehyde, formed by aldol fission at the end of sequence V. It will be seen that in 3,4-dimethylcatechol the methyl groups are so placed as to allow both alternatives. Accordingly, it is found that some strains of *P. putida* can fully utilize 2,3-xylenol (2,3-dimethylphenol) and also 3,4-xylenol because each substrate is hydroxylated to give 3,4-dimethylcatechol prior to ring fission.

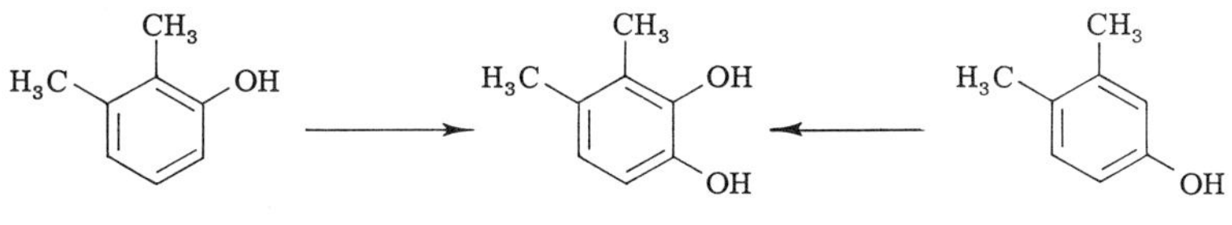

However, the enzymes of these reaction sequences are not capable of degrading xylenols that contain methyl groups placed 2,5- or 3,5- to the hydroxyl group. Such compounds appear only to be metabolized by those pseudomonads that utilize the gentisic pathway. Provided one methyl group is situated *meta* to hydroxyl, these organisms catalyze the following general reaction sequence to give substituted gentisic acids:

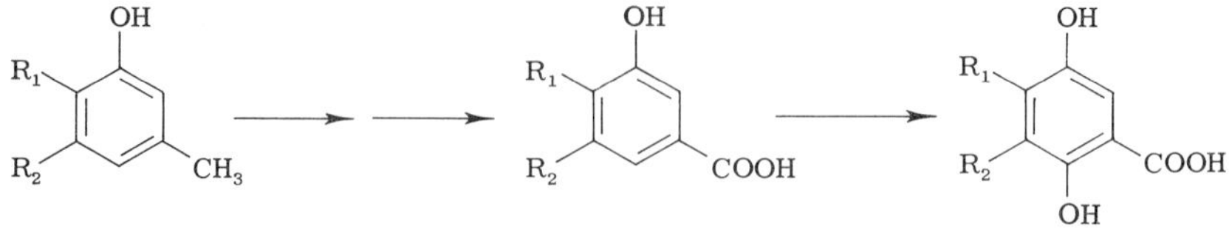

The utilization of several monohydric phenols has been investigated (Hopper and Chapman, 1971) for which R_1 = H, CH_3, or C_2H_5 and R_2 = H, or CH_3. Two metabolic routes for gentisic acid have also been studied: one of them yields fumarate, and hence L-malate, because maleylpyruvate (the product of ring fission II, Fig. 1) is converted to a reduced glutathione-dependent isomerase into fumarylpyruvate which is then hydrolyzed to fumarate and pyruvate. The alternative route does not involve isomerization, and maleylpyruvate is immediately hydrolyzed to give pyruvate and maleate, and hence D-malate (Hopper *et al.*, 1968, 1970). Of the two gentisate pathways, the one that does not involve isomerization has been found to exhibit the broadest substrate specificity of all the routes of aromatic catabolism studied to date. Accordingly, the alkyl-substituted gentisic acids formed from these monohydric phenols are converted by the organisms that use this pathway into alkyl-substituted D-malic acids. The CoA esters of these acids are then cleaved by aldolases and the products are further metabolized (Hopper *et al.*, 1971). These investigations of the microbial metabolism of cresols and xylenols illustrate the fact that some catabolic pathways for natural products can be used more effectively than others for degrading industrial wastes.

The hydrocarbon biphenyl is distributed widely enough to permit the ready isolation of microorganisms that can utilize the compound as a carbon source, and its degradation has been investigated by several groups of workers (Lunt and Evans, 1970; Catelani *et al.*, 1971; Ahmed and Focht, 1973; Ohmori *et al.*, 1973). The hydrocarbon is attacked by a dioxygenase similar to that for benzene (Fig. 6), giving *cis*-2,3-dihydro-2,3-dihydroxybiphenyl (Gibson *et al.*, 1973b) which is dehydrogenated to give 2,3-dihydroxybiphenyl. Ring fission then occurs as shown in Fig. 3,IV, so that benzoic acid is formed by hydrolysis (R = phenyl) and is further metabolized. Ohmori *et al.* (1973) have shown that 4-chlorobenzoic acid is formed from *p*-chlorobiphenyl; from some cultures γ-benzoylbutyric acid was isolated. However, extensive substitution of biphenyl with chlorine blocks the reactions described and, as is well known, polychlorinated biphenyls are extremely persistent pollutants.

C. Industrial Products That Resist Biodegradation

Chlorinated compounds have been used extensively for pest control and have often proved to be persistent. I shall choose these and other pesticides for discussing general approaches to these problems. In the past, these compounds were selected for agricultural use on the basis of their effectiveness against target organisms and their low cost of manufacture. On the whole, the use of synthetic pesticides has been amply justified by their contribution to the success of modern agriculture in meeting human needs, and by their value in the fight against such age-old scourges of mankind as malaria and sleeping sickness. During the last decade, a great deal of research has been necessary in order to determine the environmental fates of pesticides, or compounds formed from them by microbial action. These slow and incomplete degradations may be viewed as failures of microbes when confronted by novel, nonnatural products of laboratory syntheses. The literature relating to these partial degradations is extensive but fragmentary and uncoordinated; and no details will be given in this survey unless they are germane to discussion. Several excellent reviews of the area are available (Alexander, 1967; Kearney and Kaufman, 1972; Bollag, 1972).

I shall focus this discussion by posing two questions. First, what effects do structural modifications have upon biodegradation? We shall consider the simple case of substituting a hydrogen atom by chlorine at C-4 in the readily metabolizable compound, benzoic acid. The second question stems from recognizing the fact that biochemists routinely adopt the approved scientific attitude, that progress is most rapid when an investigation is confined in scope and its variables are reduced in number. They use pure cultures of microorganisms, usually chosen for abundant overnight growth, and they investigate the metabolism of single substrates. It may be mentioned in passing that recent observations in my laboratory confirm those of Gray and Thornton (1928) which have not received the attention they merit, namely that many types of widely distributed and slow growing organisms are probably responsible for the turn-over of large quantities of aromatic material in nature, although their time cycles are ill adjusted to those of the laboratory. These considerations give rise to the second question: how is the degradation of a compound affected by additions of other compounds, and by the presence of other bacterial strains? For these are the conditions encountered in nature, if not in the laboratory.

In considering the first question, we saw that the benzene nucleus is attacked initially by an electrophilic form of oxygen, and we might therefore expect that substituents such as chlorine, which withdraw electrons from carbon would render the nucleus less susceptible to attack. Moreover, the more chlorine atoms there are, the greater the deactivation; but probably more important, the greater is the probability that chlorine-containing metabolites, should they be formed, will resist attack by subsequent enzymes. Alexander and Lustigman (1966) and MacRae and Alexander (1965) have indeed shown that the presence of halogen atoms greatly reduced biodegradability, and have investigated various other effects of molecular structure upon persistence.

However, the question I have posed, while reflecting a natural and common line of thought, is not entirely appropriate. The capabilities of an organism are determined by its evolutionary past, and it may not have encountered 4-chlorobenzoate previously. There is no reason why it should respond to this compound as it would to benzoate; moreover, its enzymes are far from being simple catalysts. These observations prompt the following reflections.

1. If chlorine is eliminated at an early stage of degradation, it is probable that its place will be taken by hydroxyl, giving 4-hydroxybenzoate. As described previously, every one of the enzymes employed for metabolizing this compound to Krebs cycle intermediates will be different from its counterpart used for benzoic acid. A separate suite of enzymes will be induced. Neither 4-chlorobenzoate nor 4-hydroxybenzoate will be treated as benzoate, slightly modified in structure, but rather as a separate essentially unrelated aromatic substrate.

2. Predictions based exclusively on chemical reactivity for nonenzymic systems are not necessarily valid when the reaction is enzyme catalyzed. For example, fumarase catalyzes the hydration of fluorofumarase to give α-fluoromalate, from which H^+ and F^- ions are immediately expelled to give oxaloacetate (Clarke *et al.*, 1968). This hydration is predicted on purely chemical grounds by Markownikoff's rule. However, chlorofumarate and other halogenofumarates are hydrated the other way around, yielding β-substituted malates, and the halogen is not expelled. The intervention of the enzyme gives an unexpected result—one that cannot be predicted from electron density distributions for molecules not bound to this enzyme (Teipel *et al.*, 1968). Purely chemical considerations might also be evoked to account for the observations that, whereas the benzene nucleus of 4-methylcatechol is oxidatively cleaved between C-2 and C-3, meta fission of 4-carboxycatechol (protocatechuate) by pseudomonads occurs between C-4 and C-5 (Dagley, 1971). However, such considerations would not explain why a dioxygenase recently isolated from *Bacillus circulans* (Crawford, 1975) differs from those from pseudomonads in cleaving the same substrate, protocatechuate, between C-2 and C-3.

3. To treat the degradation of a halogen-substituted compound exclusively in terms of the known reaction pathway for the unsubstituted parent compound is to ignore the possibility that enzymes may exist whose

natural substrates contain carbon–halogen bonds. Because such compounds do not participate in the central metabolic cycles and sequences of biochemistry, the halogen atoms must be eliminated before these points are reached in degradative routes. Will this elimination be fortuitous, that is, will the halogen be released by an enzyme that does not normally function for that purpose? Or will this be achieved by an enzyme especially evolved for elimination? Clearly, since pesticides are made by man and not by nature, the enzyme is not likely to be tailored to accommodate groups around the particular carbon–halogen linkage under consideration. However, halogenated compounds are now known to be much more widely distributed in nature than was formerly believed. Thus, although only about 30 organic compounds containing chlorine were characterized in 1968 as natural products (Fowden, 1968) a review of Siuda and De Bernardis (1973) lists more than 200 naturally occurring compounds with carbon-halogen bonds, and about 150 of them contain chlorine. Recent surveys of natural products found in marine environments indicate that many new compounds, especially those containing bromine, will shortly be added to the list (Dr. Lowell P. Hager, personal communication). It is therefore probable that many bacterial enzymes remain to be discovered that have been evolved specifically for the purpose of releasing halogens bound to carbon; and it will be interesting to evaluate their functions, if any, in pesticide degradation. Indeed, some bacterial enzymes specific for the hydrolysis of carbon–halogen bonds in halogenoacetic acids have been described by Goldman (1965, 1972a,b). Fluoroacetic acid is a natural product, synthesized by the South African shrub *Dichapetalum cymosum*, and other fluoro fatty acids have been isolated from *D. toxicarium* (ratsbane). In other instances the fluorine atoms are eliminated fortuitously from compounds during the operation of enzymes that attack the halogen-free substrates (Goldman, 1972a).

The most thorough research on herbicide degradative pathways that probably involve enzymes for the specific (or nonfortuitous) elimination of chlorine is that of the research groups of Alexander (Tiedje *et al.*, 1969; Tiedje and Alexander, 1969; Bollag *et al.*, 1968) and Evans (Evans *et al.*, 1971 a,b; Gaunt and Evans, 1971a,b). These studies relate to the microbial degradation of 4-chlorophenoxyacetate (X = H, Fig. 12), 4-chloro-2-methylphenoxyacetate (MCPA; X = CH_3, Fig. 12) and 2,4-dichlorophenoxyacetate (2,4-D; X = Cl, Fig. 12). Their findings may be summarized in the degradative pathway shown in Fig. 12, although a variation on this basic pattern occurs for 2,4-dichlorophenoxyacetate (Tiedje *et al.*, 1969). Among lines of evidence that suggest the use of specific enzymes for dechlorination may be mentioned the observations that cells grown with 2,4-D or MCPA oxidize various chlorophenols and chlorophenoxyacetic acids

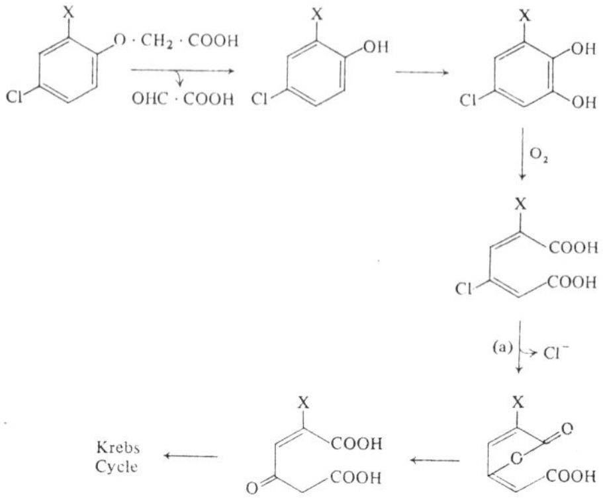

FIG. 12. Microbial degradation of substituted phenoxyacetic acids. The herbicides 4-chlorophenocyacetate, 4-chloro-2-methylphenoxyacetate (MCPA), and 2,4-dichlorophenoxyacetate (2,4-D) are represented when X = H, CH_3, or Cl, respectively.

whereas the chlorine-free compounds are poorly oxidized; and also that the lactonizing enzyme (a), Fig. 12, which was partially purified by Gaunt and Evans (1971b) from a pseudomonad grown with MCPA, appears to eliminate chlorine in a one-step reaction and differs in several respects from other lactonizing enzymes in ortho-fission pathways for substrates containing no halogen.

Bacteria that grow at the expense of chlorinated phenoxyacetic acids may be isolated quite readily from farm soils. These herbicides possess the environmental advantage of being biodegradable, although it may be noted that some degree of persistence is essential for effective pesticidal action: the compound should persist throughout a growing or breeding season and then, ideally, should be removed. Much attention has rightly been devoted in past years to the fate of pesticides more persistent than this; many empirical data have been accumulated, but it cannot be said that a coherent view of microbial activities has emerged. For future research, although it might not appear to be so immediately relevant to some environmentalists, an investigation of microbial enzymes that are used to eliminate halogens from natural products would ultimately be more profitable. The search for acceptable pesticides would be greatly helped by a better understanding of what bacteria are capable of doing, rather than of what they usually fail to do.

The second question posed earlier in this section was concerned with the effects produced by other compounds, and other bacteria strains, upon the metabolism of one particular compound by one pure strain. Soil contains a wide variety of metabolites and microorganisms with many possibilities for interaction. I shall enumerate a few of these possibilities.

1. A compound may fail to gain access to the interior of the cell in the absence of other metabolites. Thus *E. coli* characteristically fails to grow in a citrate–mineral-salts medium despite the fact that it possesses, or can elaborate, enzymes capable of metabolizing citrate. However, citrate is readily taken up and metabolized in a nutrient-rich medium providing glucose as an additional energy source (Lominski *et al.*, 1947; Vaughn *et al.*, 1950; Dagley, 1954).

2. A mutant microorganism may lack the enzyme that attacks Y, an intermediate formed in the degradation of X, and therefore will not grow at the expense of X as would the wild type. However, when glucose serves as source of carbon, X may be metabolized and Y will accumulate in the culture in amounts that permit its isolation and characterization. This experimental approach has been much used for delineating metabolic pathways (Dagley and Chapman, 1971).

3. A somewhat similar situation may occur if a structural feature prevents a compound from being attacked by an enzyme in a degradative pathway. Thus, a species of *Pseudomonas* grew with *p*-cresol (4-hydroxytoluene) as sole source of carbon but it did not grow when provided with structural analogues substitued with methyl groups at C-2 or C-3 of the nucleus (Dagley and Patel, 1957). When cells were induced to grow with *p*-cresol, however, these compounds were oxidized as far as the corresponding 4-hydroxy-methybenzoic acids, which therefore accumulated in culture fluids; but this partial oxidation of the substrate was insufficient to sustain growth. In this example further oxidation was prevented by the strict substrate specificity of 4-hydroxybenzoate 3-hydroxylase which does not tolerate substitutions at C-2 and C-3. Leadbetter and Foster (1959) also observed that certain hydrocarbons do not support the growth of *P. methanica* but are partially oxidized by cultures utilizing methane. They described this behavior as "cooxidation," a term that remains acceptable since its application is well defined. Enzymes induced by methane cooxidized *n*-butane to *n*-butanol, *n*-butyric acid and 2-butanone (Foster, 1962). Other examples of hydrocarbon cooxidation have been furnished by Raymond *et al.* (1967) and Jamison *et al.* (1969). In these examples the function of the growth substrate A was to induce enzymes that partially degraded B, a compound that did not support growth.

4. In other systems, B might be attacked only *in the presence* of A, a readily metabolized substrate: that is, prior growth with A would not in itself equip cells to oxidize B, because the function of A was not that of an inducer of the necessary enzymes. It might serve instead to generate a cosubstrate needed for the initial attack on B; or it might provide the energy needed for synthesizing the enzymes required for this attack, B serving as inducer. These considerations may well apply to the observations of Kirk *et al.* (1976)

who found that the white-rot fungi *Phanerochaete chrysosporium* and *C. versicolor* did not grow with, or decompose, lignin until a growth substrate such as cellulose or glucose was added. Decomposition of lignin ceased when the carbohydrate was exhausted. Lignin is a polymer containing many carbon–carbon and ether linkages uniting its constituent monomers, and as we have seen, these bonds are likely to be broken only when the structure is attacked by oxygenases. Further, the energy requirements of growth will be met only after substantial degradation has occurred. At first, a considerable expenditure of energy will be needed both for the synthesis of oxygenases and, in most cases, for their operation which calls for a continuous supply of NAD(P)H. Clearly, lignin is such a stable structure that for some time during the course of its degradation it will consume, rather than generate, energy; and this will be furnished by a second substrate such as cellulose or glucose.

5. We may now consider the case when a single compound is degraded only if two or more microbial species are present. Thus, complete utilization of cycloparaffins by single isolates is a very rare occurrence (Beam and Perry, 1973). However, cyclohexane is mineralized, i.e., degraded to carbon dioxide, by the complex mixture of microorganisms present in marine mud despite the fact that this cannot be accomplished by individual organisms acting alone (Beam and Perry, 1974). It appears that some organisms can oxidize the paraffin only as far as cyclohexanone, and that this compound then serves as a growth substrate for others. This is an example of commensalism, in which one type of microorganism produces a compound that it cannot utilize itself but which serves as nutrient for a second type. For this particular system it has been shown that during growth of certain microorganisms at the expense of propane as a carbon source, enzymes are used that can also oxidize cyclohexane to cyclohexanone; this compound then accumulates unless degraded by other organisms (Beam and Perry, 1974).

6. In other instances when two organisms are required to metabolize one compound present in a particular medium it has been found (Dr. P. J. Chapman, personal communication) that one organism can elaborate enzymes for degrading the compound but cannot grow at its expense because it is nutritionally exacting. Some growth factor, provided by the second organism, is required.

7. Finally, we may note that a compound may be extensively degraded by a single microbial strain, but during degradation it may give rise to a toxic product that prevents growth from occurring. Thus, 5-fluoronicotinic acid was oxidized almost as extensively as the growth substrate by nicotinate-grown pseudomonads (Behrman and Stanier, 1957b); however, the formation of fluoroacetate and subsequent "lethal biosynthesis" of fluorocitrate (Peters, 1957) prevented the utilization of 5-fluoronicotinate for growth.

In 1970, Horvath and Alexander observed that *Arthrobacter* could not grow with 3-chlorobenzoate, but that washed suspensions of cells grown with benzoate converted 3-chlorobenzoate almost quantitatively into 4-chlorocatechol. Similar observations were made with benzoate-grown *Brevibacterium* which formed 3,5-dichlorocatechol from 2,3,6-trichlorobenzoate. These findings are similar to those noted in item 3, and the term "cometabolism" was proposed to describe them. Unfortunately, subsequent workers have applied this term without discrimination to systems encompassed by any, or all, of the descriptions given in items 1 through 7. Thus, it may be observed that glucose, when added to activated sludge, promotes the faster disappearance of a particular compound; and this may be cited as an example of "the phenomenon of cometabolism." In view of the vast number of consequences, biochemical and microbiological, that would follow this addition the explanation is scarcely more informative than the statement that Beethoven's Fifth Symphony is due to the phenomenon of sound.

VI. Potentially Harmful Microbial Activities

Microorganisms occasionally form toxic compounds when their enzymes metabolize substrates that are not usually encountered in a natural habitat. Thus, Wood *et al.* (1968) showed that anaerobic bacteria can methylate Hg^{2+} ions by means of the methylcobalamin enzyme system that normally produces methane. The highly toxic compound methylmercury (CH_3-Hg^+) is thereby synthesized and later tends to concentrate in food chains; this can occur whenever industrial effluents containing traces of mercury are discharged into water where methane-forming bacteria are present in sediments. The accumulation of methylmercury in fish has caused concern in the United States and Canada. Fortunately, other bacteria can reduce methylmercury to Hg° and methane; and there are others that catalyze an NADH-dependent reduction of Hg^{2+} to Hg° which is much less toxic than methylmercury; while a third microbial reaction forms dimethylmercury from methylmercury. Dimethylmercury, being volatile, may escape to the atmosphere and decompose to give Hg° which can re-enter the aqueous environment. Accordingly, a cycle may be constructed from which a steady state concentration of methylmercury in sediments may be predicted. Another cycle has been described in which methylarsines are formed by microbial action (Wood, 1974).

In the case of methylmercury formation, bacterial enzymes were involved in the synthesis of a toxic material. Likewise, when a pesticide or organic pollutant is degraded, bacterial action may not be beneficial unless complete

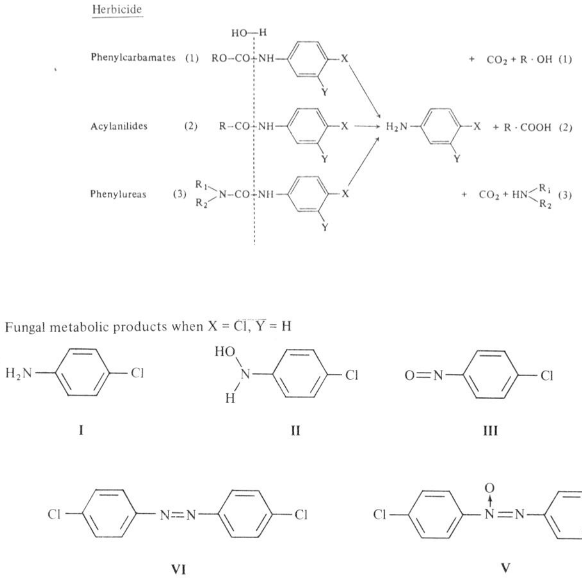

FIG. 13. Enzymatic hydrolysis of some herbicides, with examples of products formed from 4-chloroaniline(I) by fungal metabolism.

mineralization occurs. Persistent intermediate degradative products may be formed that are more harmful than the original compound. This may be illustrated with the aniline-based herbicides, for which the general structures of some families (acylanilides, phenylcarbamates, phenylureas) are given in Fig. 13. Microbial attack upon these compounds is initiated, as shown, by acylamidases that have been isolated and characterized from various bacteria, fungi and higher plants (for references, see Blake and Kaufman, 1975). The aniline-based herbicides comprise about one-third of the registered herbicides in today's pesticide market, and the degradation of propanil [3,4-dichloropropionanilide; $R = C_2H_5$ in Fig. 13 (2), and $X = Y = Cl$] has received particular attention. Concern about its use was aroused when Bartha and Pramer (1967) reported that soil treated with 2,000 ppm propanil (a very high application) produced a major metabolite subsequently identified as 3,3′, 4,4′-tetrachloroazobenzene, formed by the condensation of two molecules of 3,4-dichloraniline arising from propanil by hydrolysis. Other herbicides in this category, and various dichloroanilines and monochloroanilines, have also been shown to be transformed into chlorine-substituted azobenzenes in soil (Bartha *et al.*, 1968). These compounds are structurally related to known potent carcinogens, and studies of their mode of formation from 4-chloroaniline and 3,4-dichloroaniline provide instructive examples of how chemically reactive intermediates can be furnished by microbial action.

Kaufman *et al.* (1973) showed that the soil fungus *Fusarium oxysporum* Schlecht degraded 4-chloroaniline [(I), Fig. 13] to the following products, which were isolated and characterized: (II) 4-chlorophenylhydroxylamine; (III) 4-chloronitrosobenzene; (IV) 4-chloronitrobenzene; (V) 4,4′-dichloroazoxybenzene; and (VI) 4,4′-dichloroazobenzene. It appeared likely that microbial activation of the amino group of (I) by hydroxylation to give (II) and then (III) by oxidation, facilitated the condensation of (I) with (III) to give (VI) and of (II) with (III) to give (V). Other possibilities were the formation of (V) from two molecules of (II), and of (VI) by the reduction of (V). The isolation of 4-chloronitrobenzene (IV) was of interest since the reduction of nitroaryl compounds to arylamines has been observed for *F. oxysporum* (Madhosingh, 1961) and other microorganisms (Chacko *et al.*, 1966). Accordingly, although it was shown in the experiments of Kaufman *et al.* (1973), that (IV) accumulated by oxidation of (III), nevertheless reduction of (IV) to (I), via (III) and (II), also appears to take place under the appropriate conditions; and indeed, these authors obtained evidence that the reactions for metabolism of 4-chloroaniline were reversible. Increasing attention has been given in recent years to the fate of nitro groups present in pesticides and their degradative products, because dinitroaniline herbicides consitute a rapidly expanding category. These herbicides contain nitro groups at C−2 and C−6 of the anilinine molecule, with an alkyl or sulfonyl group at C−4, and with one or two hydrogens of the amino group replaced with alkyl and other groups. The behavior and fate of these compounds has been reviewed by Helling (1976). They are metabolized by soil microbes more rapidly under anaerobic than aerobic conditions, and it appears that attack is initiated by reduction of one nitro to an amino group with the necessary electrons and protons supplied through anaerobic degradation of organic matter in soil.

Other biochemical properties of pesticides that have aroused public concern have already been reviewed (Dagley, 1972, 1975b). What is perhaps less generally appreciated is the scope of the research now required of a manufacturer before a new pesticide can be registered and marketed. Thus, the Federal Environmental Pesticide Control Act (1972) is implemented through the Environmental Protection Agency (EPA), and this body has developed a comprehensive set of guidelines containing protocols for a great variety of laboratory and field studies. These give the requirements necessary for the registration of a pesticide, and direct the manufacturer to develop registration data in the following areas: product efficacy; hazards to humans and domestic animals, fish, and wildlife; and environmental chemistry. Thus, before registration, the manufacturer of a new pesticide will have undertaken animal studies including (for a particular animal species) a balance study for the applied pesticide, a tissue residue study, and metabolite characterization to assist the identification and, as far as possible, the interpretation of metabolic pathways. He will have undertaken plant metabolism studies including those for uptake of the free and bound pesticide residues, for distribution of the pesticide in the plant and its concentration therein; and also for metabolite characterization and interpretation of the metabolic pathway. The manufacturer is, of course, also required to give a thorough evaluation of hazards, involving numerous toxicological tests. Residue tolerances and reentry data must be furnished. Hazards to humans and domestic animals must be evaluated through toxic, teratogenic, mutagenic, and oncogenic effects. Hazards to fish and wildlife must be evaluated through acute and subacute toxicity testing.

It is therefore evident that when a manufacturer has satisfied all the safeguards which the EPA rightly insists upon, his new product will have been very thoroughly and expensively investigated. Without claiming that increased research and development costs are due entirely to investigations of this type, it may be observed that during the period 1961–1970, 19 new pesticides were introduced at an average cost of $4.1 million and in the period 1971 to the present, three have been introduced averaging $13.6 million each after 8–10 years of research and testing. The emergence of unifying principles and concepts might reduce the enormous labor and cost that are involved in the procedure of treating each new compound as though it has biological properties that are unknown and unpredictable. Industry might then be able to develop compounds for use with minor, or speciality, crops instead of concentrating exclusively on those that might be profitable only for major crops. There is a strong argument to be made in favor of developing a broad conceptual framework, based on microbial metabolism, that will facilitate informed discussion about environmental problems. The strength of the argument, as seen from the figures quoted, lies not so much in intellectual satisfaction as in practical, economic considerations.

VII. Evolution of Microbial Degradative Pathways

In discussing problems created by persistent pesticides, a recent report by the Committee on Agriculture and the Environment (1974) of the National Academy of Sciences points out that microbes have had at their disposal a time span of billions of years in which to acquire the ability to degrade organic natural products. Organic chemists have been contributing novel substances for only a century or so, and the biosphere has received the products of their ingenuity in significant amounts as pesticides only in the last 20 years. "How long," the report asks, "will it take for microbes to acquire the ability to destroy these compounds?" Evidently, if we are to speculate as to what microbes might be able to accomplish eventually, it would be helpful to have a clear picture as to how they came to acquire their present capabilities. Evolution in the microbial world was the subject of a recent symposium; however, as Hartley (1974) concludes in his contribution: "We are a long way from the systematic molecular anatomy that must underlie any firm opinions about how these molecules [microbial enzymes] evolved." Excellent summaries of speculations relating to these questions are also to be found in the reviews of Clarke (1974) and Hegeman and Rosenberg (1970).

Ornston and Parke (1976) have summarized two general experimental approaches to questions concerning the evolution of microbial catabolic pathways. One, the mechanistic approach, gives an indication of the types of changes that might have occurred during evolution by observing mutational responses of organisms when challenged by novel substrates under laboratory conditions. The mutations observed might be those that alter the specificity of a mechanism for transporting a compound into cells, or for inducing enzymes required for metabolism, or mutations might alter the substrate specificities of these enzymes. The second, or descriptive, approach seeks to learn what happened in the past when catabolic pathways evolved in a natural environment. The investigator explores possible evolutionary relationships amongst the genes for catabolic pathways that presently exist. Evolutionary affinities may be suggested by comparison of metabolic sequences, enzyme reaction mechanisms or inductive controls. In principle, the most rigorous test of theories concerning the evolution of a catabolic pathway would be to purify each enzyme of the sequence and establish its primary structure and subunit composition. Comparisons could then be made that would assist in deciding whether similarities were close enough to suggest that the pathway evolved by sequential mutations; that is, whether a duplication of the gene for enzyme 1, followed by a mutation in the duplicate gene, might have given rise to a gene coding for enzyme 2, and so on along the pathway. In some segments of catabolic pathways this would be unlikely: thus, if a dehydrogenase were followed by a kinase, the reactions are so different in mechanism as to indicate that one would not be evolved from the other. It would be more feasible to propose that the dehydrogenase, for example, was recruited from another reaction sequence, so that following a mutation, the enzyme became an efficient catalyst in the newly evolving metabolic pathway.

Investigations that take the second, or descriptive approach have been initiated for the ortho-fission pathways of catechol and protocatechuic acid

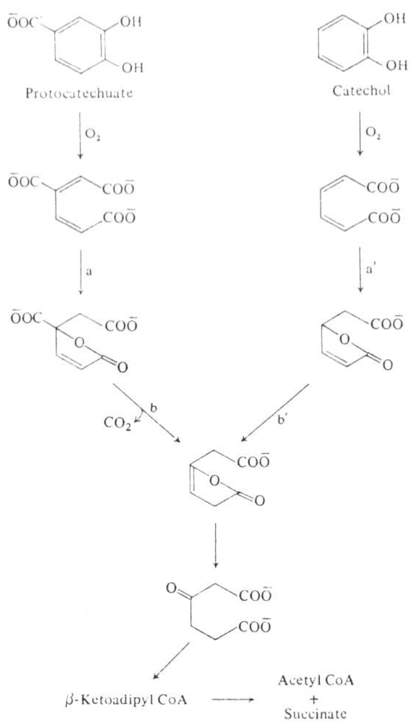

FIG. 14. *Ortho* fission pathways for protocatechuate and catechol. Enzymes referred to in the text are β-carboxymuconate lactonizing enzyme, (a); γ-carboxymuconolactone decarboxylase, (b); muconate lactonizing enzyme, (a′); and muconolactone isomerase, (b′).

(Fig. 14). These, and other, degradative sequences for aromatic compounds are well suited to evolutionary studies because although pairs of reactions from each route are of similar chemical type, they are catalyzed by separate and distinct proteins. Enzyme a that lactonizes β-carboxymuconate does not lactonize *cis*, *cis*-muconate, and enzyme b does not accept the substrate of enzyme b′. Briefly, the results of Ornston and his colleagues (Meagher and Ornston, 1973; Patel *et al.*, 1973; Parke *et al.*, 1973) suggest that in these systems at least, the enzymes of one metabolic route were evolved first and were then "recruited" by mutations to function in the second pathway. Thus enzymes a and a′ from *P. putida* had about the same molecular weight, subunit structure, crystalline form and N-terminal amino acids. Close resemblances were also found for b and b′ rather than between a and b, or a′ and b′. It is interesting to note that Stanier *et al.* (1970) observed no immunological cross-reaction between enzymes a and a′, whereas the results of Ornston and co-workers suggest extensive homology. However, a relatively few changes in an amino acid sequence might profoundly affect the way a protein chain folds and the shape it assumes, properties that would be reflected in altered immunological behavior. Antisera prepared against enzyme b′ from *P. putida* also failed to cross-react with enzyme b′ from *A. calcoaceticus*, but Patel *et al.*, (1974) have shown that the enzymes from the two genera also have very similar properties, strongly suggestive of close evolutionary homology. It may well be that the structural gene for enzyme b′ was exchanged long after the divergence of other genes that dictated the evolutional separation of the two genera. Another area of bacterial catabolism, holding promise for fruitful applications of the descriptive approach, encompasses single, individual reactions which, in the same organism, are catalyzed by more than one enzyme (Canovas and Stanier, 1967; Charnetzky and Mortlock, 1974; Koo and Adams, 1974; Wilson and Mortlock, 1973). By means of multiple enzymes, the cell gains physiological flexibility; because the biosynthesis of one enzyme, which in each case functions on a different pathway from the other(s), can be governed separately in response to nutritional needs. Such evidence as is presently available suggests that while the structural genes for multiple enzymes may have had a common ancestral gene, significant divergences in structure have occurred in the course of evolution (Koo and Adams, 1974; Patel *et al.*, 1975).

The mechanistic approach to these problems stems from the observation that some microorganisms, in addition to their ability to grow at the expense of a wide range of carbohydrates encountered in nature, are sometimes able to metabolize immediately certain novel carbohydrates which they are not likely to have encountered previously in natural habitats. These investigations were begun twenty years ago at the Prairie Regional Laboratories in Saskatoon, Saskatchewan, Canada by Simpson; elucidation of the reactions used for the metabolism of novel aldopentoses and pentitols were continued by Wood, Anderson and Mortlock and their colleagues at Michigan State University and at the University of Massachusetts; and the results of these studies have been reviewed recently by Mortlock (1976). Similarly, the ability of bacteria to utilize a series of amides which they would not normally encounter has been investigated by Clarke and her colleagues; and this area has also been reviewed (Clarke, 1974). For our present discussion, the most relevant experiments are those conducted with organisms that do not grow at the expense of a novel substrate, but acquire the ability to do so after they have undergone mutations.

We may take as an example a wild-type *Klebsiella aerogenes* that grows with ribitol but not with xylitol; this is not normally found in nature and fits the description of a novel substrate. One obstacle to growth is that xylitol is not able to derepress enzymes involved in its own metabolism; however, it does serve as substrate for ribitol dehydrogenase although catalysis is not very efficient due to the low affinity of the enzyme for this alternative substrate (K_m for ribitol is about 1 mM, that for xylitol is about 1 M). Slow growth with xylitol might therefore be possible for cells that contained an adequate level of dehydrogenase, provided this could be obtained without the necessity of derepression; for this enzyme converts xylitol into D-xylulose, a compound that is utilized efficiently by these organisms. This freedom from the constraints of inducer specificity can be achieved by isolating "constitutive mutants" that can produce large amounts of enzymes, 2% or more of the total cell protein, in the absence of any inducer. This is the most common mutation studied that permits the use of a novel substrate, whether it be a pentitol or one of a series of amides investigated by Clarke (1974). By using a variety of different amides in selection media, these workers selected additional mutants having amidases with altered substrate specificities compared with the wild type *P. aeruginosa*. Accordingly these mutants grew at the expense of amides that the parent organisms could not utilize. Wu *et al.* (1968) selected mutants of *K. aerogenes* possessing the new characteristic of being able to grow with xylitol by isolating first a mutant constitutive for ribitol dehydrogenase (RDH), and from this a second mutant that possessed an altered RDH with an increased affinity for xylitol. As a result of a third mutation, the xylitol was transported more rapidly into the cell, apparently by means of an active transport system that was produced constitutively, and although used by the wild type to transport D-arabitol, could also accommodate xylitol.

Hartley and his colleagues (Hartley, 1974; Rigby *et al.*, 1974) have taken a different approach to studying the part played by RDH in enabling *K. aerogenes* to grow with xylitol as carbon source. A strain constitutive for RDH was again used, but growth now took place in a chemostat, so that measurements of the steady state biomass and the concentration of xylitol in the effluent could be used to signal the appearance of spontaneous mutants able to utilize the carbon source more effectively. It appeared that, under these conditions, this result was achieved simply by the mutants making more and more enzyme, rather than by producing RDH with a stronger affinity for xylitol. Thus, extracts made from cells taken at appropriate time intervals from the chemostat contained, respectively, 2, 9, and 32 units of RDH/mg of extract protein, but the K_m for xylitol was about the same for each, namely 1.3 *M*. From these observations and many others made by this group, it is clear that mutations leading to increased production of enzyme occurred much more frequently than those leading to improved enzyme specificity. In this particular system it is probable that multiplication of the RDH gene occurred in some mutants, and was reflected in increased enzyme production, and that any mutations to give improved specificity would be acceptable only in "silent gene" copies. It should be mentioned that the mutant with increased specificity for xylitol, used by Wu *et al.* (1968) was obtained by powerful nitrosoguanidine mutagenesis, and not by use of a chemostat as described by Rigby *et al.*, (1974).

The experiments I have summarized suggest at least the approaches that might be made to the selection of microorganisms capable of degrading some of the novel, or unnatural, compounds that have appeared in our environment as a result of man's agricultural and technological activities. I shall conclude with some comments as to how such degradative versatility, once acquired, might be transmitted throughout bacterial populations through the agency of plasmids. These are autonomous genetic elements (sequences of deoxyribonucleotides) that exist in a wide variety of bacteria (for reviews, see Clowes, 1972; Helsinki, 1973). Although they have a stable existence apart from the bacterial chromosome, combination with chromosomal DNA has been observed, and also combination with other plasmids. They code for various functions which, although not an absolute requirement for growth, may enable the bacteria to exist in ecological niches from which they would otherwise be excluded. Of particular interest is the fact that they may sometimes confer an advantage upon a mixed population by passing from one cell into another when the cells come into contact. The investigation of plasmids coding for enzymes of catabolic pathways is a recent development, briefly summarized by Williams and Worsley (1976). I shall choose one example from bacterial aromatic catabolism to illustrate the advantages to an organism that accrue from its possession of a plasmid. *Pseudomonas putida* (*arvilla*) is able to grow at the expense of *m*- or *p*-toluic acids (3- or 4-methylbenzoic acids). Mutants may be selected that no longer grow with these substrates but are still able to grow with benzoate, and when they do so, they use the enzymes of the ortho-fission route. By contrast, the wild type uses the meta-fission enzymes for degrading the toluates and benzoate: the genes for these enzymes are accommodated upon a plasmid, the TOL plasmid, which the mutants have lost (Williams and Murray, 1974). Further, this plasmid carries genes for catabolizing the hydrocarbons toluene, *m*-xylene and *p*-xylene; they are metabolized by oxidation of a methyl group to the corresponding acid, with the alcohol and aldehyde as oxidation intermediates. Thus benzoate, *m*-toluate and *p*-toluate are formed from the respective hydrocarbons, and by simultaneous dihydroxylation and decarboxylation they give rise to catechol, 3-methylcatechol and 4-methylcatechol which serve as ring-fission substrates in meta-fission pathways as described earlier. The broad substrate specificities of these enzymes allow, apparently, one set of plasmid-borne genes to code for the enzymes needed to support growth with 12 substrates, namely, the particular aromatic hydrocarbons, alcohols, aldehydes, and carboxylic acids mentioned (Worsley and Williams, 1975). Williams and Worsley (1976) have drawn attention to the results of Audus (1960) which show that three features are characteristic of microbial utilization of herbicides and other man-made compounds: (a) a long lag phase of up to several months before any disappearance of the compound can be detected; (b) the difficulty of isolating pure cultures capable of the complete breakdown even after the onset of disappearance; (c) the inherent instability of the degradative ability in pure strains of competent bacteria, once isolated, when removed from selective media. They make the interesting suggestion that the lag phase might represent a period during which processes of plasmid exchange and recombination result in the selection of one or more organisms that finally have a battery of enzymes capable of the complete degradation. The instability of the competent isolates might therefore be a result of the responsible genes being located on an unstable and hence easily lost plasmid.

ACKNOWLEDGMENTS

I thank S. J. Flashinski and J. R. Plimmer who supplied me with some of the information used in this survey. I am also indebted to P. J. Chapman and P. W. Trudgill for information and many valuable suggestions. This work was in part supported by Public Health Service Grant ES A1 00678 from the National Institute of Environmental Health Sciences.

REFERENCES

Ahmed, M., and Focht, D. D. (1973). *Can. J. Microbiol.* **19**, 47.
Aldrich Chemical Company. (1976). *Aldrichimica Acta* **9**, No. 2.
Alexander, M. (1967). *In* "Agriculture and the Quality of our Environment" (N. C. Brady, Ed.), Publ. No. 85, pp. 331–342. Am. Assoc. Sci., Washington, D.C.
Alexander, M., and Lustigman, B. K. (1966). *J. Agric. Food Chem.* **14**, 410.
Amphlett, M. J., and Callely, A. G. (1969). *Biochem. J.* **112**, 12P.
Audus, L. J. (1960). *In* "Herbicides and the Soil" (E. K. Woodford and G. R. Sagar, eds.), pp. 1–19. Blackwell, Oxford.
Auret, B. J., Boyd, D. R., Robinson, P. M., Watson, C. G., Daly, J. W., and Jerina, D. M. (1971). *Chem. Commun.* p. 1585.
Barker, R. (1971). "Organic Chemistry of Biological Compounds," p. 273. Prentice-Hall, Englewood Cliffs, New Jersey.
Bartha, R., and Pramer, D. (1967). *Science* **156**, 1617.
Bartha, R., Linke, H. A. B., and Pramer, D. (1968). *Science* **161**, 582.
Bayly, R, C., Dagley, S., and Gibson, D, T. (1966). *Biochem. J.* **101**, 293.
Beam, H. W., and Perry, J. J. (1973). *Arch. Microbiol.* **91**, 87.
Beam, H. W., and Perry, J. J. (1974). *J. Gen. Microbiol.* **82**, 163.
Behrman, E. J., and Stanier, R. Y. (1957a). *J. Biol. Chem.* **228**, 923.
Behrman, E. J., and Stanier, R. Y. (1957b). *J. Biol. Chem.* **228**, 947.
Blake, J., and Kaufman, D. D. (1975). *Pestic. Biochem. Physiol.* **5**, 305.
Blakley, E. R. (1974). *Can. J. Biochem.* **20**, 1297.
Blumer, M. (1961). *Science* **134**, 474.
Blumer, M. (1976). *Sci. Am.* **234**, 34.
Bolin, B. (1973). *In* "Chemistry in the Environment" (C. L. Hamilton, ed.), pp. 53–61. Freeman, San Francisco, California.
Bollag, J.-M. (1972). *Crit. Rev. Microbiol.* **2**, 35.

Bollag, J.-M., Briggs, G. G., Dawson, J. E., and Alexander, M. (1968). *J. Agric. Food Chem.* **16**, 829.
Britton, L. M., Brand, J. N., and Markovetz, A. J. (1974). *Biochim. Biophys. Acta* **369**, 45.
Brown, S. A., El-Dakhakhny, M., and Steck, W. (1970). *Can. J. Biochem.* **48**, 863.
Canovas, J. L., and Stanier, R. Y. (1967). *Eur. J. Biochem.* **1**, 289.
Catelani, D., Sorlini, C., and Treccani, V. (1971). *Experientia* **27**, 1173.
Chacko, C. I., Lockwood, J. L., and Zabik, M. L. (1966). *Science* **154**, 893.
Challenger, F. (1959). "Aspects of the Organic Chemistry of Sulphur," pp. 73–114. Butterworth, London.
Chapman, P. J., and Ribbons, D. W. (1976). *J. Bacteriol.* **125**, 975.
Chapman, P. W., and Duggleby, R. C. (1967). *Biochem. J.* **103**, 7C.
Charnetzky, W. T., and Mortlock, R. P. (1974). *J. Bacteriol.* **119**, 170.
Christopherson, B. O. (1969). *Biochim. Biophys. Acta* **176**, 463.
Clarke, D. D., Nicklas, W. J., and Palumbo, J. (1968). *Arch. Biochem. Biophys.* **123**, 205.
Clarke, P. H. (1974). *Symp. Soc. Gen. Microbiol.* **24**, 183–217.
Clowes, R. C. (1972). *Bacteriol. Rev.* **36**, 361.
Committee on Agriculture and the Environment. (1974). "Productive Agriculture and a Quality Environment," p. 17. Nat. Acad. Sci., Washington, D.C.
Crawford, R. L. (1975). *J. Bacteriol.* **121**, 531.
Cripps, R. E. (1973). *Biochem. J.* **134**, 353.
Cripps, R. E. (1975). *Biochem. J.* **152**, 233.
Crosby, D. G. (1969). *J. Agric. Food Chem.* **17**, 413.
Dagley, S. (1954). *J. Gen. Microbiol.* **11**, 218.
Dagley, S. (1971). *Adv. Microb. Physiol.* **6**, 1.
Dagley, S. (1972). *In* "Degradation of Synthetic Organic Molecules in the Biosphere," pp. 336–347. Nat. Acad. Sci. Washington, D. C.
Dagley, S. (1975a). *Am. Sci.* **63**, 681.
Dagley, S. (1975b). *Essays Biochem.* **11**, 81.
Dagley, S. (1976). *Biochem. Soc. Trans.* **4**, 455.
Dagley, S. and Chapman, P. J. (1971). *Methods Microbiol.* **6A**, 217–268.
Dagley, S., and Johnson, P. A. (1963). *Biochim. Biophys. Acta* **78**, 577.
Dagley, S., and Patel, M. D. (1957). *Biochem. J.* **66**, 227.
Davies, J. I., and Evans, W. C. (1964). *Biochem. J.* **91**, 251.
Davis, J. B. (1967). "Petroleum Microbiology." Elsevier, Amsterdam.
Donoghue, N. A., and Trudgill, P. W. (1975). *Eur. J. Biochem.* **60**, 1.
Donoghue, N. A., Norris, D. B., and Trudgill, P. W. (1976). *Eur. J. Biochem.* **63**, 175.
Dutton, P. L., and Evans, W. C. (1969). *Biochem. J.* **113**, 525.
Evans, W. C. (1976). *Biochem. Soc. Trans.* **4**, 452.
Evans, W. C., Fernley, H. N., and Griffiths, E. (1965). *Biochem. J.* **95**, 819.
Evans, W. C., Smith, B. S. W., Moss, P., and Fernley, H. N. (1971a). *Biochem. J.* **122**, 509.
Evans, W. C., Smith, B. S. W., Fernley, H. N., and Davies, J. I. (1971b). *Biochem. J.* **122**, 543.
Faulkner, J. K., and Woodcock, D. (1964). *Nature (London)* **203**, 865.
Ferris, J. P., Fasco, M. J., Stylianopoulou, F. L., Jerina, D. M., Daly, J. W., and Jeffrey, A. M. (1973). *Arch. Biochem. Biophys.* **156**, 97.
Flohé, L. (1971). *Klin. Wochenschr.* **49**, 669.
Foster, J. W. (1962). *In* "Oxygenases" (O. Hayaishi, ed.), pp. 241–247. Academic Press, New York.
Fowden, L. (1968). *Proc. R. Soc. London, Ser. B* **171**, 5.
Fridovich, I. (1974). *Horiz. Biochem. Biophys.* **1**, 1.

Fujiwara, M., Golovleva, L. A., Saeki, Y., Nozaki, M., and Hayaishi, O. (1975). *J. Biol. Chem.* **250**, 4848.

Gaunt, J. K., and Evans, W. C. (1971a). *Biochem. J.* **122**, 519.

Gaunt, J. K., and Evans, W. C. (1971b). *Biochem. J.* **122**, 533.

Gauthier, J. J., and Rittenberg, S. C. (1971). *J. Biol. Chem.* **246**, 3737.

Gibson, D. T. (1972). *In* "Degradation of Synthetic Organic Molecules in the Biosphere," pp. 116–136 Nat. Acad. Sci., Washington, D.C.

Gibson, D. T., Hensley, M., Yoshida, H., and Mabry, T. J. (1970a). *Biochemistry* **9**, 1626.

Gibson, D. T., Cardini, G. E., Maseles, F. C., and Kallio, R. E. (1970b). *Biochemistry* **9**, 1631.

Gibson, D. T., Gschwendt, B., Yeh, W. K., and Kobal, V. M. (1973a). *Biochemistry* **12**, 1520.

Gibson, D. T., Roberts, R. L., Wells, M. C., and Kobal, V. M. (1973b). *Biochem. Biophys. Res. Commun.* **50**, 211.

Goldman, P. (1965). *J. Biol. Chem.* **240**, 3434.

Goldman, P. (1972a). *In* "Degradation of Synthetic Organic Molecules in the Biosphere," pp. 147–165. Nat. Acad. Sci. Washington, D.C.

Goldman, P. (1927b). *Carbon-Fluorine Compd. Chem., Biochem. Biol. Act., Ciba Found. Symp.,* pp. 335–349.

Gray, P. H. H., and Thornton, H. G. (1928). *Zentralbl Bakteriol Parasitenkd Infektionskr. (2 Abt.)* **73**, 74.

Griffin, M., and Trudgill, P. W. (1972). *Biochem. J.* **129**, 595.

Griffin, M., and Trudgill, P. W. (1976). *Eur. J. Biochem.* **63**, 199.

Gunsalus, I. C. (1972). *In* "Degradation of Synthetic Organic Molecules in the Biosphere," pp. 137–146. Nat. Acad. Sci. Washington, D.C.

Guyer, M., and Hegeman, G. D. (1969). *J. Bacteriol.* **99**, 906.

Hamilton, G. A. (1971). *Prog. Bioorg. Chem.* **1**, 83–157.

Hareland, W. A., Crawford, R. L., Chapman, P. J., and Dagley, S. (1975). *J. Bacteriol.* **121**, 272.

Harrison, P. G., Bailey, B. K., and Steck, W. (1971). *Can. J. Biochem.* **49**, 964.

Hartley, B. S. (1974). *Symp. Soc. Gen. Microbiol.* **24**, 151.

Hayaishi, O. (1964). *Proc. Plen. Sess. Int. Congr. Biochem. 6th, 1964* p. 31.

Hayaishi, O. (1966). *Bacteriol. Rev.* **30**, 720.

Hegeman, G. D., and Rosenberg, S. L. (1970). *Annu. Rev. Microbiol.* **24**, 429.

Helling, C. S. (1976). *J. Environ. Qual.* **5**, 1.

Helsinki, D. R. (1973). *Annu. Rev. Microbiol.* **27**, 437.

Hendrickson, J. B., Cram, D. J., and Hammond, G. S. (1970). "Organic Chemistry," 3rd ed., p. 698. McGraw-Hill, New York.

Higgins, P. M. (1968). *Dev. Ind. Microbiol.* **9**, 146.

Holcenberg, J. S., and Stadtman, E. R. (1969). *J. Biol. Chem.* **244**, 1194.

Holmes, P. E., and Rittenberg, S. C. (1972). *J. Biol. Chem.* **247**, 7622.

Hopper, D. J., and Chapman, P. J. (1971). *Biochem. J.* **122**, 19.

Hopper, D. J., Chapman, P. J., and Dagley, S. (1968). *Biochem. J.* **110**, 798.

Hopper, D. J., Chapman, P. J., and Dagley, S. (1970). *J. Bacteriol.* **104**, 1197.

Hopper, D. J., Chapman, P. J., and Dagley, S. (1971). *Biochem. J.* **122**, 29.

Horvath, R. S., and Alexander, M. (1970). *Can. J. Microbiol.* **16**, 1131.

Howell, L. G., Spector, T., and Massey, V. (1972). *J. Biol. Chem.* **247**, 4340.

Hunt, A. L., Hughes, D. E., and Lowenstein, J. M. (1958). *Biochem. J.* **69**, 170.

Jamison, V. W., Raymond, R. L., and Hudson, J. O. (1969). *Appl. Microbiol.* **17**, 853.

Jerina, D. M., Daly, J. W., Witkop, B., Zaltzman-Nirenberg, P., and Udenfriend, S. (1968). J. Am. Chem. Soc. **90**, 6525.

Jerina, D. M., Daly, J. M., Jeffrey, A. M., and Gibson, D. T. (1971). *Arch. Biochem. Biophys.* **142**, 394.
Kaneda, T. (1974). *Biochem. Biophys. Res. Commun.* **58**, 140.
Kaufman, D. D., Plimmer, J. R., and Klingebiel, U.I. (1973). *J. Agric. Food Chem.* **21**, 127.
Kearney, P. C., and Kaufman, D. D. (1972). *In* "Degradation of Synthetic Organic Molecules in the Biosphere," pp. 166–189. Nat. Acad. Sci., Washington, D.C.
Kellogg, E. W., and Fridovich, I. (1975). *J. Biol. Chem.* **250**, 8812.
Kirk, T. K., Connors, W. J., and Zeikus, J. G. (1976). *Appl. Environ. Microbiol.* **32**, 192.
Kobal, V., Gibson, D. T., Davis, R. E., and Garza, A. (1973). *J. Am. Chem. Soc.* **95**, 4420.
Koo, P. H., and Adams, E. (1974). *J. Biol. Chem.* **249**, 1704.
Leadbetter, E. R., and Foster, J. W. (1959). *Arch. Biochem. Biophys.* **82**, 491.
Lominski, I., Conway, N. S., Harper, E. M., and Rennie, J. B. (1947). *Nature (London)* **160**, 573.
Loomis, W. D., and Battaile, J. (1966). *Phytochemistry* **5**, 423.
Lunt, D., and Evans, W. C. (1970). *Biochem. J.* **118**, 54P.
McCord, J. M., Keele, B. B., and Fridovich, I. (1971). *Proc. Natl. Acad. Sci. U.S.A.* **68**, 1024.
MacRae, I. C, and Alexander, M. (1965). *J. Agric. Food Chem.* **13**, 722.
Madhosingh, C. (1961). *Can. J. Microbiol.* **7**, 553.
Mason, H. S., Fowlks, W. L., and Peterson, E. (1955). *J. Am. Chem. Soc.* **77**, 2914.
Meagher, R. B., and Ornston, L. N. (1973). *Biochemistry* **12**, 3523.
Michalover, J. L., and Ribbons, D. W. (1973). *Biochem. Biophys. Res. Commun.* **55**, 888.
Mortlock, R. P. (1976). *Adv. Microb. Physiol.* **13**, 1.
Nakamura, S., Ogura, Y., Yano, K., Higashi, N., and Arima, K. (1970). *Biochemistry* **9**, 3235.
Nakazawa, A., Kojima, Y., and Taniuchi, H. (1967). *Biochim. Biophys. Acta* **147**, 189.
Niemer, H., and Bucherer, H. (1961). *Hoppe-Seyler's Z. Physiol. Chem.* **326**, 9.
Niemer, H., Bucherer, H., and Kohler, H. (1959). *Hoppe-Seyler's Z. Physiol. Chem.* **317**, 238.
Norris, D. B., and Trudgill, P. W. (1976). *Eur. J. Biochem.* **63**, 193.
Ohmori, T., Ikai, T., Minoda, Y., and Yamada, K. (1973). *Agric. Biol. Chem.* **37**, 1599.
Ohta, Y., and Ribbons, D. W. (1970). *FEBS Lett.* **11**, 189.
Orf, H. W., and Dolphin, D, (1974). *Proc. Natl. Acad. Sci. U.S.A.* **71**, 2646.
Ornston, L. M., and Parke, D. (1976). *Biochem. Soc. Trans.* **4**, 468.
Parke, D., Meagher, R. B., and Ornston, L. N. (1973). *Biochemistry* **12**, 3537.
Patel, R. N., Meagher, R. B., and Ornston, L. N. (1973). *Biochemistry* **12**, 3531.
Patel, R. N., Meagher, R. B., and Ornston, L. N. (1974). *J. Biol. Chem.* **249**, 7410.
Patel, R. N., Mazumdar, S., and Ornston, L. N. (1975). *J. Biol. Chem.* **250**, 6567.
Patel, R. N., Hou, C. T., Felix, A., and Lillard, M. O. (1976). *J. Bacteriol.* **127**, 536.
Pederson, T. C., and Aust, S. D. (1975). *Biochim. Biophys. Acta* **385**, 232.
Peters, R. A. (1957). *Adv. Enzymol.* **18**, 113.
Prairie, R. L., and Talalay, P. (1963). *Biochemistry* **2**, 203.
Raymond, R. L., Jamison, V. W., and Hudson, J. O. (1967). *Appl. Microbiol.* **15**, 857.
Reiner, A. M., and Hegeman, G. D. (1971). *Biochemistry* **10**, 2530.
Rigby, P. W. J., Burleigh, B. D., and Hartley, B. S. (1974). *Nature (London)* **251**, 200.
Robinson, T. (1967). "The Organic Constituents of Higher Plants," 2nd ed., pp. 50 and 106. Burgess, Minneapolis, Minnesota.
Rotruck, J. T., Hoekstra, W. G., and Pope, A. L. (1971). *Nature (London), New Biol.* **231**, 223.
Siuda, J. F., and De Bernardis, J. F. (1973). *Lloydia* **36**, 107.
Smith, D. I., and Callely, A. G. (1975). *J. Gen. Microbiol.* **91**, 210.

Smith, R. V., Humphrey, D. W., Engel, K. O., and Rosazza, J. P. (1976). *Appl. Environ. Microbiol.* **31**, 448.
Sparnins, V. L., and Chapman, P. J. (1976). *J. Bacteriol.* **127**, 362.
Sparnins, V. L., and Dagley, S. (1975). *J. Bacteriol.* **124**, 1374.
Sparnins, V. L., Chapman, P. J., and Dagley, S. (1974). *J. Bacteriol.* **120**, 159.
Spector, T., and Massey, V. (1972). *J. Biol. Chem.* **247**, 5632.
Stanier, R. Y., Wachter, D., Gasser, C., and Wilson, A. C. (1970). *J. Bacteriol.* **105**, 351.
Strickland, S., and Massey, V. (1973). *J. Biol. Chem.* **248**, 2944.
Swisher, R. D. (1963). *J. Am. Oil Chem. Soc.* **40**, 648.
Swisher, R. D. (1964). *Chem. Eng. Prog.* **60**, 41.
Tack, B. F., Chapman, P. J., and Dagley, S. (1972). *J. Biol. Chem.* **247**, 6438.
Takemori, S., Yasua, H., Mihara, K., Suzuki, K., and Katagiri, M. (1969). *Biochim, Biophys. Acta* **191**, 58.
Teipel, J. W., Hass, G. M., and Hill, R. L. (1968). *J. Biol. Chem.* **243**, 5684.
Tiedje, J. M., and Alexander, M. (1969). *J. Agric. Food Chem.* **17**, 1080.
Tiedje, J. M., Duxbury, J. M., Alexander, M., and Dawson, J. E. (1969). *J. Agric. Food Chem.* **17**, 1021.
Trudgill, P. W. (1969). *Biochem. J.* **113**, 577.
Trudgill, P. W. (1976). *Biochem. Soc. Trans.* **4**, 458.
Trudgill, P. W., Dubus, R., and Gunsalus, I. C. (1966). *J. Biol. Chem.* **241**, 4288.
Underwood, E. J. (1962). "Trace Elements in Human and Animal Nutrition," 2nd ed., pp. 291–324. Academic Press, New York.
Vaughn, R. H., Osborne, J. T., Wedding, G. T., Tabachnick, J., Beisel, C. G., and Braxton, T. (1950). *J. Bacteriol.* **60**, 119.
Watson, G. K., Houghton, C., and Cain, R. B. (1974a). *Biochem. J.* **140**, 265.
Watson, G. K., Houghton, C., and Cain, R. B. (1974b). *Biochem. J.* **140**, 277.
White-Stevens, R. H., and Kamin, H. (1972). *J. Biol. Chem.* **247**, 2358.
White-Stevens, R. H., Kamin, H., and Gibson, Q. H. (1972). *J. Biol. Chem.* **247**, 2371.
Williams, P. A., and Murray, K. (1974). *J. Bacteriol.* **120**, 416.
Williams, P. A., and Worsley, M. J. (1976). *Biochem. Soc. Trans.* **4**, 466.
Wilson, B. L., and Mortlock, R. P. (1973). *J. Bacteriol* **113**, 1404.
Wood, J. M. (1974). *Science* **183**, 1049.
Wood, J. M., Kennedy, F. S., and Rosen, C. G. (1968). *Nature (London)* **220**, 173.
Worsley, M. J., and Williams, P. A. (1975). *J. Bacteriol.* **124**, 7.
Wu, T. T., Lin, E. C. C., and Tanaka, S. (1968). *J. Bacteriol.* **96**, 447.
Yu, C. A., and Gunsalus, I. C. (1969). *J. Biol. Chem.* **244**, 6149.
Ziffer, H., Jerina, D. M., Gibson, D. T., and Kobal, V. M. (1973). *J. Am. Chem. Soc.* **95**, 4048.

The Thermochemical Generation of Hydrogen from Water*

FELIX SCHREINER AND EVAN H. APPELMAN

Chemistry Division,
Argonne National Laboratory,
Argonne, Illinois

I. INTRODUCTION

In recent years there has been much discussion of the "hydrogen economy." The term is usually intended to refer to a future situation in which hydrogen has become a major fuel, replacing natural gas in pipelines and gasoline in automobiles and airplanes. Whether or not this state of

* Based on work supported by the U.S. Energy Research and Development Administration.

affairs actually comes about in the future, it is important to realize that today hydrogen is used almost entirely as a chemical feedstock. Of the 2×10^{12} cubic feet of hydrogen produced annually in the United States, approximately 90% is used in the production of ammonia and methanol and for the refining of petroleum. This usage pattern is unlikely to change significantly as long as natural gas and oil are available at reasonable prices. As oil and gas reserves dwindle, hydrogen will become increasingly more interesting as a substitute pipeline gas and as a mobile fuel. The use of hydrogen in a natural gas distribution network appears feasible, while its use as a mobile fuel will require the development of satisfactory means of storing the gas compactly.

Hydrogen may also have to take the place of the natural gas now used to generate process steam for industrial plants. The amount of energy used to generate industrial steam in the United States today is equal to that used by all the private automobiles in the country.

As oil becomes scarce, processes for the liquefaction of coal will become competitive, and these will substantially increase the use of hydrogen as a chemical feedstock.

In considering the hydrogen economy, it is important to bear in mind that hydrogen is a secondary energy source, or energy carrier. Hydrogen can only be made through the expenditure of primary energy. It can be produced from energy-rich substances, or it can be made by the use of primary energy to dissociate water, e.g., by electrolysis.

The principal source of hydrogen in the United States today is hydrocracking of natural gas (methane). The chemistry of this process may be represented by the equations

$$CH_4 + H_2O \rightarrow CO + 3H_2$$

$$CO + H_2O \rightarrow CO_2 + H_2$$

The process is economical largely because of the low regulated price of natural gas. If natural gas is available at \$0.75 per million Btu, the hydrogen made from it will cost around \$1.50 per million Btu. Some hydrogen is made by a similar process from oil. If crude oil costs \$9.00 per barrel, the hydrogen made from it will cost about \$3.25 per million Btu.

At this time no significant quantity of hydrogen is made from coal. However, the technology for doing so is well known. It consists of the water-gas reaction,

$$C + H_2O \rightarrow CO + H_2$$

followed by the "shift" reaction over a suitable catalyst,

$$CO + H_2O \rightarrow CO_2 + H_2$$

It is estimated that hydrogen from coal may cost as much as \$5.00 or \$6.00 per million Btu.

Production of hydrogen by electrolysis of water is more expensive than production from fossil fuels, and therefore this process is not in use. However, because of the high cost, there has been little incentive to improve electrolytic techniques, and the technology has largely been frozen into the state of the art of the 1930s. It makes little sense to burn fossil fuels to generate electricity to electrolyze water to make hydrogen for use as a fuel, because two thirds of the energy of the fossil fuel is lost in the generation step and can never be recovered. A better case can be made if the electricity comes from nuclear or solar generating plants, or if hydrogen production is used as a means of storing off-peak electrical energy. However, it seems highly unlikely that significant quantities of hydrogen can be produced economically by the decomposition of water so long as any fossil fuel is abundantly available. Methods of preparing hydrogen from water, which include both electrolysis and the thermochemical schemes to be discussed in this review, should thus be regarded as back-up technologies that will come into their own only when all fossil fuels, including coal, will have become extremely scarce.

II. The Concept of Thermochemical Hydrogen Generation

At the present time electrolysis is the only technically feasible means of obtaining hydrogen from water alone, that is, without the consumption of energy-rich substances such as coal or methane. The preparation of hydrogen by electrolysis of water may be considered to involve two steps. First, a heat engine converts heat from the burning of fuels or from a nuclear reactor into mechanical energy of a turbine, generating electricity. In the second step, this electrical energy is converted to chemical potential energy in an electrolysis cell.

The first step is restricted by the limitation on the efficiency of a heat engine imposed by the second law of thermodynamics. If the engine accepts heat at a temperature T_2 and rejects it at a temperature T_1, then the overall efficiency η, defined as work or electrical energy produced divided by heat consumed, cannot exceed $(T_2 - T_1)/T_2$. This is the efficiency that would be achieved by an ideal *Carnot* engine. Present-day electrical generation is 30–35% efficient. Future technology, employing high temperature topping cycles, may increase this to 50%. The efficiency of electrolysis is not subject to the Carnot limitation, but typical electrolysis efficiencies are only of the order of 65%. Efforts are presently underway to improve this number, and it is not unreasonable to think that efficiencies of 75–80% may be achieved in the future. Hence the conversion of heat to hydrogen via electrolysis of water can be carried out today with about 20% overall efficiency, and future improvements may increase this figure to nearly 40%.

In discussing the efficiency of energy conversion, it is worth mentioning an ambiguity that frequently arises with regard to the energy content of hydrogen. In this review we shall always consider this energy content to be the Gibbs free energy of formation of water at 298°K and 1 atm pressure, $\Delta G_f^0 = -237.2$ kJ/mole. This is the energy that can be obtained as electricity by recombining hydrogen and oxygen in an ideal fuel cell. The energy content of hydrogen is sometimes considered to be the heat of formation of liquid water, $\Delta H_f^0 = -285.8$ kJ/mole, which is the heat obtained when hydrogen is burned to form liquid water. Efficiencies of conversion expressed in terms of ΔH_f^0 will obviously be some 20% higher than those expressed in terms of ΔG_f^0. Efficiencies expressed in terms of ΔG are the ones to which the Carnot limitation applies directly.

The concept of thermochemical hydrogen generation involves the formation of chemical potential energy directly from heat energy, without requiring the intermediate production of electricity. The simplest approach to the realization of this concept is to heat water sufficiently to dissociate it into hydrogen and oxygen, and then to separate the two gases. However, to obtain one atmosphere of hydrogen from the reaction

$$H_2O \rightleftarrows H_2 + \tfrac{1}{2}O_2$$

it is necessary to work at temperatures above 3000°K. Heat sources at such temperatures are not readily obtainable. Nor is it obvious how the two gaseous products can be conveniently separated. What is desired is a multistep process making use of temperatures attainable with available heat sources, producing hydrogen and oxygen at separate stages, and regenerating all reagents except the dissociated water. One generalized representation of such a scheme is

$$2AB + 2H_2O \rightarrow 2AH + 2BOH \qquad (\Delta G \cong 0, \Delta H < 0 \text{ at } T_1) \qquad (1)$$

$$2BOH \rightarrow 2B + H_2O + \tfrac{1}{2}O_2 \qquad (\Delta G \cong 0, \Delta H > 0 \text{ at } T_2) \qquad (2)$$

$$2AH \rightarrow 2A + H_2 \qquad (\Delta G \cong 0, \Delta H > 0 \text{ at } T_3) \qquad (3)$$

$$2A + 2B \rightarrow 2AB \qquad (\Delta G \cong 0, \Delta H < 0 \text{ at } T_4) \qquad (4)$$

where T_2 and T_3 are substantially higher than T_1 and T_4. This cycle will absorb heat at the high temperatures T_2 and T_3 and reject a smaller amount of heat at the low temperatures T_1 and T_4. The excess of heat absorbed over heat rejected is converted to chemical potential energy of the hydrogen and oxygen released. Hence we can see that we are dealing here with a heat engine that produces chemical potential energy instead of mechanical energy. All that is needed to make the thermochemical process strictly analogous to the conventional steam turbogenerator is the addition of a fuel cell in which hydrogen and oxygen can be recombined to form water and produce electrical energy. The combination of thermochemical process and fuel cell is exactly equivalent to the turbogenerator: both convert heat into electricity. As a heat engine, a thermochemical cycle is still constrained by the thermodynamic *Carnot* limitation on its efficiency, i.e., this efficiency cannot exceed $(T_3 - T_1)/T_3$ if T_3 and T_1 are respectively the highest and lowest temperatures of the cycle. We shall subsequently see that any real cycle is subject to even more severe constraints.

It is perhaps worth emphasizing that whatever heat is not converted into chemical energy must be discharged at a low temperature. The thermochemical water-splitting process must therefore involve at least one exothermic reaction that proceeds at a relatively low temperature. This is usually the reaction by means of which the water enters the cycle.

The energy requirements for the thermochemical decomposition of water were first investigated by Funk and Reinstrom (1966). These authors considered a two-step process for the decomposition of water, consisting of the reaction of water with a metal to liberate hydrogen and to form an oxide, followed by the thermal decomposition of the oxide:

$$A + H_2O \rightarrow AO + H_2$$

$$AO \rightarrow A + \tfrac{1}{2}O_2$$

They concluded that water decomposition by such a two-step process was not possible, and it is intuitively clear why it is not. If the reaction of water with a metal proceeds to form an oxide, the metal–oxygen bond in the oxide will be at least as strong as the oxygen–hydrogen bond in water, and the oxide will only decompose at a temperature comparable to that required for the direct thermal dissociation of water itself. The following principle makes it possible for a compound to be dissociated by a multistep process at a much lower temperature than that required for direct thermal dissociation: The compound to be dissociated is brought to reaction with a very weakly bonded chemical species in such a way that two new species are formed with bond energies roughly the average of the high- and low-energy bonds of the original reactants. While the temperature for the dissociation is thereby drastically reduced, the number of molecules that have to be dissociated is doubled for each reaction step in which such an averaging of high and low bond energies takes place.

From this general consideration it follows that a water-splitting cycle must have at least the four distinguishable steps that are illustrated in the sequence of reactions (1) to (4) and that may be identified as

(1) The reaction in which water is absorbed.
(2) The reaction in which the oxygen from the water is given off.
(3) The reaction in which hydrogen is liberated.
(4) The reaction in which the weakly bonded reagent is regenerated.

A real cycle will frequently require several additional steps, and often the species identified as A and B in reactions (1)–(4) will actually represent complex mixtures of substances.

We have seen that a thermochemical cycle is thermodynamically equivalent to the combination of heat engine, turbogenerator, and electrolyzer as a means of obtaining hydrogen from water. There is no fundamental reason why one approach should be more efficient than the other. The choice will ultimately have to be made on practical grounds.

A steam turbine is limited to maximum temperatures around 800°K. Advanced topping cycles, such as gas turbines and magnetohydrodynamic generators, will be able to utilize higher temperatures, but these technologies are not yet perfected. We may reasonably hope that a thermochemical cycle will be able to accept heat at temperatures considerably higher than 800°K, and it could therefore have a higher efficiency than a steam turbine-electrolyzer combination. The present goal for thermochemical cycles is the use of heat at 1000–1100°K, which can be delivered by a high temperature gas-cooled nuclear reactor. Many chemical processes presently operate at temperatures exceeding 1000°K. The production of steel, for example, requires much higher temperatures.

There is also reason to hope that a thermochemical hydrogen plant will have economies of scale over a generator-electrolyzer combination. Electrolysis tends to be a modular operation showing little improvement in economy as the installation gets larger. A thermochemical plant, like other chemical plants, may be expected to become progressively more economical to operate as it grows larger. Doubling the size of an electrolysis plant may increase its cost by perhaps a factor of 1.9, while doubling the size of a thermochemical plant might only increase its cost by a factor of 1.5.

III. General Thermodynamic Considerations

As we have already pointed out, a water decomposition cycle is analogous to a heat engine in that it converts heat into the chemical potential energy residing in elemental hydrogen and oxygen. There are, however, features that set the thermochemical cycle apart from the conventional heat engine and that impose additional restrictions with regard to the energy-conversion efficiency. These restrictions stem from the thermodynamic characteristics of the reactions of the cycle, and thus have their origins in the chemical properties of the elements and compounds involved.

For maximum efficiency each reaction of a thermochemical cycle should be carried out at a temperature at which its standard Gibbs free energy, ΔG^0, vanishes. To understand what this means physically, let us consider the endothermic reaction (3). To be useful, the hydrogen produced by this reaction must be at about 1 atm pressure. If the reaction is carried out at a temperature where its ΔG^0 is substantially positive, hydrogen will be evolved at a very low pressure, and it will be necessary to do work to compress it to one atmosphere. Inasmuch as some three units of heat are required to produce one unit of work, this requirement of external work input will substantially degrade the efficiency of the process. On the other hand, if the reaction is carried out at a temperature such that ΔG is substantially negative, the hydrogen will be obtained at a very high pressure. In principle, this hydrogen could be brought to atmospheric pressure by passage through a turbine that performed useful work. In a real process, however, it will probably not be possible to recover this work, and the efficiency of the process will be reduced. In practice, of course, it may be desirable to deliver hydrogen at pressures above atmospheric (e.g., around 100 atm), but this will require only a comparatively small increase of the reaction temperature. Hence, we see that the optimum condition is where the reactions proceed without either requiring or generating useful work, which is just the condition where $\Delta G = 0$. The temperature at which ΔG vanishes is a characteristic parameter of each reaction, which we may call the equilibrium temperature of the reaction.

If external work is not to be required, the temperature of the primary heat source must be as high or higher than the equilibrium temperature of any endothermic reaction in the cycle. Going back to our general water decomposition scheme, reactions (1)–(4), we may expect at least two endothermic reactions for even the simplest case. The equilibrium temperatures of these two reactions are dictated by their respective chemical characteristics and will in general not be equal. As long as there is only one heat source for the process, it must have a temperature high enough to drive the reaction characterized by the highest equilibrium temperature. Hence, there will be at least one other endothermic reaction in the cycle that has a lower equilibrium temperature but that still must receive its heat from the high temperature heat source. This is equivalent to running this second reaction at the higher temperature and has the consequence that the efficiency of the thermochemical process will be less than the Carnot efficiency calculated from the temperature of the primary source.

A similar problem arises if the cycle contains two exothermic reactions. Again, these will in general have two different equilibrium temperatures, and this will usually mean that a portion of the heat rejected by the cycle will be rejected at a temperature higher than that of the lowest temperature reaction. As in the case of the two endothermic reactions, this will have the effect of reducing the "ideal" efficiency of the cycle below that calculated by applying the Carnot limitation to the highest and lowest reaction temperatures.

We may conclude from the foregoing that the thermochemical cycle has less flexibility than a conventional heat engine with regard to the temperatures at which it can operate reversibly, i.e., under equilibrium conditions. The Carnot efficiency calculated from the highest and lowest temperatures cannot ordinarily be achieved, even under conditions of complete reversibility.

A convenient way of assessing the ideal efficiency of thermochemical cycles makes use of a diagram of entropy versus temperature, a method first applied in this context by Abraham and Schreiner (1974). Such a diagram for our general four-reaction scheme is shown in Fig. 1. The entropy changes depicted in the diagram refer to the entropy changes of all reactants and products of the cycle, including the water itself. A closed loop is obtained by including the entropy of formation of water at ambient temperature.

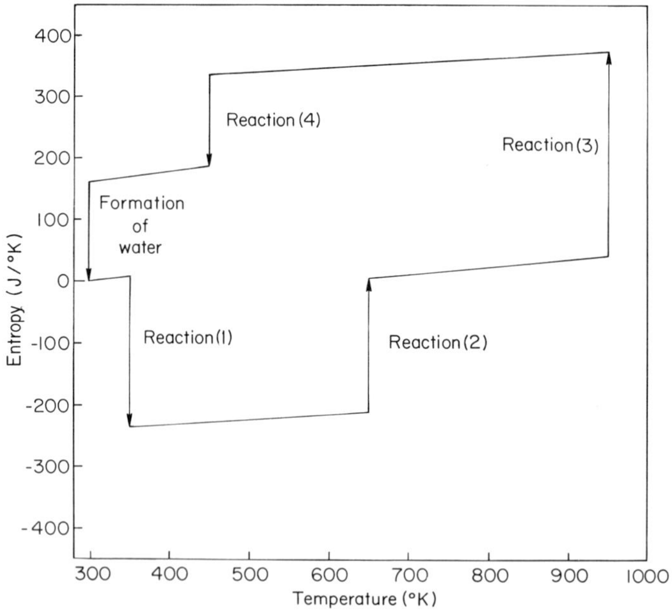

FIG. 1. Diagram of entropy *vs* temperature for a generalized four-step cycle for the thermochemical decomposition of water.

Each vertical arrow represents the entropy of a reaction plotted at its equilibrium temperature, i.e., the temperature at which the standard Gibbs energy of the reaction is zero. The only exception to this is the entropy of formation of water, which is plotted at ambient temperature to close the cycle. The lines connecting the arrows give the entropy changes during heating and cooling of reactants and products. An upward pointing arrow implies an endothermic reaction, a downward pointing arrow implies an exothermic reaction.

As a consequence of plotting all reaction entropies at equilibrium temperatures with the exception of the entropy of formation of water, the area enclosed by the loop equals the standard Gibbs energy of formation of water at ambient temperature. The heat involved in each reaction is the product of its ΔS (the length of the arrow) and its equilibrium temperature. Hence, the ideal efficiency of the process is approximately the area enclosed by the loop divided by the sum of the $T\Delta S$ terms for the endothermic processes. Actually a small correction must be included to account for heat effects in going from one reaction to another. This is reflected in the fact that in general the lines connecting the reactions are not exactly horizontal.

Diagrams of this type provide a means for the preliminary screening of proposed thermochemical cycles. A minimum thermodynamic requirement for an acceptable cycle is that it can be represented as a closed cycle on an entropy–temperature diagram.

IV. Practical Requirements for a Thermochemical Cycle

In the preceding section we have recognized the thermodynamic characteristics of an optimum thermochemical cycle: The heat-absorbing reactions should take place at a temperature as close as possible to the source temperature, i.e., they should have ΔG values near zero at the source temperature, while the heat-evolving reactions should take place as close as possible to ambient temperature. Such a situation, however, cannot be realized in practice and, in fact, may not be desirable. If heat is being supplied by a gas-cooled nuclear reactor, the heat will be provided in the form of the sensible heat content of a stream of helium. The temperature at which the heat is available will obviously decrease as heat is extracted from the helium stream. Hence, to make the best use of the helium as it cools, it will be advantageous for the thermochemical process to accept heat over a range of temperatures.

A number of other features are clearly desirable in a practical thermochemical process. The process should involve as few sequential operations as possible in order to minimize plant complexity. The individual reactions should proceed rapidly to produce the desired products, without side reactions and without attack on reaction vessels and piping. The speed of the reactions will obviously determine the throughput per unit time for a given inventory of reagents and thus will affect the size of plant needed for any specified rate of hydrogen production. A large inventory of reagents is undesirable in that it will increase the mass of material that has to be cycled from one temperature to another and will correspondingly increase the heat losses that are invariably associated with such cycling. Reactions in which one solid material is formed from another are generally relatively slow and should be avoided if possible. For example, the reaction

$$Ag + HCl \rightarrow AgCl + \tfrac{1}{2}H_2$$

slows down markedly as a coating of silver chloride forms on the metallic silver. Desirable reactions will have either all of the reactants or all of the products in fluid states. Solid reagents and products are also less desirable because of the difficulties that may be involved in transporting them and in incorporating them into a continuous process.

Reaction products should be readily separable from each other and from reactants. Incomplete separations will require the recycling of material, adding complexity to the process and increasing heat losses. Separations may take advantage of phase differences or may make use of "scrubbing" techniques. Often it will be advantageous for reactions to go nearly to completion in order to facilitate the separation of products from reactants.

Ideal reagents for a thermochemical cycle should be inexpensive and should not be so corrosive as to require the use of expensive materials for construction. Nor should the reagents constitute a serious hazard in the event that they are inadvertently released into the environment.

Needless to say, no real thermochemical process is likely to possess all of the optimal characteristics that we have enumerated. For any proposed cycle we must decide, first, whether the thermodynamics are favorable, second, whether the reactions will actually go as written, and finally, whether the balance of practical advantages and disadvantages is such as to make the cycle an attractive prospect. Until now, most of the effort in this field has been put into a search for sets of reactions that have favorable thermodynamics. Much less work has been done to verify that reactions do, in fact, proceed in the way we expect, and no cycles have been developed to the point of practical assessment. This should not be too surprising. It is fairly easy to write down a set of reactions that comes close to meeting the thermodynamic criteria for a thermochemical cycle. It is something else again to determine experimentally how these reactions actually take place, and it is harder yet to assess the cycle as a complete practical entity.

Later in this article we will discuss a number of cycles that have been proposed, and we will try to indicate the extent to which they have been evaluated.

V. General Aspects of Real Water Decomposition Processes

The number of thermochemical water decomposition cycles that has been published is fairly large. In their review on the subject, Bamberger and Richardson (1976) list 72 different cycles and point out that there are some 20 more that remain undisclosed for proprietary reasons. Bamberger and Richardson classify these 72 cycles according to the principal chemical species involved in the water decomposition. There are other aspects under which water-splitting cycles can be classified, such as the number of independent reaction steps or the maximum temperature required. The chemical classification is probably the most useful scheme since it is generally applicable and emphasizes natural relationships among the cycles.

Obviously the variety of proposed reaction sequences owes its existence to the fact that many chemists and chemical engineers have independently addressed themselves to the problem of thermochemical water splitting. The development of the basic original patterns incorporated in the cycles relied probably entirely on the general knowledge and intuition of the individual investigator. Efforts have been made to utilize computers to generate water-splitting cycles, but the computer's role appears to be limited to the identification of new examples of cycles that fit a predetermined pattern. It has not heretofore been possible to program computers to search out entirely new patterns for cycles.

There is no simple and straightforward procedure according to which reactions can be combined to form a water-splitting cycle. There are, however, a number of rules and principles that apply to the concept of thermochemical water splitting and that can be used profitably in the search for new cycles.

At this point it should be emphasized that there is no need to restrict water decomposition processes to exclusively thermochemical operation. A combination of thermochemical reactions and operations requiring the expenditure of useful work may, in fact, be superior in terms of overall efficiency, as will be discussed later. Cycles in which thermochemical steps are combined with electrolysis or with other steps requiring work are generally referred to as hybrid cycles. Two of the processes discussed later fall into this category. One does well to keep in mind that from a practical standpoint the best process is the one that requires the least cost for the decomposition of water. The first step toward this end is to design the highest overall energy efficiency into the cycle. It is obvious, on the other hand, that the amount of useful work supplied must be kept to a minimum since such work will in general be derived from heat with an efficiency of about 30%.

In beginning to look for suitable reactions, the chemist soon recognizes that the crucial step of a water decomposition process is the one in which hydrogen is liberated. It is assumed here that the decomposition process follows the general pattern that we have described earlier Eqs (1)–(4). In that case the hydrogen will be the product of a high-temperature endothermic reaction. It is possible to design a water-splitting cycle around a low temperature reaction for the production of hydrogen, but this will lead to the formation of a hydrogen-free species of stability equal to or greater than that of water. Such a species will in turn require its own decomposition cycle.

The high-temperature endothermic reaction for the formation of hydrogen must be compatible with the low temperature reaction of the cycle in which the water is absorbed. Because of this requirement, there arises a severe restriction with regard to the elements that can be used as hydrogen carriers. Most elements, including almost all metals, have a higher affinity to oxygen than to hydrogen, and it is therefore not possible to use such elements to achieve a separation of the hydrogen from water. One concludes that the elements that hold promise as hydrogen carriers are limited to the following: C, N, S, Se, Br, and I. With the exception of ammonia, the hydrides of all of these elements have in one way or another been incorporated into proposed water decomposition cycles.

Once a decision has been made as to which hydride to use in the cycle, the next step is to find a reaction in which this hydride is formed from water. This reaction must occur at or near ambient temperature to satisfy the general energetic requirements that we have outlined. It should lead to products that can be separated with relative ease, so that the hydride and the oxygen-carrying compound can be processed in separate high temperature reactions. Finally, the oxygen-carrying compound from this reaction must be suitable as reactant for the endothermic high temperature reaction that evolves O_2. This latter condition can usually be satisfied without great difficulty because there are many oxides and salts of oxyacids that give off oxygen when heated.

As a final step in setting up the chemistry of a water-splitting cycle, it may be necessary to include reactions in which the original reactants are regenerated.

The four-step basic scheme for a water decomposition process that we have presented in Eqs. (1)–(4) incorporates the necessary chemistry. It is designed around the principle of replacing a single strong bond between atoms by two bonds of less energy, thereby permitting thermal dissociation at lower temperatures. Both endothermic reactions in this scheme are seen as simple dissociations of binary compounds. As was shown earlier, the laws of thermodynamics dictate a minimum amount of heat that has to be absorbed at the high temperature. The minimum heat requirement at 1000°K is

$$\frac{237.2}{(T_3 - T_1)/T_3} = \frac{237.2}{0.7} = 339 \text{ kJ/mole of water decomposed}$$

It is not possible to absorb this large a quantity of heat at 1000°K in two endothermic reactions, one of which is the dissociation of an oxide, and the other the dissociation of a hydride. Hence a process of this simple type cannot achieve the dissociation of water at temperatures below 1000°K. It is necessary to consider more complex endothermic high temperature reactions which can absorb the necessary heat by virtue of larger entropy changes. Both the hydrogen-evolving and the oxygen-evolving reactions must therefore be more complex than simple dissociation reactions.

We can see from this discussion that the maximum temperature at which heat can be provided to a cycle severely restricts the choice of reaction schemes. In some cases, this has led to the proposal of cycles operating at temperatures considerably in excess of 1000°K; one such proposal even includes the design of a special reactor to provide the necessary ultra-high temperature heat.

VI. Description of Selected Water-Splitting Cycles

In this article we will not attempt to describe all the different schemes that have been proposed for the thermochemical generation of hydrogen. For this the reader is referred to the comprehensive listing in the report published by Bamberger and Richardson (1976).

The water decomposition cycles that we will discuss in this section are meant to give an idea of the different approaches used to solve the problem, of the variation in the chemistry among the cycles, and of the present state of progress in the area.

The following cycles will be included here:

(1) Three cycles proposed by a group of researchers working at the Euratom Laboratory in Ispra, Italy, and designated respectively Mark 1, Mark 9 and Mark 14.

(2) The zinc selenide cycle of Lawrence Livermore Laboratory.
(3) The barium chromate cycle of Oak Ridge National Laboratory.
(4) The iodine-sulfur dioxide cycle of General Atomic Company.
(5) The ANL-4 iodine-mercury cycle of Argonne National Laboratory.
(6) The sulfur dioxide electrolysis process of Westinghouse Electric Corporation.

A. The Ispra Cycles (Marchetti, 1973)

1. *Mark 1*

The Mark 1 cycle, invented by De Beni in 1969, evolved from studies directed at the obtaining of chemical process heat from nuclear reactors. With its publication a period of dormancy in the search for thermochemical water decomposition processes was replaced by one of widespread and intense activity. At present Mark 1 is mainly of historical interest as the first of a series of processes developed at Ispra. The reactions of the Mark 1 cycle are

$$CaBr_2 + 2H_2O \xrightarrow{T = 1005°K} Ca(OH)_2 + 2HBr$$

$$Hg + 2HBr \xrightarrow{T = 525°K} HgBr_2 + H_2$$

$$HgBr_2 + Ca(OH)_2 \xrightarrow{T = 475°K} CaBr_2 + HgO + H_2O$$

$$HgO \xrightarrow{T = 825°K} Hg + \tfrac{1}{2}O_2$$

The overall result of this sequence of reactions is easily seen to be the decomposition of one mole of water into oxygen and hydrogen.

In laboratory studies it was found that at the temperatures specified some of the reactions had unfavorable kinetics or unsatisfactory yields. As a consequence, improved cycles were developed, including those dubbed Mark 9 and Mark 14.

2. *Mark 9 and Mark 14*

Mark 9 and 14 are members of the family of iron chloride cycles. The iron chloride cycles make use of the thermal conversion of iron(III) chloride into iron(II) chloride to produce a reducing species capable of releasing hydrogen from water. Mark 9 proposed to achieve the decomposition of water by the following three reactions:

$$6FeCl_2 + 8H_2O \xrightarrow{T = 925°K} 2Fe_3O_4 + 12HCl + 2H_2$$

$$2Fe_3O_4 + 3Cl_2 + 12HCl \xrightarrow{T = 450°K} 6FeCl_3 + 6H_2O + O_2$$

$$6FeCl_3 \xrightarrow{T = 695°K} 6FeCl_2 + 3Cl_2$$

Unfortunately, the 450°K reaction did not proceed as indicated at the specified temperature, and the scheme had to be modified. The modified version was called Mark 14 and consisted of the reactions:

$$9FeCl_2 + 12H_2O \xrightarrow{T = 925°K} 3Fe_3O_4 + 18HCl + 3H_2$$

$$3Fe_3O_4 + 1\tfrac{3}{2}Cl_2 \xrightarrow{T = 505°K} 4Fe_2O_3 + FeCl_3$$

$$4Fe_2O_3 + 24HCl \xrightarrow{T = 505°K} 8FeCl_3 + 12H_2O$$

$$9FeCl_3 \xrightarrow{T = 705°K} 9FeCl_2 + \tfrac{9}{2}Cl_2$$

$$3H_2O + 3Cl_2 \xrightarrow{T = 1075°K} 6HCl + \tfrac{3}{2}O_2$$

The last step of this cycle is the reverse of the well-known Deacon reaction, and it has been employed in a number of other cycles.

After the principal equilibrium and kinetic data had been assembled, this cycle was subjected to an engineering assessment to establish material flows and heat requirements (Hoogstoel *et al.*, 1976). It was found that the overall thermal efficiency of the cycle would be substantially less than 30%, and that mass flow would have to be quite large. Unless these difficulties can be overcome, the engineering analysis leaves little hope for Mark 14 to become practical.

B. The Barium Chromate Cycle of Oak Ridge National Laboratory

In 1975 Bamberger *et al.*, published the following water decomposition cycle:

$$Cr_2O_3(s) + 4Ba(OH)_2(l) \xrightarrow{T = 925°K} 2Ba_2CrO_4(s) + 3H_2O(g) + H_2(g)$$

$$2BaCrO_4(s) + Ba(OH)_2(l) \xrightarrow{T = 1125°K} Ba_3(CrO_4)_2(s) + H_2O(g) + \tfrac{1}{2}O_2$$

$$2Ba_2CrO_4(s) + Ba_3(CrO_4)_2(s) + 5H_2O(l) \xrightarrow{T = 375°K}$$
$$Cr_2O_3 + 2BaCrO_4(s) + 5Ba(OH)_2(aq)$$

The authors report that this cycle has favorable reaction kinetics and that it has the attractive feature of not requiring the separation of multicomponent gas mixtures. The latter advantage, however, may be partially offset by the fact that solid reactants will have to be moved about in the operation of the cycle.

The first reaction represents a remarkable approach to the problem of generating hydrogen in a thermochemical cycle. The authors state that significant reaction rates can be obtained in the 725°–925°K range. No estimate of the thermal efficiency of this chromate cycle is available inasmuch as the necessary thermochemical data are lacking.

C. The Zinc Selenide Cycle of Lawrence Livermore Laboratory

An original version of a cycle employing the thermal dissociation of hydrogen selenide to release hydrogen was proposed by Dreyfuss and Krikorian in 1975. The initial scheme was later modified to consist of the following reactions (Pearson *et al.*, 1976):

$$2ZnO + Se(l) + SO_2 \xrightarrow{T = 770°K} ZnSe + ZnSO_4$$

$$ZnSe + ZnSO_4 + 2HCl(g) \xrightarrow{T = 350°K} ZnSO_4(aq) + ZnCl_2(aq) + H_2Se$$

$$ZnSO_4(aq) + ZnCl_2(aq) + SO_3 + 2H_2O(g) + H_2O(l) \xrightarrow{T = 350°K} 2ZnSO_4 \cdot H_2O + 2HCl(g)$$

$$2ZnSO_4 \cdot H_2O \xrightarrow{T = 600°K} 2ZnSO_4 + 2H_2O$$

$$2ZnSO_4 \xrightarrow{T = 1100°K} 2ZnO + SO_2 + SO_3(g) + \tfrac{1}{2}O_2$$

$$H_2Se \xrightarrow{550°K} Se(l) + H_2$$

The maximum temperature for this water-splitting cycle is 1100°K, the temperature required to decompose zinc sulfate. Work on this cycle has not yet advanced to the point at which an evaluation of its merits has been published.

D. The ANL-4 Cycle of Argonne National Laboratory

One of the cycles that relies on hydrogen iodide as the hydrogen carrying species is the scheme dubbed ANL-4, which was developed at Argonne National Laboratory (Abraham *et al.*, 1976). There are four principal reaction steps in this cycle:

$$2H_2O + 2NH_3 + 2CO_2 + 2KI(aq) \xrightarrow{T = 300°K} 2KHCO_3 + 2NH_4I(aq) \quad (5)$$

$$2KHCO_3 \xrightarrow{T = 425°K} K_2CO_3 + H_2O + CO_2 \quad (6)$$

$$2NH_4I + Hg \xrightarrow{T = 650°K} 2NH_3 + HgI_2(g) + H_2 \quad (7)$$

$$K_2CO_3 + HgI_2 \xrightarrow{T = 1000°K} 2KI(l) + Hg + CO_2 + \tfrac{1}{2}O_2 \quad (8)$$

The high temperature reactions (7) and (8) can take place at 600° and

825°K, respectively, but reaction (8) in particular is better carried out at about 1000°K in order that the K_2CO_3 may be present as a solution in molten KI. From a chemical standpoint the reactions of ANL-4 are remarkably "clean," i.e., they are not encumbered by adverse side reactions. The first reaction, the precipitation of $KHCO_3$ from a solution of KI, must be conducted in a mixture of water and an organic solvent such as acetone or isopropanol. In a pure aqueous solution NH_4HCO_3 is less soluble than $KHCO_3$ and is precipitated instead.

The presence of mercury in this cycle is considered a liability, especially so because it is released at the same time that the oxygen is given off. A working plant will probably vent the oxygen, and it will therefore be necessary to remove the mercury very thoroughly in order to avoid contaminating the environment.

Inasmuch as the authors of this article are intimately familiar with the cycle ANL-4, it is chosen here to demonstrate how some of the general features of water-splitting procedures discussed earlier have been incorporated into an actual sequence of reactions.

The water that enters the cycle in reaction (5) is, in fact, split at once by virtue of its capability of dissociating into ions. The hydrogen becomes attached to the ammonia, and the oxygen finds itself part of the hydrogen carbonate ion. Since the potassium hydrogen carbonate is precipitated from the solution, there is at once a physical separation of the two constituent elements of the water. The potassium salt is obtained in a rather pure form, and as a solid it can be used directly in the next reaction, which is the calcination of $KHCO_3$ to form potassium carbonate. The hydrogen in the $KHCO_3$ is evolved as water and is returned to step (5).

The ammonium iodide carries with it the hydrogen, but it can be obtained from the solution only by a certain expenditure of energy. Simple evaporation of the solvent, however, must be ruled out. At an expected concentration of 10 mol%, 20 moles of solvent would have to be vaporized for every 2 moles of NH_4I recovered, or every mole of hydrogen produced. The amount of heat that would have to be furnished for the evaporation would make the process quite uneconomical in terms of thermal efficiency. There are, however, other options for the recovery of the ammonium iodide. These are techniques that have been applied to water desalination and include multistage regenerative distillation of the solvent, distillation by vapor compression, and perhaps even hyperfiltration. In water desalination, however, no solid phase is involved, and application of these techniques to the separation of a solid product has never been demonstrated. Vapor compression may be the most promising approach. It involves the use of a pump to evaporate solvent and compress the vapor so that it can be condensed at a temperature slightly higher than that of the evaporator. The heat evolved on condensation can then be used to provide the heat needed for evaporation. This type of evaporation obviously requires the expenditure of work to drive the compressor. Depending on the efficiency of the compressor, however, this work should be only some 30% more than the thermodynamic energy required to separate the solute reversibly from the solvent. This amount of energy will be a great deal less than that needed for the irreversible evaporation of large volumes of solvent.

The hydrogen is released through the reaction of ammonium iodide with mercury at 650°K. This reaction is not a simple dissociation. Closer analysis shows that it is best considered as a combination of two processes: the evaporation and simultaneous dissociation of ammonium iodide to form NH_3 and HI, followed by the subsequent reaction of the hydrogen iodide with the mercury vapor to produce hydrogen.

The overall heat balance of reaction (7) shows it to be an endothermic reaction. It is therefore one of the steps of the cycle during which heat is absorbed from the primary source. When one looks at the reaction as proceeding in two stages, however, only the dissociative evaporation of the ammonium iodide turns out to be an endothermic reaction. The reaction of hydrogen iodide with mercury is, in fact, strongly exothermic. The amount of heat released from it must be used elsewhere in the cycle in order to maintain overall energy efficiency. In magnitude this amount of heat is approximately equal to the heat required for the calcination reaction (6), and this latter reaction is therefore the most appropriate one to be driven by the heat from the hydrogen iodide-mercury reaction.

The highest temperature is needed to carry out the reaction of potassium carbonate with mercuric iodide, which results in the release of the second principal product of the cycle, the oxygen. This is also an endothermic reaction, and it is therefore also a point at which primary heat is absorbed by the cycle. Again, this is more complex than a simple thermal dissociation. The mercuric iodide in this case serves to bring about the decomposition of potassium carbonate and the release of oxygen at the desired temperature. It is interesting to observe that if the potassium were replaced by sodium, the corresponding reaction would take place only at a considerably higher temperature. It is the relatively low stability of K_2CO_3 with respect to KI that lets this reaction proceed at an acceptably low temperature.

While the chemistry of ANL-4 is described by the four-reaction sequence (5)–(8), this sequence of reactions can be misleading with regard to the practical energy efficiency of the cycle. In order to evaluate the energy efficiency of ANL-4 it is necessary to know exactly during which operations heat must be absorbed from the primary source. These stages can be identified only if a general flow diagram is drawn up that shows the individual steps of the process as they are supposed to take place in practice. Such a flow diagram for ANL-4 shows that there are three operations during which heat must be absorbed from the primary source. These operations are

(1) The high temperature reaction (8) at 1000°K, including the heat of solution of K_2CO_3 in molten KI.
(2) The evaporation of liquid mercuric iodide at 900°K.
(3) The evaporation of ammonium iodide at 650°K.

To the amount of heat directly absorbed from the source in these three operations there has to be added the amount of heat necessary to generate work for the vapor compression step during which ammonium iodide is recovered from solution.

For convenience, the respective amounts of heat are compiled in Table I. The sum of these heats constitutes the total heat requirement of the cycle ANL-4 for the production of 1 mole of hydrogen from water. The efficiency of the cycle, as expressed by the ratio of Gibbs energy of formation of water to total heat requirement, is 37%. In the opinion of the authors it is this comparatively high conversion efficiency at the relatively low maximum temperature of 1000°K, along with the straightforward chemistry without side reactions, that makes the cycle an attractive one with which to demonstrate the thermochemical concept. The use of mechanical work for the vapor compression stage of the process constitutes a slight deviation from the ideal of a purely thermochemical cycle. ANL-4 in its present form must therefore be classified as a hybrid process.

TABLE I

HEAT INPUT TO ANL-4 CYCLE FOR THE GENERATION OF 1 MOLE OF HYDROGEN

Operation	Temperature (°K)	Heat (J)
Dissolution of K_2CO_3 in melt	1000	27,600[a]
Heat of reaction (8): $K_2CO_3 + HgI_2 \rightarrow 2KI + Hg + CO_2 + \frac{1}{2}O_2$	1000	169,500
Heat of evaporation of HgI_2	900	59,800
Heat of evaporation of $2NH_4I$	650	176,800
Heat equivalent of the work of vapor compression	—	204,900[b]
Total heat to be supplied from the primary heat source	1000	638,600

Efficiency:

$$\frac{\Delta G_{f\mathrm{H_2O}}(298)}{\text{total heat}} = \frac{237200}{638600} = 0.37$$

[a]The heat required to dissolve K_2CO_3 in a salt mixture of composition $3KI \cdot 2K_2CO_3$ was estimated from the heat of fusion of K_2CO_3.

[b]Assuming an electrically driven compressor operating at 75% efficiency and using electricity produced with 30% efficiency.

E. The Sulfur Dioxide–Iodine Cycle of General Atomic Company*

A cycle based on the Bunsen reaction of sulfur dioxide with iodine in an aqueous medium has been proposed by General Atomic Company. There are only three fundamental reactions for this cycle:

$$2H_2O + I_2 + SO_2 \xrightarrow{T = 300°K} H_2SO_4 + 2HI(aq)$$
$$H_2SO_4 \xrightarrow{T = 1300°K} H_2O + SO_2 + \tfrac{1}{2}O_2$$
$$2HI \xrightarrow{T = 500°K} H_2 + I_2$$

This scheme, although it looks enticingly simple on paper, is nonetheless very difficult to carry out in practice with any degree of energy efficiency. The first reaction occurs very rapidly in a aqueous suspension of iodine. Problems arise when the concentration of HI begins to build up, because hydriodic acid is capable of reducing sulfuric acid to elemental sulfur and even to sulfide.

Under certain conditions of concentration and SO_2 pressure, it has been found that the system spontaneously separates into two liquid phases, one which contains most of the sulfuric acid and water, and one which is of higher density and consists mainly of aqueous HI. This behavior appears to facilitate the separation of the two acids.

The sulfuric acid must be concentrated before it can be decomposed in the second reaction. This decomposition requires a temperature of about 1300°K, which constitutes an inconveniently high maximum temperature for the cycle. We will refer to this problem again when we discuss the Westinghouse sulfur cycle.

In the General Atomic Company cycle the hydrogen is evolved by the dissociation of gaseous, anhydrous hydrogen iodide at a comparatively low temperature. While hydrogen iodide is partially dissociated under conditions of thermodynamic equilibrium, at low temperatures (up to 700°K) the rate of equilibration is slow and would have to be accelerated catalytically in order to be useful for a hydrogen production scheme. It is obvious from what we have said here that further fundamental experimentation is necessary in order to find an effective way of utilizing the sulfur dioxide-iodine reaction sequence.

F. The Westinghouse Hybrid Sulfur Process†

Our final example of a water-splitting scheme is the Westinghouse sulfur process. It is important because scientists and engineers at Westinghouse

* Russell *et al.* (1976).
† Brecher *et al.* (1976).

Electric Corporation have drawn up a comprehensive plan for a hydrogen-generating plant that utilizes this process.

The Westinghouse scheme is designed to be a hybrid process in which the hydrogen is produced electrolytically while the oxygen is released in a thermal decomposition reaction. The following two reactions describe the chemistry of the cycle.

$$2H_2O + SO_2 \xrightarrow[T \geqq 300°K]{\text{electrolysis}} H_2 + H_2SO_4$$

$$H_2SO_4 \xrightarrow{T = 1140°K} H_2O + SO_2 + \tfrac{1}{2}O_2$$

The electrolysis reaction in effect makes use of sulfurous acid as an anodic depolarizer to render the operating voltage substantially lower than that of ordinary water electrolysis. The high Gibbs energy of formation of water, which determines the reversible cell voltage for the electrolysis in the absence of a depolarizer, is diminished here by the Gibbs energy of formation of aqueous sulfuric acid from SO_2. From known standard potentials we may calculate the reversible voltage for the electrolysis of a sulfurous acid solution to be 0.17 V. Actual experimental data obtained in the Westinghouse laboratories show the cell voltage to vary between 0.6 and 0.8 V for current densities up to 0.1 A/cm^2, with higher values at higher current densities. The electrolyte was 50 wt.% H_2SO_4 at a temperature of 30°C. Higher temperatures lead to lower cell voltages.

The sulfuric acid decomposition reaction requires very high temperatures. It also requires subsequent concentration of the ca. 40 wt.% acid that leaves the electrolyzer cells. These are problems that this process has in common with the SO_2–I_2 cycle of General Atomics. The highest temperature of the process is the temperature at which sulfur trioxide decomposes. Depending on the extent of decomposition and other process variables, this temperature will be in the 1100°–1300°K range. Such high temperatures are not reached by present-day nuclear heat sources. The Westinghouse process is therefore dependent on the development of a new type of reactor, the so-called VHTR or "very high temperature reactor." It is believed that the technology exists on which to base the development of such a machine, which will deliver nuclear heat at the temperature of 1144°K required for the process.

The Westinghouse plan for a hydrogen production plant makes use of a VHTR with a capacity of 3345 MW of nuclear heat. The plant will produce 10,120,000 standard cubic meters of 99.9%-pure hydrogen per day. This corresponds to the production of 4869 moles of hydrogen per second, and from this number we obtain an overall efficiency of 237,200 (4869)/ 3,345,000.000 = 34.5%.

Of all the water-splitting processes discussed here, the Westinghouse process is at the most advanced stage with respect to a realization of the concept in an actual hydrogen-producing plant. If the process can be made to function as well as predicted, it will be a promising competitor to the conventional electrolytic production of hydrogen.

G. Additional Remarks

Our foregoing description of several proposed thermochemical water decomposition cycles has concentrated on the basic chemical reactions that characterize each cycle. In many cases this is all the information that is presently available. But we must also recognize that the successful combination of certain chemical reactions to form a water-splitting cycle is only the first step toward the goal of developing a large scale process. The final judgment as to whether a proposed process is viable is determined by the cost of the hydrogen that can be obtained from it. Implicit in the cost figure are the operating expenses and the capital investment in the physical plant. By attempting to design a process to have the highest possible overall energy efficiency, a favorable base for low operating costs is obtained. But there are other factors influencing the operating cost, among them the cost of chemicals and feedstock used in the process. While in principle only water enters the plant and hydrogen and oxygen leave it, in reality there will be certain losses of the reactants used in the various steps. These losses must be made up, and the cost of makeup may be substantial in the case of rare and expensive materials. Without detailed knowledge of actual chemical plant conditions it is difficult to give a general quantitative estimate of such losses. Looking again at the ANL-4 cycle it is obvious that losses of both mercury and iodine must be kept to the lowest attainable level, not only because of the cost of these elements but also because a plant producing, say, 10 million standard cubic meters of hydrogen per day is in danger of causing unacceptable environmental contamination if its losses exceed the level of parts per million. However, it appears not to be beyond present engineering capability to reach such low loss levels, as is evidenced by the present practice of reducing noxious emissions from industrial plants. Actually it is very likely that losses of organic solvent will be more substantial than those of mercury and iodine.

As to the cost of the plant itself, there is little to go on until a specific design has been worked out by chemical engineers. Only the Westinghouse process is currently at a stage where plant costs have been assessed. However, one comment may be made: The use of fairly expensive and rare chemicals need not impose an unbearable financial burden on a process. If, for instance, we assume that with ANL-4 hydrogen can be generated at the rate of 1 mole/sec with an inventory of 100 moles of mercury at \$2.50/mole, we see that the cost of the mercury will be 250/641.1 = \$0.39 per installed kW thermal.

A nuclear reactor alone will cost at least \$100 per thermal kilowatt of installed capacity, and a thermochemical plant will probably be at least as expensive as the reactor that powers it. Therefore, if the inventory of expensive chemicals can be maintained at the level indicated, which implies cycling of the inventory once every 100 sec, the cost of even a chemical as expensive as mercury will contribute negligibly to the overall plant cost.

It is of interest that two of the most promising cycles, Argonne's ANL-4 and the Westinghouse cycle, both use a significant amount of work in addition to heat. The Argonne cycle proposes to use the work of vapor compression to separate ammonium iodide from a solution, while the Westinghouse cycle uses the electrical equivalent of work to obtain hydrogen from a solution of sulfurous acid. It may be that such hybrids will always offer an advantage over purely thermal cycles.

We have already noted as a limitation on thermochemical processes the need to work with the characteristics of real materials, which fix the energetics of various reaction steps as well as the temperatures at which they take place. In principle, it seems that it might be possible to vary the energetics of the reactions in a cycle by carrying them out in solvent systems or multicomponent melts. An example of this approach is the use of a mixed solvent system in the Argonne ANL-4 cycle to precipitate $KHCO_3$, whereas in a purely aqueous medium only NH_4HCO_3 precipitates. One might hope to be able to use this approach to "fine tune" the energetics of a cycle in order to optimize its efficiency. So far, however, no cycle has actually made use of such fine tuning, and it is not possible to say whether the concept will really prove useful.

VII. Conclusions

In this review we have discussed the rationale of the concept of thermochemical generation of hydrogen and have examined a number of cycles that have been proposed. It is apparent that at the present time we cannot say whether this method of hydrogen production will ever become a viable technology. On the other hand, there is no basis on which the thermochemical approach can be categorically rejected. There is no alternative to careful evaluation of candidate cycles: first, to establish that the chemistry really works, and then to carry out detailed engineering and economic analyses designed to demonstrate or refute the practical viability of the process. It seems fairly evident that most of the cycles now under discussion will not survive such close scrutiny. Many of them can already be seen to face severe obstacles that could well turn out to be insurmountable. We may hope, however, that one or more of these schemes, or possibly others that have yet to be put forth, will ultimately work well enough to provide a basis for the development of a useful technology for the thermochemical generation of hydrogen.

References

Abraham, B. M., and Schreiner, F. (1974). *Ind. Eng. Chem., Fundam.* **13**, 305.

Abraham, B. M., Schreiner, F., Attia, A., and Jeannotte, A. (1976). *Proc. Thermochem. Generation Hydrogen, Div. Fuel Chem. Am. Chem. Soc., 1976.* Vol. 21/3, p. 31.

Bamberger, C. E., and Richardson, D. M. (1976). *Cryogenics* **16**, 197.

Bamberger, C. E., Richardson, D. M., and Bredig, M. A. (1975). *Science* **189**, 715.

Brecher, L. E., Spewock, S., and Warde, C. J. (1976). *Proc. World Hydrogen Energy Conf., 1st, 1976.* Vol. I, p. 9A-1.

Dreyfuss, R. M., and Krikorian, O. H. (1975). *Lawrence Livermore Lab.* [*Rep.*] **UCRL-51741**.

Funk, J. E., and Reinstrom, R. M. (1966). *Ind. Eng. Chem., Process Des. Dev.* **5**, 336.

Hoogstoel, M., Goossens, W. R. A., Francesconi, A., and Baetslé, L. H. (1976). Paper presented at World Hydrogen Energy Conf., 1st. 1976.

Marchetti, C. (1973). *Chem. Econ. & Eng. Rev.* **5**, 7.

Pearson, R. K., Krikorian, O. H., Elson, R. E., Condit, R. M., and Dreyfuss, R. M. (1976). *Lawrence Livermore Lab.* [*Rep.*] **UCRL-77859**.

Russell, J. L., McCorkle, K. H., Norman, J. H., Porter, J. T., Roemer, T. S., Schuster, J. R., and Sharp, R. S. (1976). *Proc. World Hydrogen Energy Conf., 1st, 1976.* Vol. I, pp. 1A-105.

Theoretical and Practical Aspects of Immobilized Enzymes

J. Konecny

Pharmaceutical Division,
CIBA-GEIGY Ltd.,
Basel, Switzerland

I. Purpose and Scope

Enzymes have intrigued biologists, chemists, and physicists for many decades. These compact globular-shaped proteins with molecular weights ranging from ten thousand to several million catalyze all the processes in living cells and display outstanding levels of activity and specificity. Thus a molecule like the cephalosporin shown in Fig. 1 can be selectively hydrolyzed at the indicated positions by three different enzymes, while hydrogen and hydroxide ions attack all these bonds. Although the maximum activities displayed vary enormously from protein to protein, reaction rates corresponding to the turnover of 100–10,000 substrate molecules per molecule of enzyme per second may be regarded as typical. That these remarkable properties are necessary becomes evident when we consider the reproduction of a bacterium like *E. coli*: The process of cell doubling involves, in round figures, 10^6 catalyst macromolecules of 10^3 different kinds, all confined to a space of 10^{-12} ml, the time required being only 30 minutes when 20 simple amino acids are available as nutrients (Watson, 1965; Maaløe and Kjeldgaard, 1966).

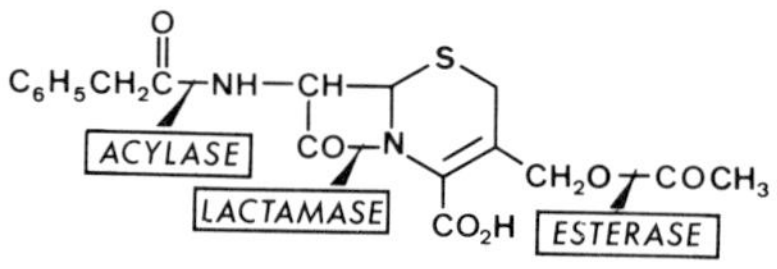

FIG. 1. Selective cleavage of a cephalosporin molecule by three different enzymes.

The term immobilized enzymes designates enzymes which have been confined to insoluble carriers either by chemical or physical attachment or some form of entrapment. As heterogeneous catalysts they differ in many respects from their soluble counterparts and they are of theoretical and practical interest for several reasons.

Many of the 2000 or more enzymes which have been characterized and catalogued (Barman, 1969, 1974) are localized in biological membranes in living cells. Our knowledge of their kinetic properties is derived largely from studies of the isolated species in solution, and thus it is pertinent to ask how immobilization and the proximity of other enzymes modifies their catalytic and regulatory behavior. Owing to the complexity of biological membranes (Capaldi and Green, 1972), synthetic matrices were used as substitutes in the pioneering work of Katchalski and co-workers (1971).

Actually, much of the current work is motivated by interest in immobilized enzymes as catalysts in their own right and in their possible applications. These lie mainly in the area of natural products and related substances and extend from analytical to industrial uses, such as the resolution of D,L-amino acids into the stereoisomers, the isomerization of glucose to fructose, and chemical modifications of penicillins. Heterogeneous catalysts have the advantage that they are easy to recover for repeated use and can also be employed for long periods of time in continuous processes provided that they have adequate mechanical properties and an adequate life. This long term use is crucial because most enzymes are too expensive to be discarded after a single application on a large preparative scale.

Obviously, technological and economic factors such as the yields and costs of immobilized activity and the mechanical properties of the carriers are an integral part of the framework within which the applied research is conducted. However, the dividing line between fundamental and applied work is very diffuse. Knowledge of the structure of enzymes, of the factors which determine their stability and specificity and of the regulation of their biosynthesis are of vital interest in the technological realm. On the other hand; the mathematical and experimental tools developed by chemical engineers and physical chemists for handling heterogeneous catalysts are germain to most investigations in this field.

The present work deals with both aspects of immobilized enzymes. It addresses itself mainly to the chemist interested in the preparation and properties of these novel catalysts and the factors which determine their utility. Since the comprehensive monograph by Zaborsky (1973) covers all literature available prior to December 1971, a detailed discussion of standard methods would be superfluous. An attempt has therefore been made to present an overview from the perspective of early 1977. For the sake of completeness, relevant properties of enzymes are reviewed in the next section. Their isolation and purification are also outlined because many of the stationary phases employed in their column chromatography are used as carriers for their immobilization, and because it is an integral part of enzyme engineering.

II. Enzymes

Enzymes are linear chains of the general structure $H(-NH \cdot CHR \cdot CO-)_nOH$ and are formed by the head-to-tail polymerization of the 20 L-amino acids listed in Table I, the sequence being specified by the gene. The molecular weights of these chains, which are folded into compact molecules, range from 10 to 80 thousand, larger enzymes being invariably composed of two or more globular subunits held together by physical forces and sometimes covalently (Darnall and Klotz, 1975; Wold, 1971). The subunits of such complexes often carry out different reactions or have regulatory functions whereby the activity of the "working" part of the molecule is modified

TABLE I

THE SIDE CHAINS R OF THE 20 ESSENTIAL AMINO ACIDS

Hydrocarbon type		Inert heteroatom type	
Alanine	$-CH_3$	Methionine	$-CH_2CH_2SCH_3$
Valine	$-CH(CH_3)_2$	Tryptophan	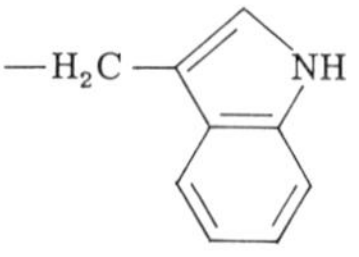
Leucine	$-CH_2CH(CH_3)_2$		
Isoleucine	$-CH(C_2H_5)(CH_3)$		
Phenylalanine	$-CH_2C_6H_5$		
Charged type		Amide type	
Aspartate	$-CH_2CO_2^-$	Asparagine	$-CH_2CONH_2$
Glutamate	$-CH_2CH_2CO_2^-$	Glutamine	$-CH_2CH_2CONH_2$
Arginine	$-CH_2CH_2CH_2NHC(NH_2)_2^{\oplus}$		
Lysine	$-CH_2CH_2CH_2CH_2NH_3^+$		
Hydroxylic type		Miscellaneous	
Serine	$-CH_2OH$	Glycine	$-H$
Threonine	$-CH(OH)CH_3$	Cysteine	$-CH_2SH$
Tyrosine	$-CH_2C_6H_4OH$	Histidine	
		Proline[a]	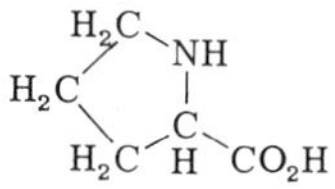

[a]Structure of the amino acid.

according to the concentration levels of products, substrates, or other metabolites.

The globular structures are stabilized in their compact conformation by van der Waals forces and hydrogen bonding and frequently by internal $-S-S-$ bridges formed from two adjacent $-CH_2SH$ residues of the amino acid cysteine (Fig. 2). As an example of a small enzyme, consisting of 129 amino acids, the CPK model of chicken lysozyme (Dickerson and Geis, 1969) is shown in Fig. 3. The cleft on the upper left hand side is the active site, where the reaction with the substrate occurs. The geometrical complimentarity between the catalyst and the substrate is evident from the second model, showing the enzyme-substrate complex. The high activity and the

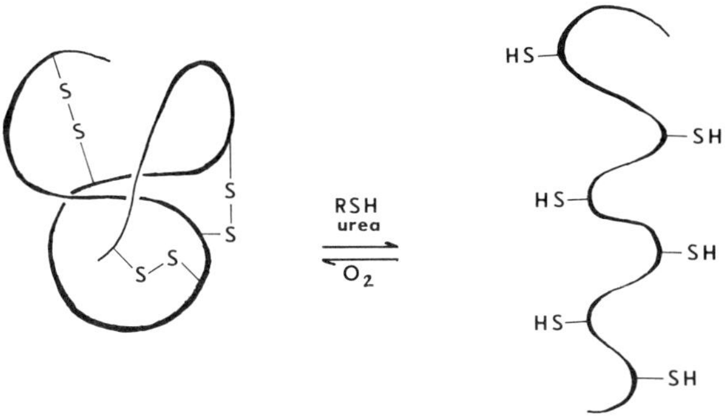

FIG. 2. Schematic illustration of internal —S—S— bridges and of the unfolding of the protein chain under the influence of a reducing and a structure-breaking reagent. Small proteins sometimes recover their native conformation on mild oxidation.

specificity of enzymes is due to the geometry of the active site and the ideal arrangement of the reactive groups, an outcome of millions of years of evolution.

The chain of lysozyme is folded in such a manner that all charged polar side chains of the amino acids, as well as uncharged polar groups with one or two exceptions, are on the external surface. A great majority of the non-

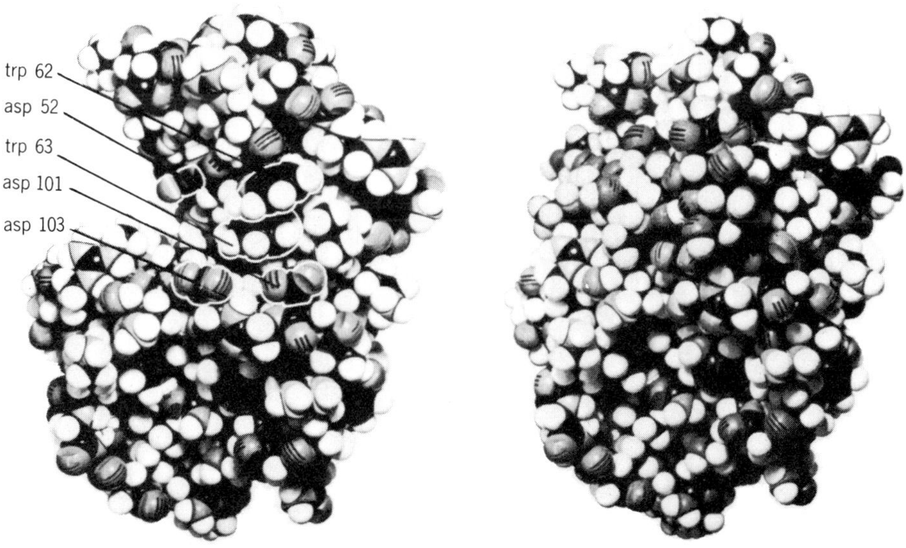

FIG. 3. The CPK model of lysozyme. Left: Enzyme without substrate, showing the active site crevice. Right: Enzyme-substrate complex. [From Dickerson and Geis (1969), by permission.]

polar hydrophobic groups are buried in the interior. Since some of the external polar side chains are acids and bases with various pK values (Tanford and Hauenstein, 1956), the net charge of proteins in solution depends on pH. The isoelectric pH at which the net charge is zero varies from species to species.

Clearly, the activity of the protein is lost when it denatures, i.e., when it loses its natural conformation. The unfolding of the chain and the dissociation of the complexes into subunits occurs more or less readily under the influence of heat, solvents, high concentrations of denaturing agents like urea, or a more or less drastic departure from the physiological pH. For practical purposes, losses of activity associated with major conformational changes are usually irreversible. Mild conditions must therefore be employed for the isolation and immobilization of proteins.

The activity of enzymes depends on temperature, pH, substrate concentration, and sometimes on the concentrations of other substances such as reaction products, inhibitors, and effectors. The influence of pH may usually be represented by a bell-shaped curve, as illustrated in Fig. 4. Curves of this type may be represented by the equation

$$v = \frac{v_{\max}}{1 + [K_a/(\mathrm{H}^+)] + [(\mathrm{H}^+)/K_b]} \tag{1}$$

where v is the reaction rate at a given activity of hydrogen ions (H^+), $v_{\max}$ is the observed rate maximum, and K_a and K_b are constants.

The dependence on the substrate concentration S which is most frequently

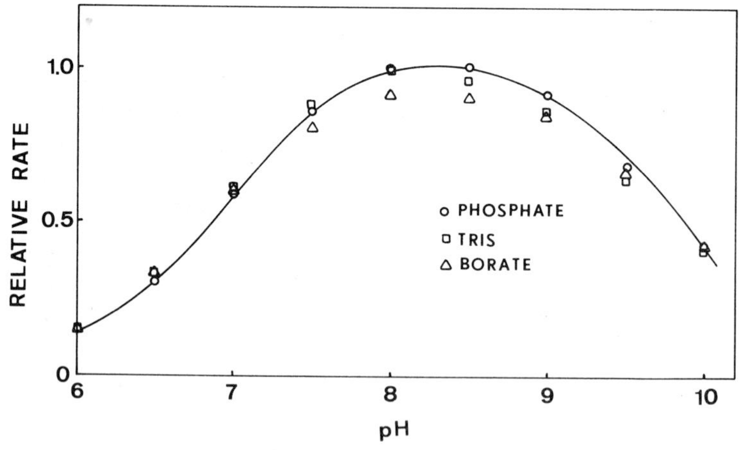

FIG. 4. pH-activity curve for the hydrolysis of (2-methoxy ethyl) acetate by a *B. subtilis* esterase as measured in three different 10-m*M* buffers.

encountered corresponds to the Michaelis-Menten kinetics, i.e., Eq. (2).

$$v = -\frac{dS}{dt} = \frac{kE_tS}{K_m + S} = \frac{v_mS}{K_m + S} \tag{2}$$

where E_t is the (total) concentration of the enzyme, K_m is the so-called Michaelis-Menten constant characteristic of the reaction, temperature, and pH, and v_m or kE_t represent the maximum reaction rate, observed when the substrate concentration $S \gg K_m$. The rate v is seen to be zero order in S when $S \gg K_m$, and first order when $S \ll K_m$.

This equation follows directly from the reaction mechanism

$$\mathrm{E + S} \underset{k_{-1}}{\overset{k_1}{\rightleftharpoons}} \mathrm{ES}$$

$$\mathrm{ES} \xrightarrow{k_2} \mathrm{E + products}$$

when we apply the steady state condition $dES/dt = 0$ and bear in mind that usually $S \gg E_t = ES + E$ under the conditions of the measurements. A plot of Eq. (2) in the dimensionless form

$$\frac{v}{v_m} = \frac{S/K_m}{1 + S/K_m} \tag{3}$$

is shown in Fig. 5. When $S = K_m$, then $v/v_m = \frac{1}{2}$.

Kinetics of enzymes subject to product inhibition obey more complex rate laws (Barman, 1969; Bernhard, 1968), such as Eq. (4)

$$-\frac{dS}{dt} = \frac{kE_tS}{(S + K_m)(1 + P/K_i)} \tag{4}$$

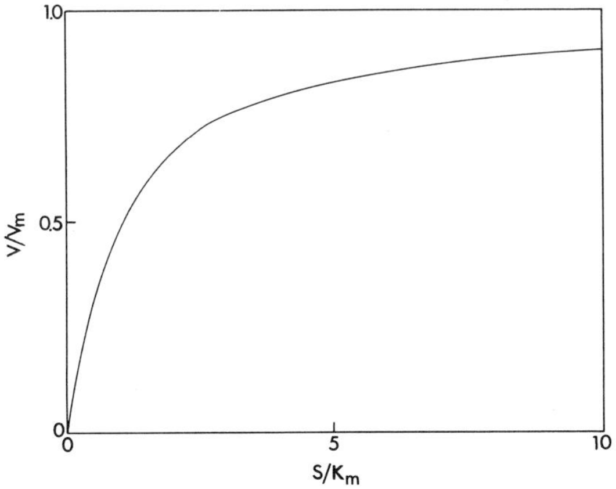

FIG. 5. A dimensionless plot of the Michaelis-Menten equation.

where P is the concentration of the product and K_i the inhibition constant.

The values of K_m are quite frequently in the 0.1–10.0 mM range but may be substantially smaller or larger depending on the enzyme and substrate. The most commonly employed and, ironically, the least reliable method (Eisenthal and Cornish-Bowden, 1974) for determining K_m and v_{max} is the linear "Lineweaver-Burke" plot of $1/v$ against $1/S$, the intercept with the ordinate giving $1/v_m$ and slope K_m/v_m.

A great many enzymatic reactions involve the participation of one or another of a small number of low molecular weight coenzymes, such as ATP (adenosine triphosphate, the "fuel" of living cells), NADH and NAD^+ (reduced and oxidized forms of nicotinamide adenine dinucleotide), and pyridoxal phosphate (Fig. 6). The enzyme brings the substrate and coenzyme into a proper juxtaposition for the reaction in its active site.

Two fundamental distinctions in regard to the mode of action of coenzymes must be made: One class of coenzymes, C, are co-catalysts which are regenerated in their original form upon each turnover of the substrate S

$$E + S + C \longrightarrow E + P + C$$

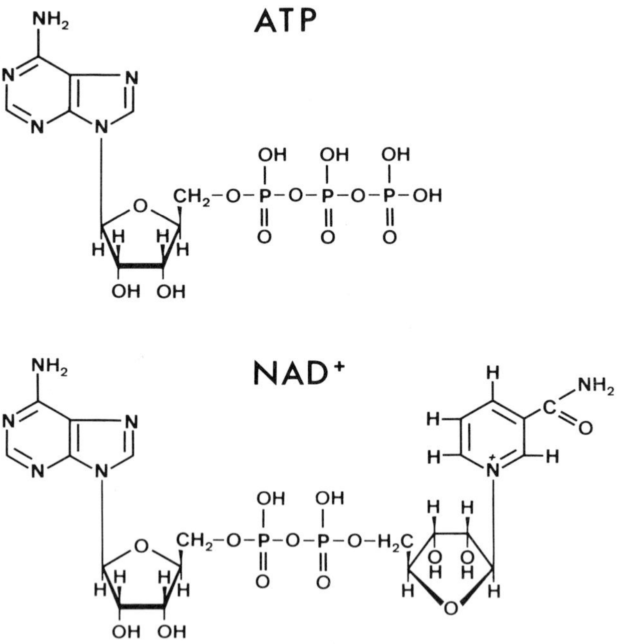

FIG. 6. Structures of the coenzymes ATP and NAD^+.

An example is pyridoxal phosphate which participates in a number of transaminations, racemizations, decarboxylations, eliminations, and condensations. The second class, to which ATP and NADH belong, are transformed in the process and must be regenerated by a reaction with another substrate $S'H_2$:

$$E + NADH + S + H^+ \longrightarrow E + NAD^+ + SH_2$$

$$E' + NAD^+ + S'H_2 \longrightarrow E' + NADH + S' + H^+$$

Enzymes fall broadly into six major classes which form the basis of their international classification (Barman, 1969): oxido-reductases, transferases, hydrolases, lyases, isomerases, and ligases. Of these groups lyases catalyze the addition of groups to double bonds, or the reverse reactions, while ligases (or "synthetases") catalyze the condensation of two molecules, the reaction involving the cleavage of a pyrophosphate bond of adenosine triphosphate or a similar triphosphate. Thus the numbers 1.1.1.1 designate an oxido-reductase acting on a CHOH group with NAD^+ or $NADP^+$ as a receptor, the last figure being the serial number of the enzyme in its subclass.

The specificity of enzymes, emphasized earlier, is high but not absolute. Thus urease, which plays a central role in metabolism, recognizes only urea as a substrate. Furthermore, enzymes necessarily display a very high specificity with respect to their coenzymes. There are, however, proteins which attack with comparable efficiency a particular chemical group of structurally unrelated substrates; furthermore, proteases like carboxypeptidase A and chymotrypsin hydrolyze both amide and carboxylic ester bonds. Whether an unnatural substrate will be transformed depends obviously on its structure and the architecture of the active site. Proteins of the same subclass but from different microorganisms commonly differ in specificity and structure. Even different strains of the same microorganism may produce different enzymes. In fact, not only the quantity but also the activity and specificity of a given protein can sometimes be altered by exposing the organism to mutagens and selection pressure (Carlile and Skehel, 1974).

Finally, a few remarks concerning the measurement and definition of enzyme concentrations are desirable. Many studies are carried out with crude preparations, the molecular weights and activities of the pure species being unknown. It is therefore necessary to use the initial rate of a specified reaction catalyzed by the enzyme as a measure of concentration. The quantity of enzyme which transforms 1 μmole of substrate per minute under these conditions is defined as one international unit (IU). All values in this article are expressed in these units.

Needless to say, screening of microorganisms for a particular catalytic activity, and then finding or developing strains which contain adequate levels of the protein, is time consuming. The "Enzyme Handbook" (Barman, 1969) and Volume X of "Techniques of Chemistry" (Jones *et al.*, 1976), as well as the recent volume on microbial transformations by Kieslich (1976) are particularly useful sources of information concerning known enzymes, their sources, properties and actual and potential uses.

III. SOURCES AND ISOLATION

Of the 150 enzymes available commercially a number are obtained from animal organs and plants. However, microorganisms are the preferred source because they reproduce fast and can be grown under controlled laboratory or industrial conditions in stirred tanks or in surface cultures. In particular, growth in stirred suspensions has the advantage that the level of inducible proteins in the cell or those susceptible to repression can be influenced by the composition of the medium. Furthermore, owing to the fast reproduction of microorganisms, the powerful methods of mutation followed by screening for better producers can be applied efficiently. Genetic and environmental manipulation may increase the concentration of an enzyme by as much as two orders of magnitude in favorable cases (Pardee, 1969; Demain, 1972), and the ambitious goal of converting bacteria into factories for the synthesis of useful proteins has been receiving increasing attention in recent years (Bernardi, 1976). However, levels of the order of 10% on total protein (Rigby *et al.*, 1974) are still an exception, and a desired enzyme usually represents only a fraction of a percent of total protein.

The isolation of an intracellular enzyme in pure form generally involves six or more steps (Jakoby, 1971). The first of these is the disruption of the concentrated cell sludge, usually by mechanical means such as grinding, followed by removal of the cell debris by centrifugation or filtration. The more or less homogeneous cell extract is then subjected to fractionation which usually involves various combinations of the following steps: Precipitation with salts, solvents, and polyionic solutes such as polyethylene imine; adsorption on and elution from ion exchangers such as hydroxyapatite, CM and DEAE cellulose or CM and DEAE Sephadex[1] carried out batchwise or on columns (Peterson, 1970); and also gel chromatography on Sephadex, agarose, and polyacrylamide beads (Fischer, 1971), as well as on porous glass. The properties utilized are differences in solubility, charge, and molec-

[1] Sephadex and agarose are polysaccharide gels, marketed as fine beads. The CM and DEAE derivatives carry carboxymethyl ($-OCH_2CO_2H$) and diethyl aminoethyl [$-CH_2CH_2N(C_2H_5)_2$] ion-exchange groups.

ular weight. A relatively new and powerful method is affinity chromatography which exploits the high affinity of enzymes for their specific inhibitors, effectors, or coenzymes (Jakoby and Wilchek, 1974). The stationary phase is prepared by the covalent binding of the coenzyme or inhibitor via a long spacer molecule to a carrier, such as agarose. The preparation of such adsorbents and the problems associated with their use, recently reviewed by Dean and Harvey (1975), have several points in common with the immobilization of enzymes on carriers of this type.

As an example of the purification procedure, the isolation of penicillin acylase from a high-producing strain of *E. coli* (Kutzbach and Rauenbusch, 1974) is outlined in Table II. The crude extract is seen to contain about 1% of the desired protein in this case.

TABLE II

PURIFICATION OF PENICILLIN ACYLASE

Step	Cumulative yield (%)	Purity (%)
1. Sludge of disrupted cells	100	—
2. Centrifuged supernatant	72	1.0
3. Ammonium sulfate precipitation	65	2.2
4. SE-Sephadex column chromatography	37	28
5. DEAE-Sephadex column chromatography	35	51
6. Ammonium sulfate precipitation	25	100

Much effort has been devoted to the rationalization and scale up of these procedures in recent years and considerable progress has been made (Atkinson, 1973, 1974; Spencer, 1974). Nevertheless, highly enriched intracellular enzymes are expensive reagents and the degree of purity required for immobilization is one of the crucial points bearing on the choice of the method in the technological realm.

Finally, there is a narrow group of predominantly hydrolytic proteins, which already have major applications, especially in the food and textile industries (Konecny, 1975; Weetall, 1975). These so-called extracellular enzymes are excreted by microbial cells into their environment, where they degrade macromolecules like starch and proteins and thus convert them into digestible nutrients. They are produced on a large industrial scale by high-producing mutants of selected microorganisms and are relatively easy to isolate from the fermentation medium, as well as being correspondingly cheap.

IV. Immobilized Enzymes

A. Overview

There are four principal immobilization methods, shown schematically in Fig. 7, and also variations and combinations of the methods which will be described later on. The list of materials which have been employed as enzyme carriers with varying degrees of success is extensive, and only a few examples are listed in Table III. They range from soft gels containing more than 90% water to oxide-coated steel and represent a wide range of morphologies, as is shown schematically in Fig. 8. The carrier configurations employed include particles with diameters extending from a few microns to several millimeters, membranes, nylon sponge, polyurethane foam, fibres, and polystyrene and nylon tubing. While some of the carriers and configurations are chosen with a view on basic research and analytical or medical applications, a large fraction represents the search for cheap materials and methods leading to catalysts suitable for long-term use on a preparative and technological scale.

It is very difficult to evaluate the large body of published data because one or another of the following pieces of information is frequently missing: the size of the particles under the conditions of their use, the activity per unit volume of carrier (in addition to the usually specified activity per unit dry weight[2]), the morphology and mechanical properties of the immobilized preparation, its operational stability, and also the reason for the large losses of activity which sometimes accompany the immobilization. These may be due to destruction of the catalyst but may also reflect severe diffusion limitations, as will be shown later.

One of the few general statements which can be made is that there is no universal method and carrier, and that the choice is determined by the

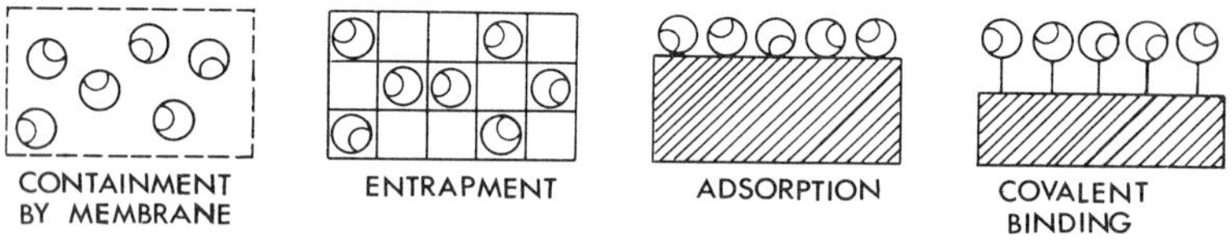

FIG. 7. The principal immobilization methods.

[2]Activity per unit dry weight is a better defined quantity. However, for the assessment of the merits of a carrier and for practical applications the activity per unit volume is more relevant. Together with the permissible packing density of the particles it determines the activity per unit reactor volume in packed-bed reactors and the ratio of catalyst/solution volume in stirred suspensions.

TABLE III

MORPHOLOGIES AND PROPERTIES OF CARRIERS

Type	Materials	Dimensional stability	Porosity
A. Nonporous or microporous	Glass, silica, steel, minerals, molecular sieves, highly cross-linked polymers, nylon	High	Pores, if any, two small to accomodate macromolecules
B. Microencapsulated	Nitrocellulose, cellulose triacetate, nylon	Low to moderate	Small pores, diameter presumably < 35 Å
C. Moderately cross-linked polymers	Mainly hydrophilic polymers: polyacrylamide, polyvinyl pyrrolidone, 2-hydroxy-ethyl methacrylate polymers and copolymers	Low to moderate	Wide pore-size distribution, small pores, diameter presumably < 35 Å
D. Polymers with a low degree of cross-linking	Polyacrylamide, Sephadex G-50 to G-200, DEAE Sephadex A-50, and other weakly cross-linked polymers including sulfonated polystyrenes	Low	Wide pore-size distribution, large pores
E. Macroporous	Controlled-pore carriers: glass, alumina, silica, silica/alumina, macroreticular polystyrene, and some forms of cellulose	High	Narrow pore-size distribution, 200–1000 Å pores
F. Bonded carriers	Various polymers formed on or bonded to a hard porous or nonporous core such as silica, pumice, brick powder	Moderate to high	Variable

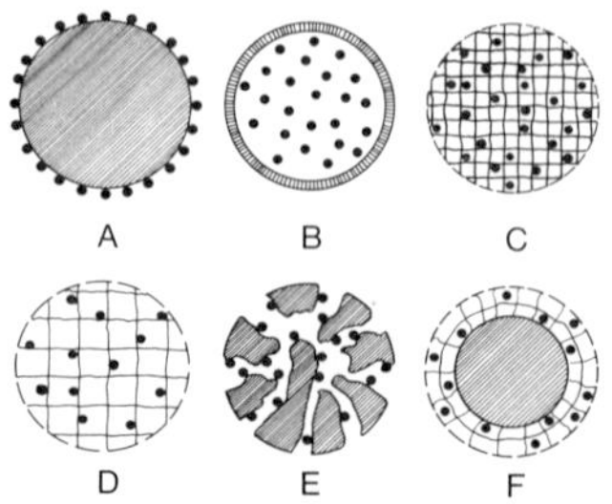

FIG. 8. Carrier morphologies.

intended application of the catalyst and usually involves a compromise between a number of desirable criteria.

In this chapter we shall outline some of the advantages and limitations of the immobilization methods and carriers. We shall also briefly discuss the practical implications of the mechanical properties of the catalysts and of their particle size. The levels of activity which have been achieved will be discussed in the next chapter in conjunction with specific carriers and methods.

B. METHODS

The first of the methods, namely containment of enzyme solutions by membranes, involves no chemical reagents and may therefore be classified as mild and widely applicable. This is particularly true of confinement by prefabricated semipermeable devices such as hollow fibers. Microencapsulation involves short exposure of droplets of aqueous protein to water-immiscible solvents, which many enzymes survive unscathed. Entrapment, which is carried out by polymerizing an aqueous solution containing the enzyme and a monomer like acrylamide, is also a mild physical method and is widely applicable. However, it does involve the exposure of the enzyme to an organic solvent—the monomer—which may inactivate it slowly (Ngo and Laidler, 1975a), and also to the reagents which initiate the polymerization by generating free radicals. All three methods can utilize quite concentrated protein solutions and have the important advantage that very crude enzyme preparations can therefore be used. The capacity of the carriers is not restricted by such factors as the number of reactive groups available for the chemical binding of the catalysts. Furthermore most of the materials used as carriers are very cheap.

All three techniques have one disadvantage in common. Since the protein is immobilized physically by the polymer network or the membrane, the mesh or the pores must be small, the estimated diameters being 35 Å in the case of microencapsulated enzymes (Chang, 1975) and 35 Å or less for typical entrapment gels (White and Dorion, 1961). The diffusion of substrates through these pores is a slow process, and its rate decreases with the increasing dimensions of the molecules and the decreasing size of the pores (White and Dorion, 1961; Mogensen and Vieth, 1973; Wadiak and Carbonell, 1975). It becomes the rate-limiting factor when the catalyst particles are too large or the concentration of the immobilized enzyme is too high or the concentration of the substrate too low.[3] These effects become apparent when a polymer containing an entrapped enzyme is broken up into progressively smaller particles, as shown by the data of Hinberg *et al.* (1974b) in Fig. 9, or when the concentration of enzyme in a microencapsulated solution is increased beyond a certain point, as shown by the plot of the data reported for penicillin acylase by Marconi *et al.* (1973) in Fig. 10.

Similar effects, discussed in detail in Section VI,D, are observed when other methods and carriers are employed, but the limitations on activity imposed by diffusion are less severe when the pores of the carrier are large. The use of such carriers is, of course, predicated by the requirement that the enzyme be held in place by physical or chemical binding.

Of the two methods which permit the use of such carriers, adsorption on hydrophilic ion exchangers of the type used in protein chromatography has the advantage that is simple and mild and undetrimental to most enzymes. It has the further advantage that the carrier can be regenerated for reuse by

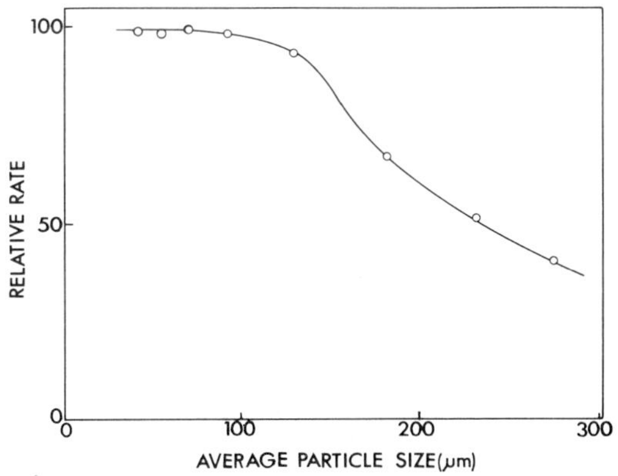

FIG. 9. Activity of entrapped β-galactosidase as a function of the size of the particles.

[3] The severity of diffusion limitations at low substrate concentrations depends also on the value of the Michaelis-Menten constant. The diffusion phenomena are discussed in detail in Section VI, D.

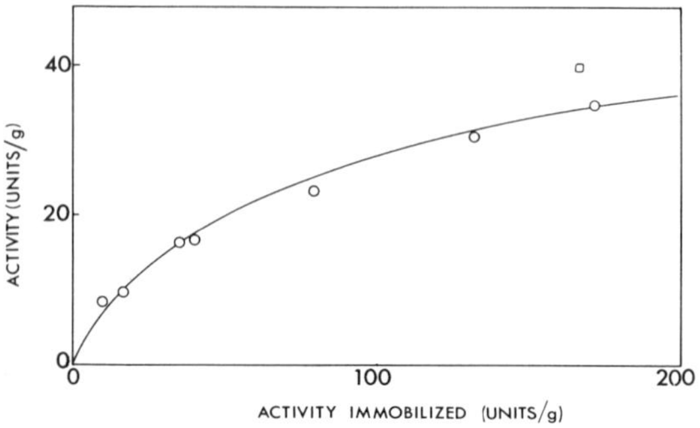

FIG. 10. Relationship between the activity of immobilized penicillin acylase and the quantity of enzyme per gram of dry carrier. Data for crude (○) and purified (□) enzyme.

the simple expedient of washing with a concentrated salt solution when the catalyst has outlived its useful life. The data in Fig. 11, taken from our work (Konecny, 1975), show the activity of a bacterial esterase adsorbed on DEAE cellulose, a presumably macroporous carrier, as a function of the increasing quantity of the adsorbed enzyme. As is apparent from the results diffusion limitations are low even at high activity levels. However, the utility of the method is restricted by the fact that only some enzymes are adsorbed and that the binding is rarely strong enough to permit the use of the

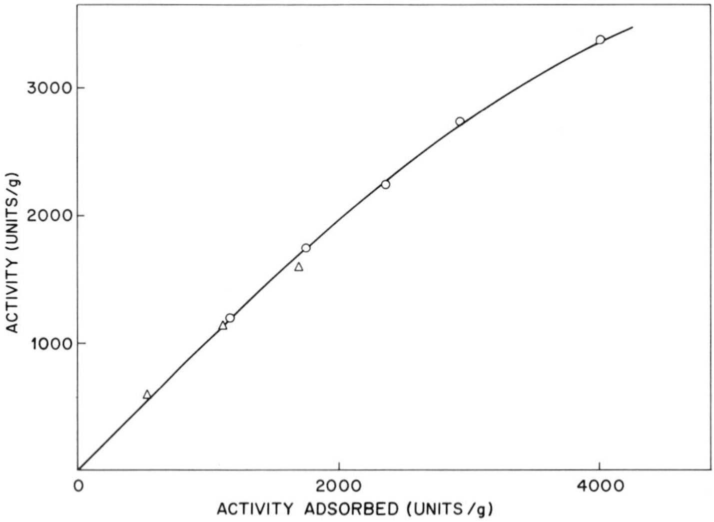

FIG. 11. Activity of an esterase adsorbed on 1 g DEAE cellulose (dry weight) as a function of the quantity of adsorbed enzyme. For conditions, see Section V,E.

adsorbed proteins as catalysts. New developments, which will be discussed later, promise to extend the applicability of the method.

There are no such restrictions when covalent binding is used, which exploits the reactivity of the exposed protein side chains, e.g., of the amino group of lysine, to couple enzymes chemically to carriers bearing, for example, reactive aldehyde groups. However, the method involves the risk that essential protein groups will undergo reaction and that the activity of the enzyme will be partly or completely lost in the process. The compiled data for chymotrypsin bound to various carriers by various methods (Lilly *et al.*, 1973) show a wide span of yields of immobilized activity, including complete loss of catalytic function. When inactivation is not extensive, the maximum activity levels which can be obtained by this method depend particularly on the number of the accessible reactive groups in the carrier which, in the case of macroporous solids, depends in turn on the available internal surface area (Weetall and Havewala, 1972). Since foreign proteins compete for the reactive sites, the maximum activity levels depend on the purity of the enzyme preparation and also on the activity of the pure species.

The covalent method is not restricted to the reaction of enzymes with activated carriers. Among the alternatives is the preparation of soluble vinyl derivatives $E-(CH=CH_2)_n$ which can subsequently be copolymerized with vinyl monomers to give three-dimensional networks which, at least in principle, can have any desired porosity. The advantage of using preformed reactive carriers is that the enzyme is necessarily excluded from regions of low porosity and is therefore relatively well accessible even to high molecular weight substrates.

C. Utility

Immobilized enzymes have the advantage over their soluble counterparts in that they are suited for repeated on continuous use. In order to be useful on a larger preparative or industrial scale, they have to meet various requirements.

In the first place their effective cost, determined by the cost of one unit of immobilized activity and the operational half-life of the catalyst must be tolerable. It must be less than that of the free enzyme if a process utilizing the free protein is economically feasible, unless some other advantages such as higher yields or savings in labor costs (Chibata *et al.*, 1972) accrue from their use.

In the second place the process must be technologically feasible. While problems associated with mechanically inferior catalysts may pass unnoticed or are easily solved in work with milligram to gram quantities of material, they often become apparent even on a modest preparative scale. Four quantities are particularly germane in this context: particle size, the dimensional stability (deformability) of the particles, the mechanical strength of the membranes of microencapsulated enzymes, and resistance to abrasion.

Clearly, particles which rapidly disintegrate into fines—which are then difficult to recover or retain—are ill-suited for long term use in stirred suspensions. Likewise, diffusion limitations which are associated with the use of relatively thick membranes can be reduced only to a certain point without running into the problem of inadequate mechanical strength.

In general there is a strong preference for packed reactor columns because this mode of operation is efficient, requires less work and permits the uninterrupted utilization of the catalyst (the activity of which declines with time) and because many of the carriers do not lend themselves well to long term use in stirred suspensions. At the same time soft, deformable particles, pose considerable, but not insurmountable problems in reactor columns.

The pressure required for a given flow rate increases with the decreasing size of the catalyst particles (Filbert, 1975) and the length of the column (Tosa *et al.*, 1971), and a dimensionally unstable material compresses to an impermeable mass when the pressure drop (pressure difference between column inlet and outlet) exceeds a critical value. These problems have been solved on the industrial scale by the use of wide and shallow, stacked columns (Davis, 1974a) and of large particles (4 mm) of the relatively soft entrapment gels (Sato *et al.*, 1975). There are, of course, limits to the levels of activity per unit volume of catalyst which can be achieved with such large particles, and other types of carriers and immobilization methods must be used when high levels of activity are called for. More extensively cross-linked entrapment gels having a higher dimensional stability would, of course, permit the use of smaller particles, but the advantages may be offset by the slower diffusion of substrate through the denser matrix.

Inorganic carriers like brick powder and pumice have excellent dimensional stability but only a relatively small internal surface (Konecny, 1977a). The development of macroporous glass and macroporous ceramics having a large internal surface area, described in the next section, is therefore a major advance in this area. Diffusion limitations become severe only at high activity levels and activities as high as 10^2–10^3 units/ml have been reported in a number of investigations utilizing these supports. The bonded phase carriers shown in Fig. 8 offer another solution to the problem of dimensional stability when polymers are used to bind enzymes. Very good results have been reported with some of these supports.

In the preceding discussion we have emphasized the relation between the carrier morphology and maximum activity levels. This should not be taken to imply that the levels of activity take precedence over other considerations. They are, admittedly, desirable but become essential only when special requirements must be met. This situation arises, for example, when the substrates are unstable like penicillins (Yamana *et al.*, 1974) and cephalosporins (Konecny *et al.*, 1973), and the desired transformation must therefore be completed in a relatively short time in order to outpace the non-enzymatic side reactions (Konecny, 1977a,b).

D. Properties of Some Carriers

Some of the frequently used carriers are well-characterized commercial materials and have the following characteristics.

The Sephadex and polyacrylamide beads employed in gel chromatography have a wide pore-size distribution and a morphology of the type A, C, or D depending on the degree of cross-linking. The weakly cross-linked materials like Sephadex G-150 and G-200 swell extensively in water, 1 g of the dry materials giving a bed volume of about 30 ml. They are correspondingly soft gels with a limited operational range. The interested reader is referred to the literature cited in Section III and the manufacturers bulletins for further details.

Agarose (also marketed under the name Sapharose) has a polysaccharide matrix like Sephadex but higher dimensional stability than Sephadex G-200. It is the preferred carrier in affinity chromatography, and its advantages and limitations have been reviewed recently (Dean and Harvey, 1975). Its derivatization with polyfunctional reagents to increase its binding capacity will be discussed in the next section.

Cellulose, another polysaccharide, has much higher dimensional stability. The microgranular ion-exchange derivatives have a high capacity and presumably a macroporous structure (Morehead, 1971) with alternating crystalline and amorphous regions. The particles desintegrate quite rapidly into fines in stirred suspensions (Regan *et al.*, 1974). The preparation of macroporous cellulose beads has recently been described by Chen and Tsao (1976).

Outstanding carriers of type E are zirconia-clad controlled-pore glass and the new controlled-pore silica, alumina, and alumina/silica developed by Weetall and Messing at the Corning Glass Company (Messing, 1974, 1975).

The most frequently used glass consists of 0.2–0.8 mm particles having uniform 550 Å pores, an internal surface area of 80 m^2/g and, in derivatized form, about 0.13 mE of NH_2 groups/g. While the surface of the unprotected carrier is susceptible to hydrolysis even at acid pH (Weetall and Havewala,

1972; Messing, 1975), the zirconia-glad material has been used for extended periods of time at pH 9 by Weetall and Jacobson (1972) and at pH 8 by the writer.

The low diffusion limitation of carriers of this type are evident from the data reported by Pitcher (1975) for lactase immobilized on controlled-pore silica, titania, and glass, which are reproduced in Fig. 12. The results were obtained with 0.5 mm particles having about 400 Å pores and a packed density of 0.5 g dry weight/ml (SiO_2, glass) and 0.9 g/ml (TiO_2), the activity measurements being made in 5% lactose as substrate at pH 3 and 40°C (W. H. Pitcher, personal communication). The coupling efficiency is defined as the ratio of the activity of the immobilized preparation to the activity of the free enzyme consumed in the coupling process.

Bonded phase carriers utilize a hard nonporous or porous core as a physical reinforcement for a polymer matrix of type C or D. They have been prepared by diazotization of phenylene diamine in the presence of various carriers (Gray *et al.*, 1974) or the copolymerization of phenylene diamine and glutaraldehyde in the presence of materials like pumice and brick powder (Krasnobajew and Böniger, 1975) or the cross-linking with glutaraldehyde of polyethylene imine adsorbed on silica (Royer, 1974). The first named type is now available commercially from Koch-Light Laboratories. Exclusion of enzyme from the inner core of the carrier may improve its utilization as a catalyst since reaction rates in this region are most severely limited by diffusion (Horvath and Engasser, 1973).

In the next section some of the results obtained by various methods are cited. They were selected as far as possible from publications dealing with operationally stable preparations and more or less defined carriers of established utility on a larger preparative scale.

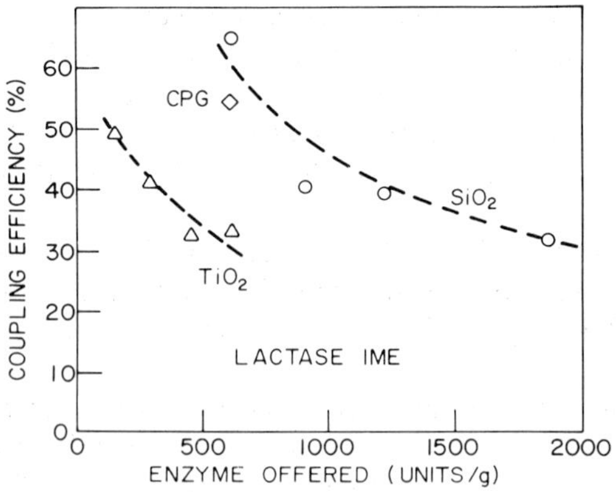

FIG. 12. Coupling efficiency of lactase on controlled-pore glass, silica, and titania.

V. Preparation

A. Containment by Membranes

1. *Hollow Fibers*

The commercially available, semipermeable hollow-fiber devices (Zaborsky, 1973) offer the simplest means for immobilizing enzyme solutions. Fibers which have been used for this purpose have typically an inside diameter of about 200 μm and an outside diameter of 250–350 μm, the bulk of the matrix having a sponge-like macroporous structure, only the inner core being a dense membrane. Devices of this type can be operated as batch or continuous reactors with the enzyme solution located inside the fibers and the substrate solution outside, or vice versa. They were first used by Rony (1971, 1972) and later by Davis (1974a) and Lewis and Middleman (1974). Fibers with a low molecular weight cut off can be employed to contain coenzymes and have been used recently by Fink and Rodwell (1975) in the study of the oxidation of alcohol to acetaldehyde by means of NAD^+ and alcohol dehydrogenase.

As a tool for studying reactions involving immobilized enzymes such reactors have the advantage that there is no risk of denaturation by chemical reagents and solvents and that there are therefore no restrictions on the use of this method. Diffusion limitations are severe. An improved theoretical model for hollow-fiber reactors has been developed by Kim and Cooney (1976).

2. *Microencapsulation*

Microencapsulation of enzymes was pioneered and perfected by Chang (1974, 1975) over a period of years, especially with a view on medical applications. In an early version of the physical method, the microcapsules were prepared as follows: Fine droplets of an aqueous enzyme solution were allowed to fall through a solution of nitrocellulose in ether/alcohol into a lower paraffin layer, the nitrocellulose adhering to the droplet solidifying to a thin membrane in the paraffin phase. The chemical method involves the polymerization of 1,6-hexane diamine and sebacyl chloride $ClOC(CH_2)_8COCl$ to nylon at the surface of the droplet. An aqueous solution containing the protein and hexamethylene diamine is dispersed in an organic phase containing an emulsifying agent. An organic solution of sebacyl chloride is then added to the stirred emulsion, a nylon membrane being formed on the surface of the water droplets where two reactive components meet. The concentration of the reagents, the composition of the solutions, stirring and the length of the reaction determine the size of the microcapsules and the properties of the membrane. On the average such spheres have a diameter of 80 μm, a membrane thickness of 0.02 μm and a pore diameter of presumably about 35 Å (Chang, 1975). Owing to the thinness of the membranes and the high surface/volume ratio the diffusion of low molecular weight substrates is presumably much faster than through hollow fibers and the fibers described in the next section, and activities as high as 140 IU/ml microcapsules have been reported (Chang, 1971). The compacting of the small particles in larger packed columns seems to be problematic (Mogensen and Vieth, 1973; Wadiak and Carbonell, 1975). Various aspects of microencapsulation were recently reviewed by Sliwka (1975).

3. *Entrapment in Fibers of Cellulose Triacetate*

This method, invented by Dinelli (1972), is a form of microencapsulation which utilizes the technology of wet spinning. The aqueous enzyme is dispersed in a methylene chloride solution of cellulose triacetate, and the emulsion is then extruded through fine orifices into toluene. Cellulose triacetate is insoluble in this medium and coagulates in the form of fibers, which contain 0.5-μm and smaller droplets of the enzyme solution dispersed in the carrier (Marconi *et al.*, 1974a). Strong threads are formed by intertwining 100–200 of these fibers, each having a diameter of about 20 μm.

Although compact beds (0.27 g dry carrier/ml bed volume) have been employed in reactor studies (Marani *et al.*, 1975), reactors with lower packing densities (< 0.1 g dry carrier/ml) in which the threads are strung parallel to the axis of a tube appear to be preferred. This design has the advantage that high flow rates are possible at low pressures and that shear forces exerted on the fibers are minimized. Among enzymes which have been successfully immobilized in this manner and which have mostly retained a major part of their initial activity for long periods of time under active conditions are tryptophan synthetase, penicillin acylase, L-aminoacylase, fumarase, and invertase (Marconi *et al.*, 1973, 1974a,b,c, 1975a), and also glucose isomerase (Giovenco *et al.*, 1973). Except for the higher values reported for invertase, enzymes immobilized with about 50% retention of activity displayed activities in the 1–30 IU/ml range, assuming a packing density of 0.1 g dry carrier per ml. The porosity of the fibers can be varied within certain limits depending on the molecular dimensions of the enzyme (Marconi *et al.*, 1974a,c).

B. Entrapment in Polymers

The trapping of enzymes in the gel matrix of cross-linked polyacrylamide was first introduced by Bernfeld and Wan (1963), and has since then been used extensively. The procedure is straightforward: A solution containing the monomer, the enzyme, and a bifunctional cross-linking agent such as bisacrylamide is polymerized by means of chemical initiators generating free radicals. Alternatively shortwave radiation may be employed (Maeda *et al.*, 1975; Kawashima and Umeda, 1975), with the advantage that low temperatures may be used. The block polymer is broken up into small particles in a Waring blendor or by forcing the material through a metal screen. It can also be cut up into membranes or slabs of well-defined dimensions for basic studies. The encagement efficiency, the permeability of the gel and also its mechanical strength depend on the composition of the reaction mixture (Degani and Miron, 1970; White and Dorion, 1961) and, no doubt, also to some extent on the nature of the monomer, which may be more or less hydrophilic than acrylamide. Thus 2-hydroxyethyl methacrylate has been employed by Hinberg *et al.* (1974a,b) and methacrylate by Johansson and Mosbach (1974a). Systematic studies of the entrapment of an L-aminoacylase, carried out by Mori *et al.* (1972) with a view on industrial applications, showed that solutions containing 15% acrylamide and 0.8% bisacrylamide gave the best results, and the same composition was subsequently used for entrapping aspartase (Tosa *et al.*, 1973), asparaginase (Mori *et al.*, 1974), and also P. *putida* and *E. coli* cells (Chibata *et al.*, 1974a,b), the latter as an alternative to the isolated aspartase. These dead entrapped cells are now in use as catalysts for the production of L-aspartic acid from ammonium fumarate (Sato *et al.*, 1975). The large, 4-mm particles have an activity of about 15 IU/ml and the oustanding operational half-life of 4 months at 37°C. Their activity is not diffusion limited in the 1 to 0.1 *M* substrate-concentration range. The aminoacylase, asparaginase, and aspartase, immobilized in 1–2.5-mm particles had activities in the 1–4 IU/ml range, their catalytic efficiency being 30–60% in relation to that of the free enzymes used in the polymerization. (For these preparations, the size of the wet particles is estimated from the values given for the dried preparations and the water content of the native gels.) The large particles reduce the pressure drop in reactor columns and perhaps have the additional advantage that the rate of bleeding of catalyst from the gel is reduced.

Instead of block polymers, gel beads can be obtained by dispersing the polymerizing solution in water-immiscible solvents. Johansson and Mosbach (1974a) have entrapped ribonuclease, trypsin, β-glucosidase, alcohol dehydrogenase and urease in this manner with various yields of immobilized activity.

C. Adsorption

Proteins are polyelectrolytes whose net charge is a function of the external pH. They are therefore bound to ion exchangers through electrostatic attraction between the oppositely charged groups. Since other ions compete for the charged sites on the carrier, adsorption can be reversed by the addition of salts, most enzymes being completely released at ionic strengths of the order of 0.5 *M* and, usually, less. As an illustration, the desorption of a bacterial esterase from five different ion exchangers as a function of increasing concentrations of sodium chloride (Konecny, 1975) is shown graphically in Fig. 13. With this particular enzyme adsorption on DEAE cellulose and Dowex 1 X-1 was strong enough at pH 8 to permit the repeated use of the adsorbed protein as catalyst at very low ionic strengths. Obviously, the utility of adsorption as an immobilization method depends on the strength of the binding of the particular enzyme at the pH of its intended use and is limited to media of low ionic strength. This implies low substrate concentrations when substrate or product are ions. When applicable, adsorption offers considerable advantages over other methods, and two of the four known industrial processes involving immobilized enzymes exploit this approach.

The first process to be realized on a production scale utilizes an L-aminoacylase adsorbed on DEAE Sephadex A-25. The activity of the catalyst, which declines from about 2.5 to 1.5 units/ml after 1 month of continuous operation, is restored by the simple expedient of adsorbing fresh enzyme on the 1-m^3 reactor column. The feasibility of reactivation permits the use of the expensive carrier.

Until recently most investigations were carried out with hydrophilic carriers like derivatized DEAE cellulose and DEAE Sephadex which rarely denature enzymes and which have long been successfully used as sta-

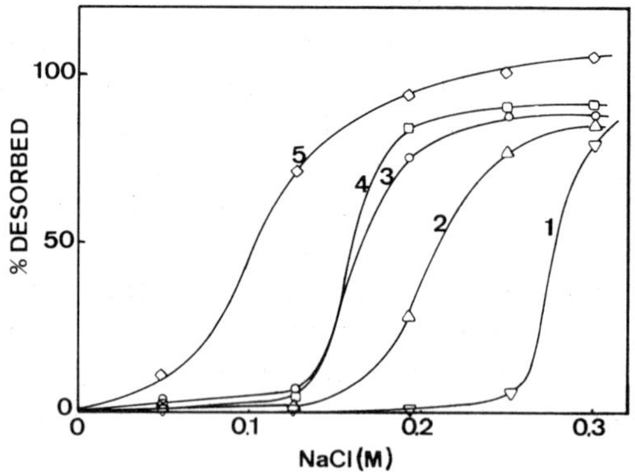

FIG. 13. Desorption of an esterase from ion exchangers as a function of ionic strength: (1) Dowex 1 X-1, (2) DEAE cellulose, (3) QAE Sephadex A-25, (4) DEAE Sephadex A-25, and (5) AE cellulose. System: 1 g wet ion exchanger 90 units of enzyme, 10 m*M* phosphate pH 8, volume 30 ml.

tionary phases in ion-exchange chromatography of proteins. However, recent observations in affinity chromatography have shown that enzymes tend to adhere quite strongly to carriers having hydrophobic (paraffin) side chains attached to a hydrophilic matrix and that they do not necessarily undergo denaturation (Shaltiel, 1974; Dean and Harvey, 1975). This adsorption is attributed to "hydrophobic bonding," i.e., the affinity which hydrophobic areas on the protein surface display for the hydrocarbons in an otherwise polar environment.[4] Recently, Peters and Fromherz (1975) have shown that malate dehydrogenase is bound strongly to negatively charged lipid films,[5] and it is known that some enzymes which are located in lipid membranes in living cells, are more stable and more active in such a partially hydrophobic environment. Preliminary studies of the adsorption of enzymes on Sepharose carrying allyl and hexyl side chains (Caldwell *et al.*, 1975) and on hydrophobic derivatives of cellulose (Butler, 1975) indicate that this approach may extend the utility of the method. Hydrophobic effects in adsorptive protein immobilization have recently been reviewed by Hofstee (1976).

Another possible way to overcome the weakness of the ionic binding between enzymes and ion exchangers is to prepare soluble derivatives of enzymes carrying dyestuff groups or polyionic chains which have a strong affinity for cellulose or ion exchangers. This approach has been tried by Vinogradova and associates (1975) and by Solomon and Levin (1974), but there is too little information so far to assess the utility of the method.

Inorganic adsorbents such as hydroxyapatite and celite which are employed in enzyme chromatography, and also bentonite, silica, alumina, and glass have been used to adsorb enzymes. Promising new materials of this type are the controlled-pore alumina, silica-alumina, silica, and titania developed by Messing (1974, 1975) and mentioned in the preceding section. They are presumably much less costly than the zirconia-clad glass and are regenerable by pyrolysis. As of this time several enzymes have been absorbed successfully after prior pretreatment of the carriers with salts of heavy metals. Thus, glucose isomerase was adsorbed on controlled-pore alumina (140–220-Å pores, internal surface 100 m^2/g, theoretical density 3.7 g/cm^2, particle size 200–700 μm) with 30% retention of activity. The immobilized

[4]The formation of the "hydrophobic bonds" is somewhat analogous to the formation of oil droplets in an emulsion of oil in water. Water molecules, which have a high affinity for each other, minimize the energetically unfavorable water/oil contacts by forcing the oil molecules together.

[5]Lipids are essential components of biological membranes. Chemically, they are long hydrocarbon chains with polar groups at one end.

enzyme had an activity of 650 IU/g and an operational half-life longer than 40 days at 60°C (Messing and Filbert, 1975).

D. Covalent Binding

Covalent methods take advantage of the exposed reactive groups on the enzyme surface to link the macromolecules to carriers by chemical bonds. Although most of the amino acid side chains listed in Table I can be derivatized (Means and Feeney, 1971), the most frequently exploited reactions are those involving primary amino groups and the phenol ring of tyrosine. The reactions are nearly invariably carried out in water and should preferably proceed at 0–25°C at mild pH to minimize the risk of denaturating the protein. The choice of the method depends on the stability of the enzyme at the pH of the contemplated coupling, the stability of the linkage at the pH of the contemplated use of the catalyst and also on the carrier. It is a matter of trial and error unless the structure of the protein is known, although some of the methods, such as those utilizing glutaraldehyde, have a wide range of applicability. Since several reactive groups are usually available on the enzyme surface, multiple attachment to the carrier may occur, especially when the backbone of the support is flexible and carries many reactive groups.

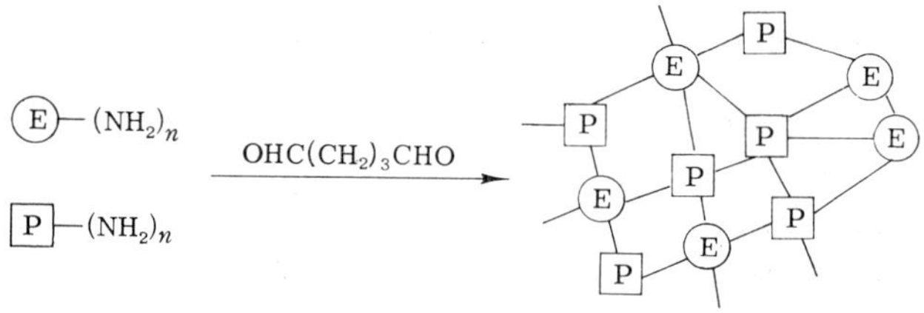

The methods described in this section involve the coupling of enzymes to preformed, reactive carriers. There are, however, several alternatives. Enzymes can be cross-linked in solution with cheap proteins or other polymers with glutaraldehyde or with one of the many other bifunctional reagents (Means and Feeney, 1971) to give insoluble aggregates. This method has been employed successfully for the preparation of enzyme membranes by Broun *et al.* (1973). Alternatively, the cross-linking reagent may be applied after prior adsorption or entrapment of the protein, or the soluble enzyme can be converted into a vinyl derivative of the type $E-(CH=CH_2)_n$ for subsequent copolymerization with a suitable monomer (Jaworek, 1974). The reader is referred to the papers by Katchalski *et al.* (1971), Goldman and associates (1971), Weetall (1975), and to Zaborsky (1973) for references to the extensive literature.

Some of the reactions employed for the coupling of enzymes are shown, the vertical lines indicating the linkage of the reactive groups to the backbone of the carrier.

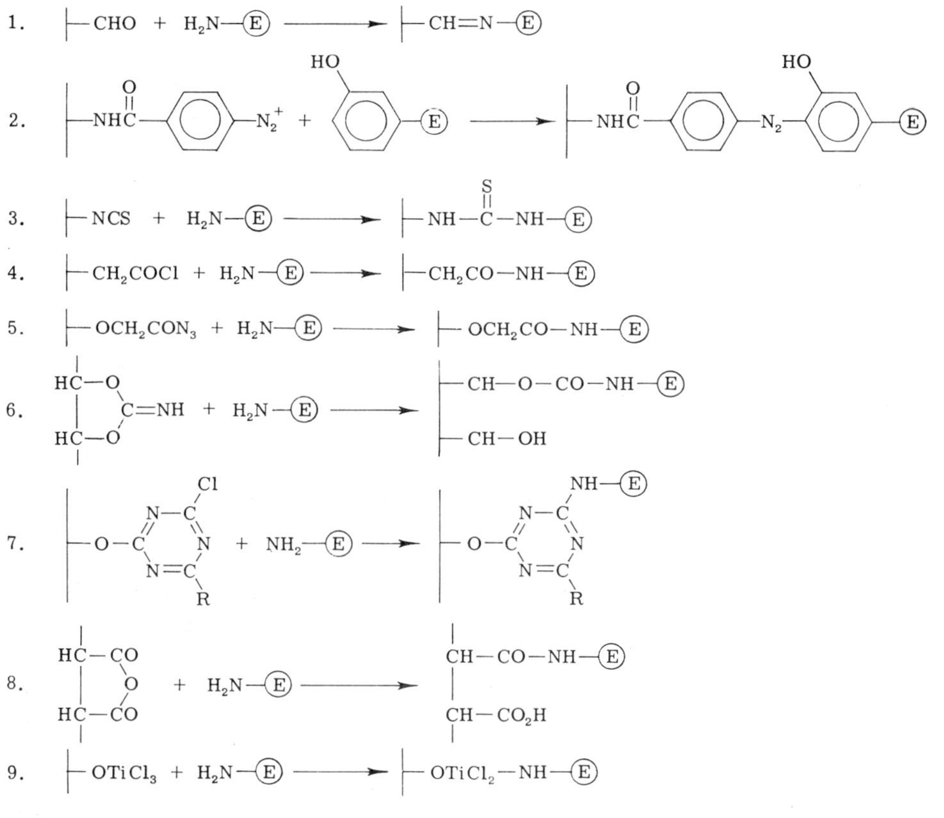

Reactive carriers of type 1–4 are obtained by straight forward derivatization of supports carrying primary amino groups, for example, the reaction of the commercially available controlled-pore alkylamino glass with glutaraldehyde[6]

$$\vdash NH_2 + OHC-(CH_2)_3-CHO \rightarrow \vdash N{=}CH-(CH_2)_3-CHO$$

and other reagents (Weetall and Filbert, 1974). The alkylamino carrier itself is prepared by treating the glass beads with an aminosilane and sub-

[6]Hardly any reagent is used as widely in enzyme immobilization as glutaraldehyde. The chemistry of the reaction, which is only imperfectly understood, has been reviewed by Quiocho (1974) and Zaborsky (1973).

sequently heating

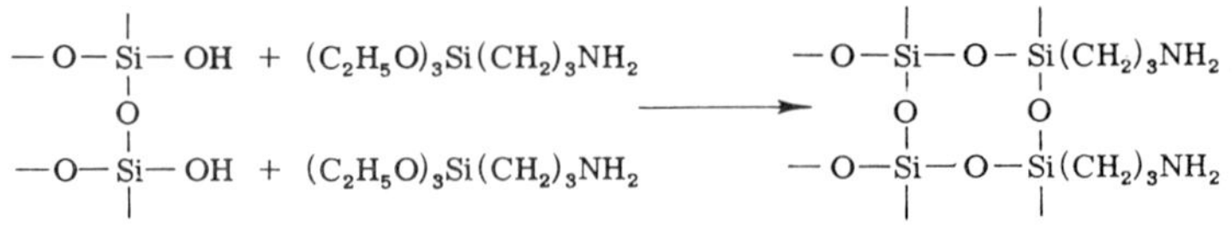

This method has been used to derivatize a variety of inorganic carriers including minerals like hornblende (Thornton *et al.*, 1975) and magnetite (Van Leemputten and Horrisberger, 1974; Dunnill and Lilly, 1974).

Carriers of type 5 are prepared from parent materials having carboxylic or amide groups, e.g., CM cellulose or polyacrylamide by conversion to hydrazides and subsequent diazotization.

$$\vdash OCH_2CONHNH_2 \longrightarrow \vdash OCH_2CON_3$$

Polysaccharides such as Sephadex, agarose and cellulose are frequently activated with cyanogen bromide

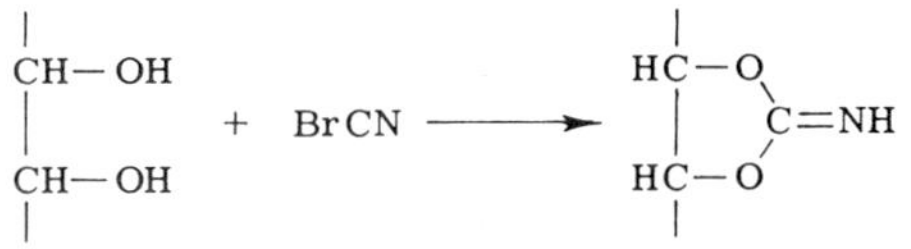

to give active carriers of type 6. Triazine $C_3N_3Cl_3$ and triazine-based dyestuffs of the type $C_3N_3Cl_2R$ have been employed for converting cellulose to derivatives of type 7.

Since the introduction of ethylene-maleic anhydride copolymers (type 8), a variety of other reactive polymers have been developed such as the commercially available Enzacryls, based on acrylamide, and Vanacryls (Brown and Joyeau, 1974a, b) prepared by copolymerization of $CH_2{=}CH{-}CO_2{-}C_6H_4CHO$ type compounds with other monomers. The metal-link and other new methods have been reviewed by Falb (1974), and the activation of nylon has been discussed by Goldstein *et al.* (1974) and Nelboeck and Jaworek (1975). The derivatization of macroreticular polystyrene beads, notable for their good mechanical properties, large pores, and a very large internal surface area, has been examined by Baum (1975).

The amount of protein which can be attached to a support obviously depends on the number of the available reactive groups. Recently polyfunctional spacers such as polylysine (Wilchek and Miron, 1974) and polyethyleneimine (Royer, 1974; Campbell *et al.*, 1975), a branched polymer containing primary, secondary and tertiary amine groups, have been introduced to increase the carrying capacity of Sepharose, silica and nylon. Some of the primary amino groups of the spacer are first bonded to the reactive carrier while the remaining free groups can be subsequently used for protein binding as shown schematically in Fig. 14. An additional advantage of this approach is that the hydrolysis of a single spacer-carrier bond does not lead to the loss of enzyme on account of the multiple attachment.

E. Synopsis

As shown in the preceding sections the reported levels of activity range from 1 to 10^3 units/ml carrier, the values for entrapped enzymes lying in the lower region while the high values for proteins immobilized on macroporous carriers are in the upper part of the range. Microencapsulated enzymes confined by a single thin membrane seem to lie in the middle while the mechanically stronger fibers containing enzyme solutions in their vacuoles are apparently less permeable.

Since the data reported in the literature refer to different enzymes, substrates, substrate concentrations and also different reaction conditions, some values determined in our laboratory with the same enzyme under the same conditions in a differential recirculation reactor (Ford *et al.*, 1972) are summarized in Table IV. All measurements were carried out at 25°C and pH 8 with 85 m*M* (2-methoxy ethyl) acetate in 10 m*M* phosphate buffer.

The values given for the adsorbed enzyme correspond to the saturation capacity of the ion exchangers, equilibrated with a crude preparation of the esterase (10% on total protein). The covalent coupling to the glutaraldehyde derivative of ZrO_2-clad aminoglass was carried out with the pure enzyme, having a specific activity of about 120 units/mg protein, a molecular weight of 150,000 and an isoelectric pH of 4.7 (Konecny, 1977a,b). The factor limiting the activity of the immobilized preparations was not substrate but buffer diffusion (see Section VI,F). The activity of the polyacrylamide particles increased by a factor of 1.4 when buffer concentration was increased four times, and by a factor of 4.5 when they were subdivided into 0.15-mm pieces.

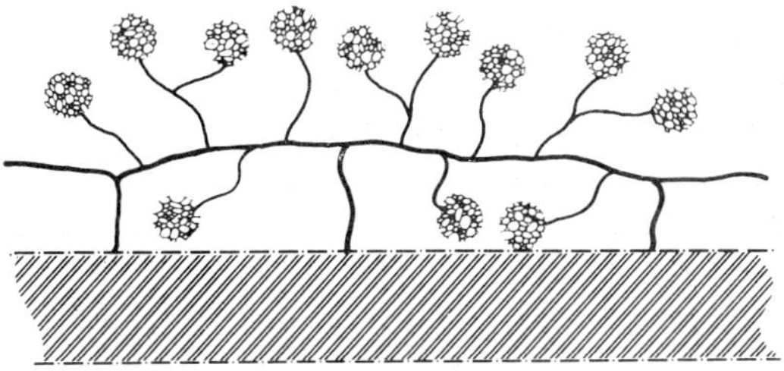

FIG. 14. Coupling of an enzyme to a carrier via a polyfunctional reagent.

TABLE IV

ACTIVITY OF VARIOUS PREPARATIONS OF AN IMMOBILIZED ESTERASE ISOLATED FROM *B. SUBTILIS*

Carrier	Method	Approximate particle diameter (mm)[a]	Activity (units/g)[b]	Activity (units/ml)[c]	Immobilization efficiency (%)[d]
Cellulose triacetate fibers	entrapment[e]	0.02	45	8	42
Polyacrylamide	entrapment[f]	2.5	50	8	14
Controlled-pore glass	covalent	0.2–1.0	200	100	75
Controlled-pore glass	covalent	0.2–1.0	400	200	52
QAE Sephadex A-25	adsorption	0.1	370	74	113
DEAE Sephadex A-25	adsorption	0.1	650	130	103
Dowex 1 X-1	adsorption	0.4	820	330	86
DEAE cellulose	adsorption	0.06	4200	670	80[g]

[a]Wet carrier.
[b]Dry carrier.
[c]Bed volume.
[d]Activity expressed/activity immobilized.
[e]See Section V, A, 3.
[f]Entrapment was carried out as described by Mori *et al.* (1972).
[g]Diameter of the fibrous particles.

VI. PROPERTIES OF IMMOBILIZED ENZYMES

A. GENERAL CONSIDERATIONS

The observed changes of catalytic properties associated with immobilization may be due to one or several of the following factors: (1) intrinsic changes of activity due to chemical modification, (2) environmental influences, analogous to solvent effects in homogeneous reactions, (3) unequal partitioning of substrate and other species (hydrogen ions, coenzyme) between the bulk of the solution and the carrier, and (4) diffusion.

That chemical binding may lead to a change of catalytic properties may be inferred from various studies which have been carried out on soluble, chemically modified proteins. Thus Simpson *et al.* (1963) observed that acetylation of carboxypeptidase A leads to a nearly complete loss of its ability to hydrolyze peptide bonds and, at the same time, to a large increase of its estereolytic activity. Bender *et al.* (1965) were able to show that this was due to changes of the K_m values for the two types of substrates. Acetylation of lysozyme (Yamasaki *et al.*, 1968) and the derivatization of pepsin with a soluble polyanion (Løwenstein, 1974) were found to displace the maxima of their pH-activity profiles, while derivatization of α-amylase with fatty acids of various chain lengths reduced the values of v_m with minor changes in the pH-activity profiles and in thermal stability (Urabe and Okada, 1972). Recently, Fries and associates (1975) showed that derivatization of liver alcohol dehydrogenase with 4-chlorobutyramide increased its activity 20-fold. Extensive succinylation of the amino groups of complex enzymes has been found to lead to dissociation into subunits, presumably as a result of the strong electrostatic repulsion between the produced $-NHCOCH_2CH_2CO_2^-$ groups (Means and Feeney, 1971).

Unequal partitioning and diffusion lead to differences between substrate concentration (and sometimes pH) at the site of the reaction and its value S, measured in the bulk of the solution. Nevertheless the observed changes of the reaction rate v with increasing substrate concentration are usually found to fit the Michaelis–Menten-type relation

$$v = \frac{v_m(\text{app.})S}{K_m(\text{app.}) + S} \tag{5}$$

over the experimental range. However, significant deviations have been observed (Ngo and Laidler, 1975a; Hinberg *et al.*, 1974b), and are, in fact, expected in diffusion-limited systems on theoretical grounds, as will be shown later.

In general, the apparent constants v_m(app.) and K_m(app.) are empirical parameters, whose values may depend on several variables such as stirring rates, particle size, and in reactions involving the generation or consumption of hydrogen ions, on buffer concentration. However, they may be related to the true molecular parameters of the reaction in well-defined systems. With very few exceptions the values of v_m(app.) have been found to be lower than the values for corresponding quantities of free enzymes. The reported values of K_m(app.) lie frequently in between 0.5 K_m and 3 K_m, where K_m is the constant for the free protein. However, much higher values are observed when diffusion limitations are very severe.

Immobilized enzymes have been found to display both higher and lower stabilities than the free species (Melrose, 1971), and the operational half-lives of a number of immobilized enzymes have been listed by Vieth and Venkatasubramanian (1974a) and Weetall (1974). For stable enzymes, values substantially higher than one month at 25°C and higher temperatures are not uncommon. In this connection the recent observations of Zaborsky (1974) on the pronounced stabilization of soluble ribonuclease, resulting from the introduction of synthetic crosslinks into the molecule by means of bifunctional reagents, are of particular interest. Apparently these artificial crosslinks oppose the unfolding of the chain in the same way as the natural, internal —S—S— bridges.

Covalent attachment of a protein to a rigid carrier backbone by several bonds may also oppose conformational changes and thus lead to enhanced thermal stability. However, other factors may have an opposite effect. At the time of this writing Lasch and associates (1976) have examined the rates of thermal denaturation of free chyomtrypsin and of chrymotrypsin covalently bound to two soluble carriers, namely detrane and a copolymer of maleic acid anhydride and acrylic acid. They followed the unfolding of the protein chain by spectroscopic methods and the loss of activity by kinetic measurements and found that the temperature-induced changes were markedly faster for the protein bound to the negatively charged copolymer. Presumably the electrostatic interactions leading to ion-pair formation between the positively charged enzyme groups and the carrier destabilize the structure. According to Jaworek (1974) multiple attachment of complex enzymes to carriers with a flexible polymer network, which expands or contracts with changes of ionic strength or temperature, may cause the subunits to be pulled apart with a concomitant loss of activity. More work is needed to shed further light on this important subject.

B. Partitioning

Let the true molecular constants characterizing the reaction of the immoblized enzyme be v_m' and K_m' and assume that the reaction is not diffusion limited. This implies that the concentration S' of the substrate in the carrier is the same throughout and corresponds to the equilibrium distribution ratio $P = S'/S$, where S represents the concentration in the solution. Substitution of the appropriate constants and of S' for S into Eq. (2) of Section II then gives

$$v = \frac{v_m' S'}{K_m' + S'} = \frac{v_m' S}{K_m'/P + S} \tag{6}$$

Comparing this expression with the empirical Eq. (5), we obtain $v_m(\text{app.}) = v_m'$ and $K_m(\text{app.}) = K_m'/P$.

Values of K_m(app.) somewhat smaller than the K_m of the free enzyme are not uncommon and are sometimes associated with preferential adsorption of substrate on the carrier (Abbott and Fukuda, 1975). A striking increase in the hydrolysis rate of glyceryl triproprionate by lipase has been observed when the enzyme was adsorbed on a hydrophobic surface which was also a good adsorbent for the substrate (Brockman *et al.*, 1973). Such observations do not, however, constitute a proof of Eq. (6) because they involve a comparison of the free and immobilized enzyme reactions, whose kinetic constants v_m and K_m are not necessarily equal. A rigorous proof would call for experiments involving changes of P by changing the composition of the carrier or the liquid phase or the nature of the substrate *without* changing K_m' or at least eliminating these changes of the Michaelis constant from the final calculation. This last approach was taken by Johansson and Mosbach (1974b) who measured the rates of oxidation of ethanol and *n*-butanol by alcohol dehydrogenase immobilized on two different carriers, namely acrylamide and methacrylate–acrylamide copolymer, the second of which is a better solvent for *n*-butanol. However, this approach also involves an assumption, namely, that K_m' behaves as a separable function f_1 (substrate) $\cdot f_2$ (carrier) of the substrate and the carrier.

Probably the best evidence derives from studies of enzymes covalently bound to electrically charged supports, i.e., to ion exchangers. Since this work has been extensively reviewed by Katchalski and associates (1971) and Goldman *et al.* (1971) only a brief outline of the analysis and the principal conclusions will be given.

The partition of a charged substrate between the solution and a charged carrier is given by the Maxwell-Boltzmann distribution

$$P = \frac{S'}{S} = \exp \frac{ze\psi}{kT} \tag{7}$$

where e is the electronic charge, ψ is the electrostatic field, and z is a positive or negative integer depending on whether the solute and the carrier have a charge of opposite or equal sign. The observed changes of K_m associated with changing ionic strength (which effects the magnitude of ψ) have been correlated with the theory.

Equation (7) applies, of course, also to the distribution of hydrogen ions, their concentration in the carrier being higher or lower than in the solution depending on whether the carrier is negatively or positively charged. The normalized pH-activity profiles of free chymotrypsin and of chymotrypsin immobilized on the negatively charged ethylenemaleic anhydride copolymer and on a positely charged polyornithyl carrier, reported by Goldstein and Katchalski (1968) are shown in Fig. 15. The observed shifts of the pH-activity profiles and of the activity maxima follow from the foregoing consideration if the rate is assumed to depend on the concentration of hydrogen ions rather than their activity. (By definition, activities and pH are the same in both phases when an equilibrium distribution is assumed.) In the case of the negatively charged carrier, for example, the concentration of hydrogen ions surrounding the enzyme is always higher than the concentration measured in the bulk of the solution. Consequently the maximum is displaced towards lower hydrogen ion concentrations as measured in the liquid.

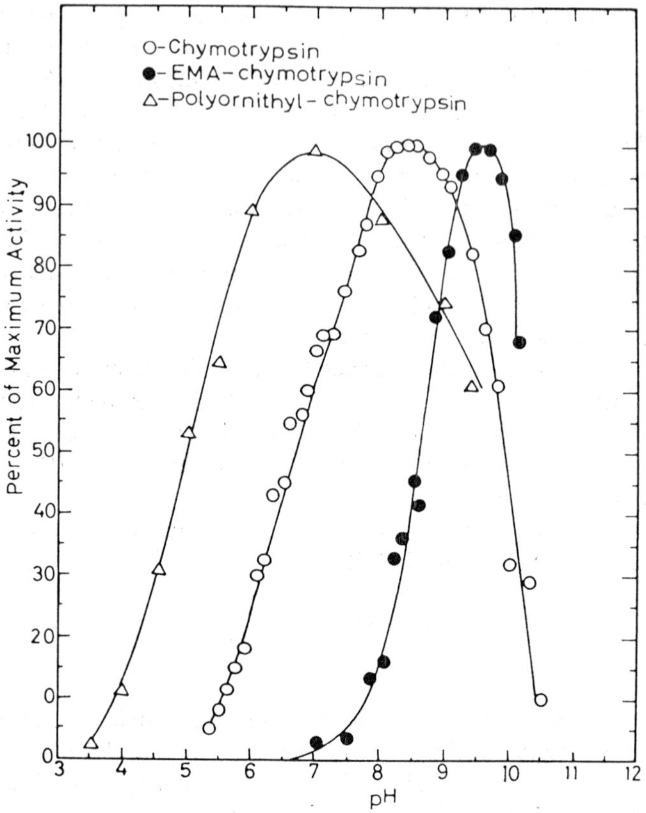

FIG. 15. pH-activity curves at low ionic strength (8 mM) for chymotrypsin, a polyanionic (EMA) derivative, and a polycationic (polyornithyl) derivative of the enzyme. Substrate: acetyl-L-tyrosine ethyl ester.

C. INTERFACIAL DIFFUSION

In all reactions involving heterogeneous catalysts a concentration gradient of substrate is established in the liquid region adjacent to the external surface of the particles where the spent solution emerging from the reaction zone and the external solution mix. Motion of the liquid in relation to the particles increases the substrate concentration in this region and reaction rates therefore depend on stirring rates in suspensions (Lilly and Sharp, 1968) and flow rates in reactor columns. The effect of flow rates on the activity of entrapped β-galactosidase, measured in a recirculation reactor by Hinberg *et al.* (1974b) as shown in Fig. 16. These changes in activity are reflected in the apparent values of the parameters of the Michaelis-Menten

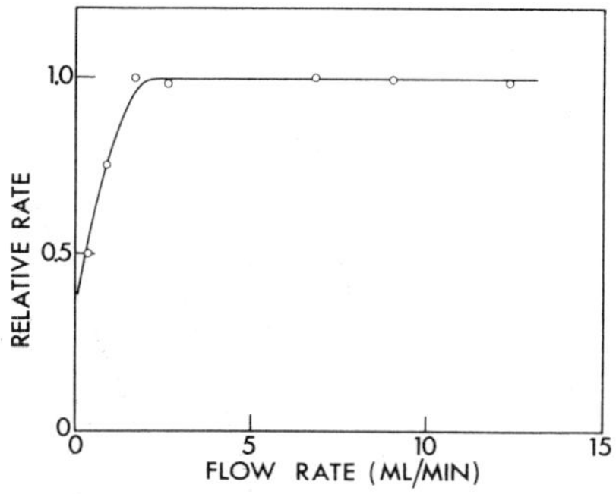

FIG. 16. Activity of entrapped β-galactosidase particles as a function of flow rate.

equation v_m(app.) and K_m(app.). When the reactions are carried out in packed columns, the results can be correlated with the flow rates.

Assuming that the reaction obeys Michaelis-Menten kinetics, Eq. (8) can be derived for the relation between the substrate concentration S_o flowing into a packed column, the fraction X of S_o that undergoes conversion in the passage, the volumetric flow rate of the solution Q and the total quantity E_c of immobilized enzyme in the column:

$$k(\text{app.})\frac{E_c}{Q} = S_oX - K_m(\text{app.})\ln(1 - X) \tag{8}$$

The results obtained by Lilly *et al.* (1966) in their study of the kinetics of columns packed with immobilized ficin are shown in Fig. 17, where the observed values of S_oX are plotted against $\ln(1 - X)$ for a series of flow rates. As is apparent from the results the predicted linear relation holds, but data for different flow rates do not coincide. The values of K_m(app.) from this and from other experiments, based on the slopes of the straight lines, are shown graphically in Fig. 18. Similar results have been reported by Tosa *et al.* (1971) for the reaction of an immobilized aminoacylase.

Recently Kobayashi and Laidler (1974a,b) have developed the theory of enzymes attached to the external surface of carriers and Ngo and Laidler (1975b) have studied the kinetics of cholinesterase attached to the interior surface of nylon tubing. In line with the theory K_m(app.) was found to be a linear function of the reciprocal cube root of the flow rate.

D. Bulk Diffusion of Substrate

Catalysis by enzymes which are imbedded in a porous matrix involves diffusion of substrate through the pores. As shown in Section IV,B, this process becomes rate determining when the pores are too small or long or

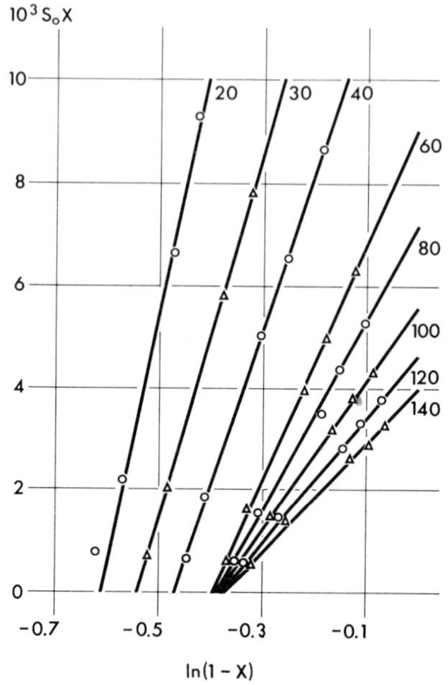

FIG. 17. Relationship between S_oX and ln $(1 - X)$ in an immobilized ficin column. The numbers above the curves indicate the flow rates Q in ml/h.

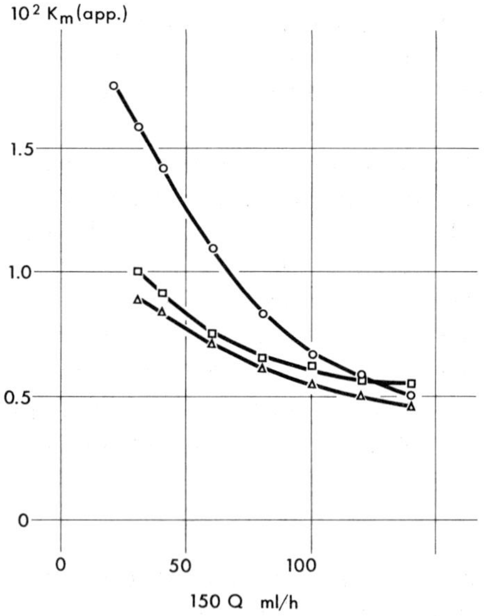

FIG. 18. Effect of the flow rate Q on K_m(app.).

few in number, or when the concentration of substrate is too low and that of the enzyme in the matrix too high.

Theoretical treatments of the effects of bulk diffusion on reaction rates have been published by Goldman *et al.* (1968), Sundaram *et al.* (1970), and other authors. We shall outline here the analysis for membrane-shaped particles as given by Sundaram and associates because the physical situation is easy to visualize, while the general conclusions are the same for other geometries such as rods and spheres (Kobayashi and Laidler, 1973). Furthermore, careful experimental studies have been carried out with this configuration.

The physical situation is illustrated schematically in Fig. 19. A membrane of cross-sectional area A and thickness l is immersed in an infinitely large volume of well-stirred substrate solution of concentration S. The enzyme is uniformly distributed in the carrier, its concentration being E_m mole/liter membrane. The true molecular constants of the reaction in the membrane are again k' and K_m' and the equilibrium distribution of substrate between the membrane and solution is $P = S'/S$. This equilibrium value is indicated by the broken line, the solid lines showing schematically the actual concentration profile of the substrate in the membrane for increasing values of E_m. The steady state concentrations decrease from the periphery to the center because the substrate undergoes conversion as it migrates through the carrier.

With regard to the corresponding profiles of the reaction rates, two extreme situations arise. When the concentration of substrate throughout the membrane $s \ll K'_m$, the reaction is first order in substrate (see Section II), and the rate and concentration profiles are therefore the same. On the other

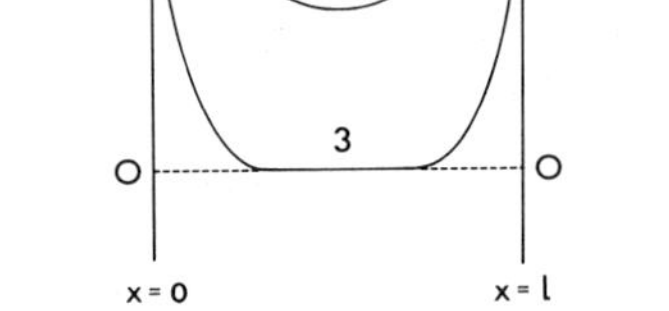

FIG. 19. Schematic representation of the steady-state concentration profiles of a substrate in a membrane of thickness l, immersed in a substrate solution of concentration S. Profiles are shown for (1) low, (2) medium, and (3) high concentrations of enzyme in the membrane.

hand, when $s \gg K'_m$ throughout the membrane, the reaction is zero order in the substrate and proceeds at the rate $k'E_m$ everywhere.

According to Fick's second law of diffusion the rate of accumulation of substrate at any given cross section parallel to the face of the membrane at any given time t is given by

$$v_d = D\left(\frac{\delta^2 s}{\delta x^2}\right)_t \tag{9}$$

where D is the diffusion constant of the substrate and s is its concentration at this cross section. When the steady state is established in the membrane, this rate must exactly balance the rate of consumption of substrate by the reaction

$$v_r = \frac{k'E_m s}{K_m' + s} \tag{10}$$

The relation to be solved is therefore

$$D\frac{d^2 s}{dx^2} = \frac{k'E_m s}{K_m' + s} \tag{11}$$

The rate of entry of substrate at each of the membrane faces, which is equal to the rate of exit of the product, is given by

$$v_t = -AD\left(\frac{ds}{dx}\right)_{x=0} = AD\left(\frac{ds}{dx}\right)_{x=l} \tag{12}$$

Equations (11) and (12) have an explicit solution for $s \gg K_m'$ and $s \ll K_m'$, while the general case can be solved only by numerical methods. The solutions for the rate of reaction catalyzed by 1 liter of membrane of thickness l are given by Eq. (13) for large s

$$v = k'E_m \tag{13}$$

and, for small s, by Eq. (14)

$$v = \frac{k'PFE_m S}{K_m'} \tag{14}$$

The function F, which is a measure of the extent to which the reaction is diffusion controlled, is given by

$$F = \frac{\tanh(\gamma l)}{\gamma l} \tag{15}$$

where

$$\gamma = \frac{1}{2}\left(\frac{k'E_m}{DK_m'}\right)^{1/2} \tag{16}$$

The theoretical plot of F against γl, together with experimental points based on rate measurements involving membranes of variable E_m and l (Ngo and Laidler, 1975a), is shown in Fig. 20. The curve is seen to be analogous to that shown in Fig. 9 in Section IV.

Thus Eq. (13) predicts that the observed rates per mole of active, immobilized enzyme will converge to the common value k' at high substrate concentrations, regardless of the thickness of the membrane and the enzyme concentration. For low substrate concentrations and a given membrane, the rates predicted by Eq. (14) are proportional to S. In this region the plot of $1/v$ versus $1/S$ is therefore a straight line with a slope inversely proportional to F. Only in the extreme case when E_m or l are small are Michaelis-Menten kinetics exactly obeyed (Ngo and Laidler, 1975a) because in this situation F approaches unity and the concentration of the substrate throughout the carrier corresponds to the equilibrium value S'. Lineweaver-Burke plots of data for acrylamide membranes containing three different concentrations of β-galactosidase, reported by Bunting and Laidler (1972) are shown in Fig. 21.

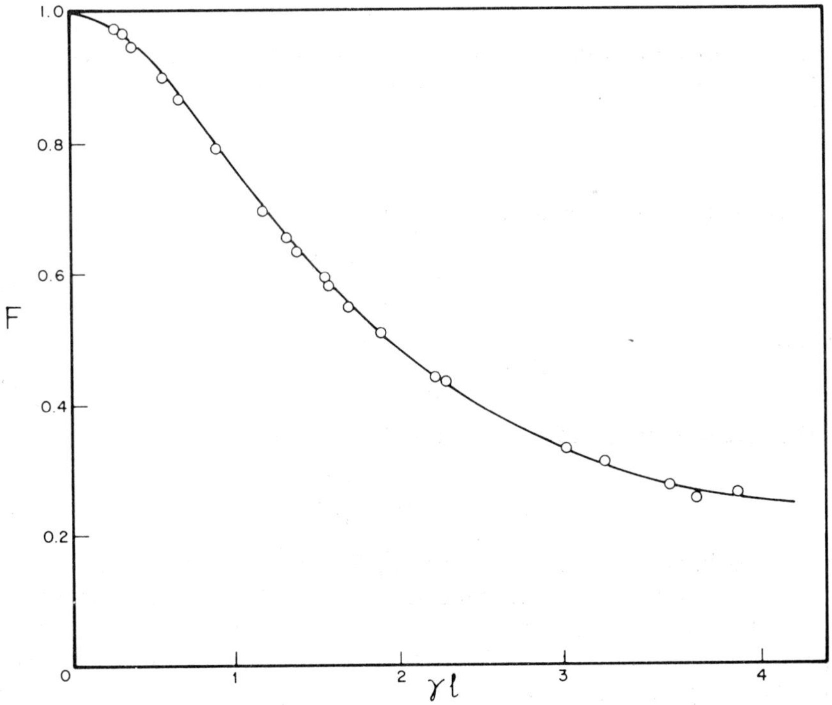

FIG. 20. Plot of the function F against γl. The line is the theoretical curve and the circles are experimental points.

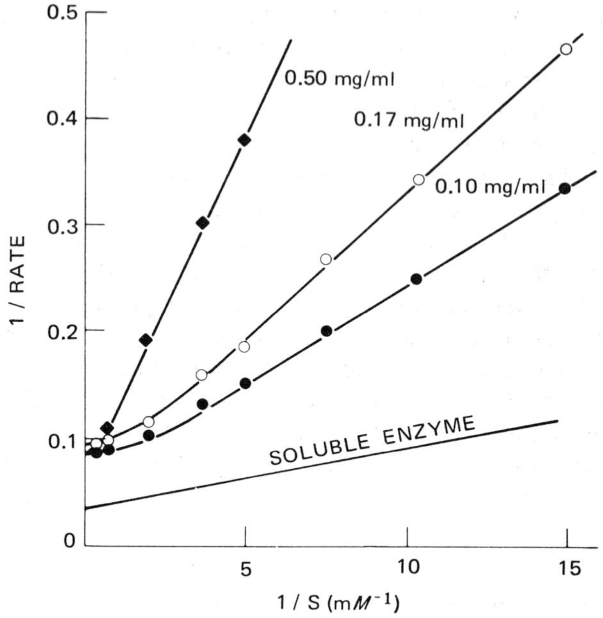

FIG. 21. Lineweaver-Burke plots for 110-μm thin slices of entrapped β-galactosidase containing three different concentrations of enzyme. Rates expressed in μmol/min per mg of enzyme.

If the curves are approximated by the equation

$$v = \frac{k(\text{app.})E_m S}{K_m(\text{app.}) + S} \tag{17}$$

where

$$k(\text{app.}) = k' \tag{18}$$

is the actual limiting value of the rate for high S, then we obtain

$$K_m(\text{app.}) = \frac{K'_m}{PF} \tag{19}$$

by comparing the limiting form of Eq. (17) for small S with Eq. (14).

Finally, for a thick membrane tanh $(\gamma l) \approx 1$ and consequently $F \approx 1/\gamma l$, where γ according to Eq. (16) is proportional to $E_m^{1/2}$. Substitution of these relations into Eq. (14) leads to the prediction that the reaction rate will be proportional to $E_m^{1/2}$, as verified experimentally by Ngo and Laidler (1975a).

E. Buffer Diffusion

Activities of enzymes depend on the pH of their environment and many reactions, such as oxidations, reductions, and ester hydrolysis involve the production or consumption of hydrogen ions. Differences between the behavior of free and immobilized proteins will therefore arise unless the pH prevailing within the carrier is identical with the pH measured in the solution.

Consider the hydrolysis of an ester substrate RX generating a relatively strong acid HX

$$RX + H_2O \longrightarrow ROH + H^+ + X^-$$

the immobilized enzyme being in a solution containing the buffer species A^- and HA, and pH being maintained constant by the addition of alkali. The basic component of the buffer A^- diffusing through the particles reacts *in situ* with the generated hydrogen ions

$$H^+ + A^- \rightarrow HA$$

and there is a net flux of A^- into and of HA out of the catalyst. Unless the external concentrations of both buffer species are high or the reaction rate is low, the internal ratio of $(HA)_i/(A^-)_i$ and the internal concentration of hydrogen ions, given by

$$(H^+)_i = K_{HA} \frac{(HA)_i}{(A^-)_i} \qquad (20)$$

will be significantly higher than the value in the external solution. Changes of external pH are therefore expected to have different effects on the rates catalyzed by the free and immobilized enzyme.

As an illustration the hydrolysis rates of an ester by a free and an entrapped esterase (Konecny and Slanicka, 1975), determined in three different 10-m*M* buffers, namely phosphate (pK 7.2), Tris (pK 8.1), and borate (pK 9.2), are shown in Fig. 22. The observed results follow from the theory of Engasser and Horvath (1974). Qualitatively, the insensitivity of the rate to external pH in alkaline phosphate, for example, may be accounted for as follows: As pH increases from 8.2 to 10.2 the external concentration of HPO_4^{2-} increases from 9 m*M* to 9.99 m*M*, while the concentration of $H_2PO_4^-$ decreases from 1 m*M* to 0.01 m*M*. The increase in external HPO_4^{2} is seen to be small. At the same time, the external concentration of $H_2PO_4^-$ is small compared to the internal concentration of the species, which derives mainly from the reaction of A^- with the generated H^+. Internal acidity is therefore insensitive to external pH changes.

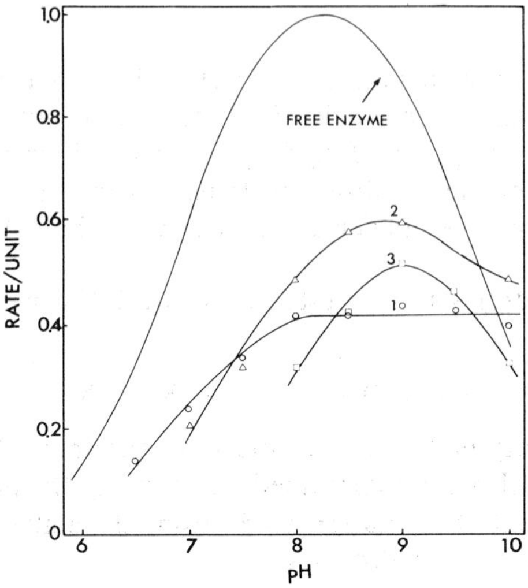

FIG. 22. pH-activity profiles for the hydrolysis of (2-methoxy ethyl) acetate by an immobilized esterase, as measured in 10 mM (1) phosphate, (2) Tris, and (3) borate buffers. Rates expressed in μmol/min per unit of enzyme immobilized.

Differences between external and internal pH lead to deviations from Michaelis–Menten kinetics (Konecny, 1977a) and may effect the operational stability of the catalyst.

F. DIFFUSION IN DUAL ENZYME SYSTEMS

Many enzymatic reactions are carried out *in vivo* by assemblies of membrane-bound enzymes, the product of one enzyme serving as a substrate for the next. In some systems, the concentration of the end product of consecutive reactions regulates the activity of the first enzyme of the series, in order to prevent overproduction. Schematically we shall consider

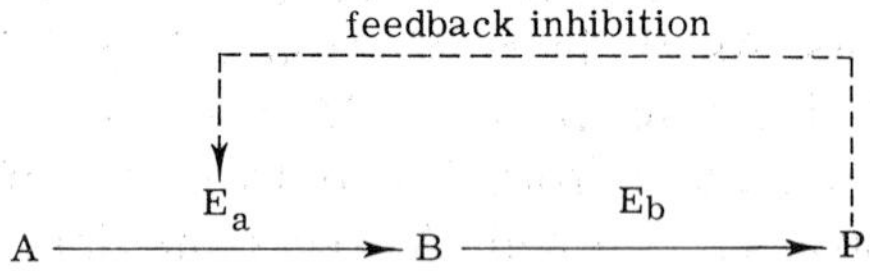

first the effects of diffusion on a system without feedback inhibition. If the two enzymes are in free form in a dilute solution, the concentration of the intermediate B has to build up to a certain level before the rate of the second reaction becomes appreciable. This point is expected to be reached earlier when the two enzymes are immobilized together, for example, by microencapsulation of their concentrated solution. The diffusion barrier represented by the membrane retards the dispersion of the intermediate B into the bulk of the solution and leads to a relatively fast build up of its concentration within the capsule, and thus to an earlier onset of the formation of P. On the other hand, when the two enzymes are immobilized in separate particles, the intermediate must diffuse out of the carrier containing enzyme E_a in order to reach enzyme E_b, and is diluted in the process. Experimental verification of these conclusions, predicted from mathematical models by Goldman and Katchalski (1971), has been reported for several systems (Mosbach *et al.*, 1974; Mosbach, 1976).

The same reasoning leads to the expectation that feedback inhibition will make itself felt sooner when the two enzymes are immobilized in the same matrix. This has been demonstrated recently in the study of the system β-glucosidase/glucose oxidase by Lecoq and co-workers (1975).

VII. Reactors

In this section we shall consider some factors bearing on the choice of reactors for the study of immobilized enzymes and their use. Of the common reactor types shown in Fig. 23 only the packed bed has been employed on an industrial scale for reasons outlined in Section IV,C. In the laboratory stirred vessels, recirculation reactors, packed beds, and in a few investigations (Charles *et al.*, 1974, 1975), fluidized beds have been used.

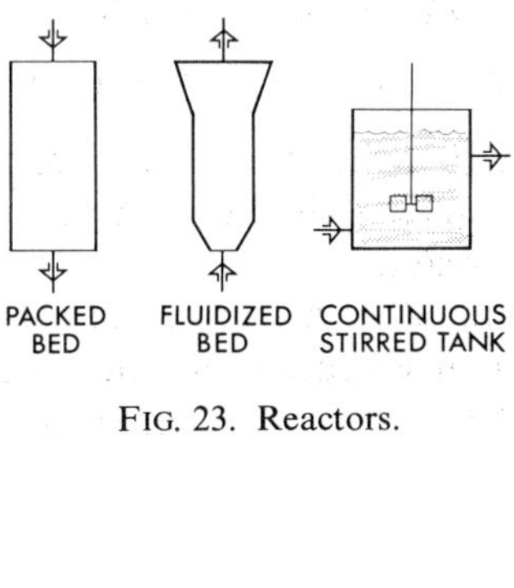

FIG. 23. Reactors.

A convenient agitated reactor for laboratory studies, which utilizes a vortex mixer in place of a stirrer and which permits monitoring of the reaction by frequent intermittent readings of optical density has been described by Widmer *et al.* (1973). As an alternative, a differential recirculation reactor (Ford *et al.*, 1972) has several advantages. It excludes the risk of disintegrating the particles by agitation and segregates the catalyst from the bulk of the solution. Moreover, the hydrodynamic conditions, which may effect the reaction rates, are well defined by two parameters, namely the volumetric flow rate and the cross-sectional area of the bed. The term "differential" implies that the quantity of catalyst in reactor A (see Fig. 23) is so small and the flow rate so high, that only infinitesimal conversion takes place as a given liquid element passes through the bed. Under the circumstances the solution in contact with the catalyst differs only negligibly from that in the vessel B. As shown earlier in Fig. 16, the rate becomes independent of the flow rate at high recirculation rates, and the time course of the reaction under these conditions is exactly the same as in a well-stirred suspension. Reactions producing acid or base can be followed by the continuous titration of the contents of vessel B, and those associated with a change of optical density by inserting a flow-through kuvette into the recirculation line. The reactor is convenient for preparative work and also for the study of kinetic parameters bearing on the scale-up, with a minimum demand on materials.

The measure of reactor efficiency is the quantity of substrate transformed per hour per unit quantity of catalyst when the initial concentration of the substrate S_0 and its desired conversion $X = (S_0 - S)/S_0$ are specified. Assuming that the reaction obeys Michaelis-Menten kinetics and that the effects of interfacial mass transfer are comparable or negligible, the performance on a stirred tank, a packed-bed reactor and a continuous stirred tank are given by Eqs. (21), (22), and (23), where k^* and K_m^* stand for k(app.) and K_m(app.), and E_c represents the total quantity of the catalyst in the reactor:

$$k^* E_c(t/V) = S_0 X - K_m^* \ln(1 - X) \tag{21}$$

$$k^* E_c/Q = S_0 X - K_m^* \ln(1 - X) \tag{22}$$

$$k^* E_c/Q = S_0 X + K_m^* X/(1 - X) \tag{23}$$

In these equations, V represents the volume of the solution in the stirred tank and t the reaction time; Q is the volumetric flow rate of the solution through the continuous reactors.

Equation (21) is obtained in a straightforward manner by substituting E_c/V for the enzyme concentration in the Michaelis-Menten equation, integrating with respect to time, and then substituting X for $(S_0 - S)/S_0$.

In comparing the first two expressions we see that the right-hand sides are identical. If we divide both equations by t, we obtain for the left-hand sides $k^* E_c / V$ and $k^* E_c / Qt$, respectively. Noting that Qt is the volume of the solution which passes through the packed bed in t hours, and that V is the volume processed by the stirred tank in the same time, we see that the efficiency of these two reactors is identical. This general conclusion, which follows from reactor theory, holds for any given rate law. The derivation of Eq. (23) is also straightforward. In a well-mixed stirred tank there are no concentration gradients and the concentration of substrate in the tank is equal to that in the effluent. When the steady state is established, the difference between the input and output of substrate per hour must be equal to its rate of consumption by the reaction, i.e.,

$$Q(S_0 - S) = V\frac{k^*(E_c/V)S}{K_m^* + S} \tag{24}$$

where V is the volume of solution in the reactor. Cancellation of V and substitution of X for S then leads to Eq. (23).[7]

When the concentration of substrate in the effluent is $S \gg K_m^*$, a reaction obeying Michaelis–Menten kinetics is effectively zero order in substrate, and the continuous stirred tank and the packed-bed reactor display equal efficiency since the reactor equations reduce to $k^* E_c / Q = S_0 X$ for both cases. However, when $S \ll K_m^*$, the reaction is first order in S and the continuous stirred tank is much less efficient because *all* enzyme is exposed to the final, low substrate concentration in this reactor. Continuous stirred tanks are potentially useful in reactions whose rates are repressed by high substrate concentrations or which require a poorly soluble gas, such as oxygen, as one of the reactants.

The theoretical efficiency of any other kind of reactor is between that of the packed bed and the continuous stirred tank. The main incentives for considering a fluidized bed, in which the catalyst is uplifted by the process stream, are the compacting of soft particles in packed beds and the danger of plugging by solid impurities in the processed solution. For a further discussion of enzyme reactors the reader should consult the papers by Pitcher (1975) and Vieth and Venkatasubramanian (1974b).

VIII. Coenzymes

The current use of free and immobilized enzymes is limited to simple processes such as hydrolysis. The impressive feats of synthesis performed

[7] When this mass balance equation is applied to an infinitesimal segment of a packed-bed reactor, and the differential equation is integrated over the length of the column, Eq. (22) is obtained.

by microbial cells are, however, the result of intricate organization and of dynamic balance in coupled sequences of reactions since many of the steps involve the recycling of coenzymes.

Actually, efficient short-term recycling of NAD^+ and ATP has been accomplished, with a very high excess of free enzymes, in analytical systems designed to determine picomole quantities of coenzymes. The analytical principle involves the measurement of the product SH_2 or S′ of the reaction generated by the catalytic cycle (see Section II), the turn over of coenzymes being as fast as 20,000 cycles per hour (Lowry and Passonneau, 1972). Large-scale preparative applications would, however, require much higher coenzyme-enzyme ratios, and raise much more difficult problems than the use of single enzymes (Wykes *et al.*, 1975; Baricos *et al.*, 1976). In view of the formidable cost of such systems the demands on catalytic efficiency and operational stability are very high. Furthermore coenzymes, whose catalogue prices range from a few to several hundred dollars per gram (NAD^+ about \$6.00/g) have to be recovered efficiently from the reaction mixture or somehow retained. We shall outline here the exploratory work which has been undertaken in this area and which has focused mainly on NAD^+ and ATP.

Adenosine triphosphate (ATP) a compound with a high free-energy content, is utilized by cells for driving reactions having un unfavorable free energy. This is accomplished by phosphorylating the substrate to a reactive intermediate, adenosine diphosphate (ADP) being formed from the triphosphate in the process. Thus if catalytic quantities of ATP are to be used in the process, ADP has to be phosphorylated to the parent compound. This has been accomplished by Marshall (1973) in a reaction involving potassium cyanate as the phosphate donor, and carbamyl phosphokinase immobilized to a highly active catalyst on controlled-pore glass. The enzyme lost 16% of its initial activity after 14 days of continuous operation at room temperature. Short-term experiments, involving the recycling of ATP by a microencapsulated dual enzyme system have been described by Campbell and Chang (1975).

Most of the work in this area has been devoted to NAD^+. In principle, various approaches may be used, as, for example:

1. The two cycling enzymes are immobilized and the coenzyme is used in its free form.
2. The regenerating enzyme and its substrate are replaced by an oxidizing (or reducing) agent capable of returning NADH (or NAD^+) to its original form; alternatively, electrochemical regeneration is employed (Coughlin and Alexander, 1975).
3. The two enzymes and coenzymes are immobilized in hollow fibers (Fink and Rodwell, 1975).

4. The coenzyme is bonded chemically to a soluble macromolecule such as polyethylene imine (Wykes *et al.*, 1972; 1975) or dextrane (Larsson and Mosbach, 1974) and this derivative together with the enzymes is contained by microencapsulation or some other semipermeable device.

The first two methods call for separation of the coenzyme from the reaction products with high requirements on quantitative recovery. There are equally high requirements on the quantitative course of the recycle reaction when chemical reagents or electrical current are used. The third possibility calls for membranes of very low porosity and correspondingly low permeability in order to retain NAD^+ (mol. wt. 663) and therefore precludes the transformation of all but very small substrate molecules. Much more porous membranes could be used if the coenzyme were bonded to a soluble macromolecule, which would prevent its escape out of the system through larger pores, and at the same time permit the co-catalyst to move freely within the confines and to interact with the enzymes. Considerable effort has been devoted to this approach.

Polyethylene imine and dextrane derivatives have been successfully prepared by three or more chemical steps, albeit with low overall yields (3%) of coenzymic activity. The coenzyme-polymer linkage in the polyethylene imine derivative was found to be susceptible to hydrolysis[8] (Wykes *et al.*, 1975), while the available information concerning the useful life of the dextrane derivative (Davies and Mosbach, 1974) is limited. At the time of this writing better results have been reported by Marconi *et al.* (1976) for a polyethylene imine derivative prepared by a different route. This preparation was microencapsulated in cellulose triacetate fibers together with high concentrations of lactate dehydrogenase and alanine dehydrogenase. This system, which displayed an activity of 5 μmole/min per gram carrier and a turn-over rate of 220 cycles/hour under test conditions, showed only a small loss of activity when used in several reaction cycles over a period of 15 days.

Another approach has been tried with NAD^+ (Mosbach, 1976) as well as with pyridoxal phosphate (Fukui *et al.*, 1975). It involves the covalent linking of the coenzymes to Sepharose, as in the preparation of stationary phases for affinity chromatography, adsorption of the enzymes on these immobilized coenzymes, followed by covalent binding of the enzyme to the coenzyme or carrier. Complexes of this type have been found to be catalytically active.

Should an increased demand for certain coenzymes develop, manufacture on a larger scale, and other developments, will no doubt reduce

[8] For the rates of decomposition of native NADH and NAD^+ as a function of temperature and pH, see Lowry and Passonneau (1972).

their price. The isolation of yeasts which leak NAD^+ into their growth medium has been reported by Sakai *et al.* (1973) and work on cheaper affinity adsorbents for isolating coenzyme A is in progress in the Chibata group in Japan. A study of synthesis of coenzyme A from simpler building stones by means of gel entrapped cells has also been described (Shimizu *et al.*, 1975). At the present time enzymatic oxidations and other transformations of steroids are being carried out industrially with whole cells. Recycling of cells as catalysts has been reported by Buckland *et al.* (1975) and studies relating to the use of live, entrapped cells in a continuous, aerated stirred reactor have recently been described by Larsson *et al.* (1976).

IX. Applications

Of the known industrial applications of immobilized enzymes, two have already been mentioned in the preceding discussion. The first of these involves the use of an adsorbed L-aminoacylase to separate D,L-methionine and other racemic amino acids into the stereoisomers (Chibata *et al.*, 1972). In this process the racemic mixture is first acetylated and then passed through the enzyme column in which only the L-isomer is hydrolyzed back to the parent acid. The produced mixture of L-methionine and D-acetylmethionine is then separated by conventional means. The continuous plant produces about 260 tons of L-methionine per year and the production costs are about 40% lower than those in a corresponding process with the free enzyme. The carrier, DEAE Sephadex, has been in continuous use for two years without significant deterioration.

The entrapped *E. coli* cells containing aspartase (Sato *et al.*, 1975) are used for the production of L-aspartic acid from ammonium fumarate

$$NH_4^+ + {}^-O_2C-CH=CH-CO_2^- \longrightarrow {}^-O_2C-\underset{\oplus NH_3}{\underset{|}{CH_2}}-CH_2-CO_2^-$$

A process of major industrial importance is the conversion of penicillin G to 6-aminopenicillanic acid, an intermediate for the production of other penicillins.

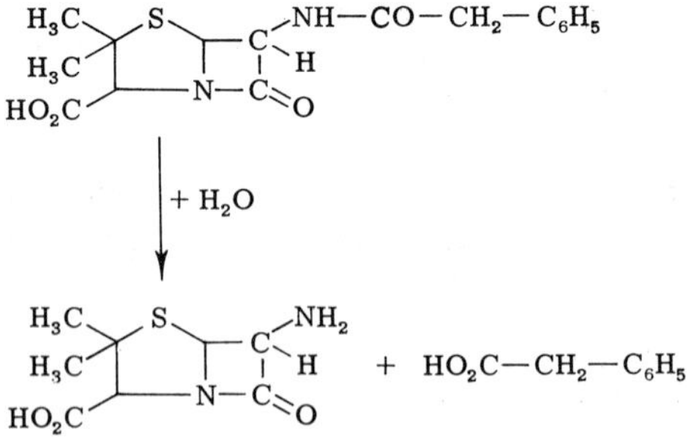

The immobilization of penicillin acylase, which catalyzes this reaction, has been described by Marconi *et al.* (1973), Self and co-workers (1969), and other authors, as well as in a number of patents. Chibata and co-workers (1975) and Sato *et al.* (1976) obtained quite high yields of 6-aminopenicillanic acid in a continuous laboratory process utilizing a column packed with gel-entrapped *E. coli* cells. The reverse reaction, namely, the acylation of the free amino group of penicillin and cephalosporin derivatives with esters of the side chain precursors, has also been described (Marconi *et al.*, 1975b; Fujii *et al.*, 1976). These reverse reactions of penicillin acylases involve the formation of an acyl-enzyme intermediate which transfers the acyl radical to the attacking amino group, or water, or another nucleophile.

In terms of bulk, by far the major users of enzymes are the food industries (Wolnak and Associates, 1972). In 1971 about 600,000 tons of glucose were produced from starch in the United States, alone. A highly active immobilized glucoamylase, producing about 1900 μmole glucose/min/ml catalyst from liquefied starch has been prepared by the covalent binding of the enzyme to zirconia-clad controlled-pore glass (Weetall, 1973), but so far this process has not been realized. Glycoamylase is cheap, the carrier expensive and the half-life of the catalyst at the desired operational temperature of 60°C is only 19 days. Nevertheless the process may be an economically attractive alternative to the free-enzyme route when new production plants are designed (Pitcher and Weetall, 1975). Pilot plant studies with a 30-liter reactor packed with glucoamylase immobilized on controlled-pore silica, and having a production capacity of 1000 lbs/day have recently been reported (Lee *et al.*, 1975).

Another important process in food industry is the enzymatic hydrolysis of sucrose to invert sugar, a mixture of glucose and fructose. An alternative route utilizing a cheaper raw material is the isomerization of glucose to an equilibrium mixture of the two sugars. The successful realization of a process utilizing glucose isomerase adsorbed on DEAE cellulose has been described by Davis (1974b) and Hermelstein (1975).

By virtue of their specificity enzymes are particularly well suited for the determination of various substances in body fluids, fermentation broths and other complex mixtures and are in fact finding increasing applications in clinical analysis and fermentation and pharmaceutical industry (Bergmeyer, 1974). There are incentives for substituting immobilized proteins for the free enzymes also in this area (Bowers and Carr, 1976).

Automated analyses, such as the continuous flow method developed by Technicon, are used extensively in routine analyses. The autoanalyzers can handle large numbers of samples but often consume appreciable quantities of reagents. As an alternative to the use of free enzymes Hornby and co-workers have prepared coils of nylon tubing with enzymes covalently bound to their inner surface, and have successfully used them for a very large number of determinations of penicillin (Hornby *et al.*, 1974), glucose (Campbell *et al.*, 1975), and other substances.

Other potential analytical applications of immobilized enzymes include enzyme electrodes, studied intensively by Guilbault (1974), and the enzyme thermistor, developed by Mosbach and co-workers (1975). Enzyme electrodes are constructed, e.g., by coating a suitable sensor, such as glass, platinum, oxygen, or other electrode by a thin layer of gel containing the enzyme in entrapped or, alternatively, in cross-linked form (Tran-Minh and Broun, 1975). A urea-sensing electrode may be constructed by dipping an ammonia-sensitive electrode into a polymerizing solution containing urease and then rotating it in air until the solution solidifies to a stiff gel. When this sensor is immersed in a solution of urea, the latter diffuses into the gel where it is hydrolyzed to carbon dioxide and ammonia. The electrode then responds to the steady-state concentration of ammonia which is established at the glass/gel interface after a certain time, the response in millivolts being a linear function of the external urea concentration within a certain range. [The theoretical analysis, given by Tran-Minh and Broun (1975) follows the basic considerations outlined in Section VI,D.] When an amino acid oxidase, which catalyzes the reaction

$$\underset{\displaystyle NH_2}{R{-}\underset{|}{C}H{-}CO_2H} + O_2 + H^+ + H_2O \longrightarrow \underset{\displaystyle O}{R{-}\underset{||}{C}{-}CO_2H} + NH_4^+ + H_2O_2$$

is substituted for the urease, the electrode can be used for the determination of amino acids. Utilizing the same principles oxygen electrodes modified with a membrane containing a mixture of glucose oxidase and catalase can be constructed for the measurement of glucose. Two commercial electrodes, utilizing a different design, are now manufactured by the Yellow Springs Instrument and Hoffmann–La Roche companies. An alternative approach, which has distinct advantages in certain applications, involves the use of enzyme microcolumns, the electrode detecting the product being located downstream (Johansson and Ogren, 1976; Rusling *et al.*, 1976).

Since a majority of chemical reactions are associated with the consumption or generation of heat, microcalorimeters (Schmidt *et al.*, 1976) and other thermal detectors can also be utilized as sensors. A particularly attractive device is the enzyme thermistor developed by Mosbach and associates (1975), which is essentially a microcolumn packed with the enzyme catalyst and furnished with a thermistor at the outlet. At sufficiently low flow rates of the solution to be analyzed through the column, complete transformation of substrate takes place and the adiabatic temperature rise associated with the reaction is directly proportional to the substrate concentration.

In this writer's laboratory, quantitative alkalimetric titrations of isolated cephalosporins have been carried out in a small recirculation reactor utilizing highly active preparations of the *B. subtilis* esterase immobilized on controlled-pore glass (Konecny, 1977b).

Potential applications of immobilized enzymes in the biomedical area have been studied intensively especially by Chang (1974, 1975) and co-workers. The possible remedy of inborn enzyme defficiencies, the suppression of certain tumors by implanted, microencapsulated asparaginase and the construction of extracorporeal shunts as alternatives to artificial kidneys for the removal of urea from blood have been examined.

In biological research immobilization of the subunits of complex enzymes is being used to prevent their reaggregation to the native protein state in studies of their properties and biological function (Chan *et al.*, 1973a,b; Fell and White, 1975). Adsorption of organelles such as mitochondria on hydrophobic carriers is now also being exploited in investigations of these subcellular organs. The ability of higher organisms to combat foreign macromolecules, toxins, viruses, and bacteria depends on the production of highly specific antibody proteins which recognize particular molecular configurations on the surface of these bodies and which complex with them at these sites. Immobilization of the antibodies on suitable carriers provides a powerful tool for isolating the toxins (Jakoby and Wilchek, 1974), and the reverse procedure can, of course, be used to isolate specific antibodies.

For further discussion of enzyme applications the reader is referred to Suckling (1974), Weetall (1975, p. 201), Jones *et al.* (1976), and Mosbach (1976). The preparation of pharmaceutical compounds by immobilized enzymes and cells has been reviewed by Abbott (1976).

X. Summary

Enzyme engineering has reached the stage where many enzymes can be immobilized on various supports with a fair to high retention of their activity. The properties of these novel heterogeneous catalysts are well understood in relation to mass transfer. On the other hand, our understanding of the intrinsic changes of catalytic properties associated with chemical modification and physical forces exerted by the carrier on the enzyme is still fragmentary.

In the relatively short time in which the technological and economic requirements on the catalysts became fully appreciated, several immobilized enzymes have found industrial applications. A number of developments in the analytical and separation areas show considerable promise and several are being exploited.

The growing awareness of the potential of enzymes as specific catalysts, the availability of convenient immobilization methods and carriers together with further advances in bio-technology and microbiology will extend their range of applications in the future.

ACKNOWLEDGMENTS

The writer is indebted to Mr. M. Sieber for the illustrations, to Mrs. E. Funk for transforming his handwriting into a legible manuscript, to Dr. J. Nüesch for reading it, and to his wife for her patience.

REFERENCES

Abbott, B. J. (1976). *Adv. Appl. Microbiol.* **20**, 203.
Abbott, B. J., and Fukuda, D. S. (1975). *Antimicrob. Agents & Chemother.* **8**, 282.
Atkinson, A. (1973). *Process Biochem.* **8**, [8] 9.
Atkinson, A. (1974). *Rep. Prog. Appl. Chem.* **59**, 375.
Baricos, W. H., Chambers, R. P., and Cohen, W. (1976). *Anal. Lett.* **9**, 257.
Barman, T. E. (1969). "Enzyme Handbook," Vols. 1 and 2. Springer-Verlag, Berlin and New York.
Barman, T. E. (1974). "Enzyme Handbook," Suppl 1. Springer-Verlag, Berlin and New York.
Baum, G. (1975). *Biotechnol. Bioeng.* **17**, 253.
Bender, M. L., Whitaker, J. R., and Menger, F. (1965). *Proc. Natl. Acad. Sci. U.S.A.* **53**, 711.
Bergmeyer, H. U. ed. (1974). "Methods of Enzymatic Analysis." Academic Press, New York.
Bernardi, G. (1976). *Nature (London)* **259**, 173.
Bernfeld, P., and Wan, J. (1963). *Science* **142**, 678.
Bernhard, S. (1968). "The Structure and Function of Enzymes." Benjamin, New York.
Bowers, L. D., and Carr, P. W. (1976). *Anal. Chem.* **48**, 544A.
Brockman, H. L., Law, J. H., and Kezdy, F. J. (1973). *J. Biol. Chem.* **248**, 4965.
Broun, G., Thomas, D., Domuraro, D., Berjonneau, A. M., and Guillon, C. (1973). *Biotechnol. Bioeng.* **15**, 359.
Brown, E., and Joyeau, R. (1974a). *Polymer* **15**, 546.
Brown, E., and Joyeau, R. (1974b). *Makromol. Chem.* **175**, 1961.
Buckland, B. C., Dunnill, P., and Lilly, M. D. (1975). *Biotechnol. Bioeng.* **17**, 815.
Bunting, P. S., and Laidler, K. J. (1972). *Biochemistry* **11**, 4477.
Butler, L. G. (1975). *Arch. Biochem. Biophys.* **171**, 645.
Caldwell, K. D., Axen, R., and Porath, J. (1975). *Biotechnol. Bioeng.* **17**, 613.
Campbell, J., and Chang, T. M. S. (1975). *Biochim. Biophys. Acta* **397**, 101.
Campbell, J., Hornby, W. E., and Morris, D. L. (1975). *Biochim. Biophys. Acta* **384**, 307.
Capaldi, R. A., and Green, D. E. (1972). *FEBS Lett.* **25**, 205.
Carlile, M. J., and Skehel, J. J. (1974). "Evolution in the Microbial World." Cambridge Univ. Press, London.
Chan, W. W., Mort, J. S., Chong, D. K. K., and MacDonald, P. D. M. (1973a). *J. Biol. Chem.* **248**, 2778.
Chan, W. W., Schutt, H., and Brand, K. (1973b). *Eur. J. Biochem.* **40**, 553.
Chang, T. M. S. (1971). *Nature (London)* **229**, 117.
Chang, T. M. S. (1974). *In* "Insolubilized Enzymes" (M. Salmona, C. Saronio, and S. Garattini, eds.), p. 15. Raven, New York.

Chang, T. M. S. (1975). *Chem. Technol.* **5**, 80.
Charles, M., Coughlin, R. W., Allen, B. R., Paruchuri, E. K., and Hasselberger, F. X. (1974). *Adv. Exp. Med. Biol.* **42**, 213.
Charles, M., Coughlin, R. W., Paruchuri, E. K., Allen, B. R., and Hasselberger, F. S. (1975). *Biotechnol. Bioeng.* **17**, 203.
Chen, L. F., and Tsao, G. T. (1976). *Biotechnol. Bioeng.* **18**, 1507.
Chibata, I., Tosa, T., Sato, T., Mori, T., and Matsuo, Y. (1972). *Ferment. Technol. Today, Proc. Ferment Symp. Y + h, 1972* p. 383.
Chibata, I., Tosa, T., Mori, T., and Yamamoto, K. (1974a). *In* "Enzyme Engineering 2" (E. K. Pye and L. B. Wingard, Jr., eds.), p. 303. Plenum, New York.
Chibata, I., Tosa, T., and Sato, T. (1974b). *Appl. Microbiol.* **27**, 878.
Chibata, I., Tosa, T., Sato, T., Mori, T., and Yamamoto, K. (1975). *In* "Immobilized Enzyme Technology" (H. H. Weetall and S. Suzuki, eds.), p. 111. Plenum, New York.
Coughlin, R. W., and Alexander, B. F. (1975). *Biotechnol. Bioeng.* **17**, 1379.
Darnall, D. W., and Klotz, I. M. (1975). *Arch. Biochem. Biophys.* **166**, 651.
Davies, P., and Mosbach, K. (1974). *Biochim. Biophys. Acta* **370**, 329.
Davis, J. C. (1974a). *Biotechnol. Bioeng.* **16**, 1113.
Davis, J. C. (1974b). *Chem. Eng.* **81**, 52.
Dean, P. D. G., and Harvey, M. J. (1975). *Process Biochem.* **10**, [7] 5.
Degani, Y., and Miron, T. (1970). *Biochim. Biophys. Acta* **212**, 362.
Demain, A. L. (1972). *In* "Enzyme Engineering" (L. B. Wingard, Jr., ed.), p. 21. Wiley (Interscience), New York.
Dickerson, R. E., and Geis, I. (1969). "The Structure and Action of Proteins." Harper, New york.
Dinelli, D. (1972). *Process Biochem.* **7**, [8] 9.
Dunnill, P., and Lilly, M. D. (1974). *Biotechnol. Bioeng.* **16**, 987.
Eisenthal, R., and Cornish-Bowden, A. (1974). *Biochem. J.* **139**, 715.
Engrasser, J., and Horvath, C. (1974). *Biochim. Biophys. Acta* **358**, 178.
Falb, R. D. (1974). *In* "Enzyme Engineering 2" (E. K. Pye and L. B. Wingard, Jr., eds.), p. 67. Plenum, New York.
Fell, D. A., and White, C. J. B. (1975). *Biochem. Biophys. Res. Commun.* **67**, 1013.
Filbert, A. M. (1975). *In* "Immobilized Enzymes for Industrial Reactors" (R. A. Messing, ed.), p. 39. Academic Press, New York.
Fink, D. J., and Rodwell, D. W. (1975). *Biotechnol. Bioeng.* **17**, 1029.
Fischer, L. (1971). "An Introduction to Gel Chromatography." North-Holland Publ., Amsterdam.
Ford, J. R., Lambert, A. H., Cohen, W., and Chambers, R. P. (1972). *In* "Enzyme Engineering" (L. B. Wingard, Jr., ed.), p. 267. Wiley (Interscience), New York.
Fries, R. W., Bohlken, D. P., Blankley, R. T., and Plapp, B. V. (1975). *Biochemistry* **14**, 5233.
Fujii, T., Matsumoto, K., and Watanabe, T. (1976). *Proc. Biochem.* **11** [8], 21.
Fukui, S., Ikeda, S., Fujimura, M., Yamada, H., and Kumagai, H. (1975). *Eur. J. Appl. Microbiol.* **1**, 25.
Giovenco, S., Morisi, F., and Pansolli, P. (1973). *FEBS Lett.* **36**, 57.
Goldman, R., and Katchalski, E. (1971). *J. Theor. Biol.* **32**, 243.
Goldman, R., Kedem, O., and Katchalski, E. (1968). *Biochemistry* **7**, 4518.
Goldman, R., Goldstein, L., and Katchalski, E. (1971). *In* "Biochemical Aspects of Reactions on Solid Supports" (G. R. Stark, ed.), p. 1. Academic Press, New York.
Goldstein, L., and Katchalski, E. (1968). *Z. Anal. Chem.* **243**, 375.
Goldstein, L., Freeman, A., and Sokolovsky, M. (1974). *In* "Enzyme Engineering 2" (E. K. Pye and L. B. Wingard, Jr., eds), p. 97. Plenum, New York.

Gray, C. J., Livingstone, C. M., Jones, C. M., and Barker, S. A. (1974). *Biochim. Biophys. Acta* **341**, 457.
Guilbault, G. G. (1974). *In* "Insolubilized Enzymes" (M. Salmona, C. Saronio, and S. Garattini, eds.), p. 199. Raven, New York.
Hermelstein, N. H. (1975). *Food Technol.* **29**, 20.
Hinberg, I., Kapoulas, A., Korus, R., and O'Driscoll, K. (1974a). *Biotechnol. Bioeng.* **16**, 159.
Hinberg, I., Korus, R., and O'Driscoll, K. F. (1974b). *Biotechnol. Bioeng.* **16**, 943.
Hofstee, B. H. J. (1976). *J. Macromol. Sci. Chem.* **A10**, 111.
Hornby, W. E., Campbell, J., Inman, D. J., and Morris, D. L. (1974). *In* "Enzyme Engineering 2" (E. K. Pye and L. B. Wingard, Jr., eds.), p. 401, Plenum, New York.
Horvath, C., and Engasser, J.-M. (1973). *Ind. Eng. Chem., Fundam.* **12**, 229.
Jakoby, W. B., ed. (1971). "Methods in Enzymology," Vol. 22. Academic Press, New York.
Jakoby, W. B., and Wilchek, eds. M., (1974). "Methods in Enzymology," Vol. 34. Academic Press, New York.
Jaworek, D. (1974). *In* "Insolubilized Enzymes" (C. M. Salmona, C. Saronio, and S. Garattini, eds.), p. 65. Raven, New York.
Johansson, A., and Mosbach, K. (1974a). *Biochim. Biophys. Acta* **370**, 339.
Johansson, A., and Mosbach, K. (1974b). *Biochim. Biophys. Acta* **370**, 348.
Johansson, G., and Ogren, L. (1976). *Anal. Chim. Acta* **84**, 23.
Jones, J. B., Sih, C. J., and Perlman, D. (1976). "Applications of Biochemical Systems in Organic Chemistry." Wiley, New York.
Katchalski, E., Silman, I., and Goldman, R. (1971). *Adv. Enzymol.* **34**, 445.
Kawashima, K., and Umeda, K. (1975). *Biotechnol. Bioeng.* **17**, 599.
Kieslich, K. (1976). "Microbial Transformations of Non-Steroid Cyclic Compounds." Georg Thieme, Stuttgart.
Kim, S. -S., and Cooney, D. O. (1976). *Chem. Eng. Sci.* **31**, 289.
Kobayashi, T., and Laidler, K. J. (1973). *Biochim. Biophys. Acta* **302**, 1.
Kobayashi, T., and Laidler, K. J. (1974a). *Biotechnol. Bioeng.* **16**, 77.
Kobayashi, T., and Laidler, K. J. (1974b). *Biotechnol. Bioeng.* **16**, 99.
Konecny, J. (1975). *Chimia* **29**, 95.
Konecny, J. (1977a). *Biochim. Biophys. Acta* (in press).
Konecny, J. (1977b). *In* "Enzyme Engineering 3" (E. K. Pye and H. H. Weetall, eds.). Plenum, New York. (In press.)
Konecny, J., and Slanicka, J. (1975). *Biochim. Biophys. Acta* **403**, 573.
Konecny, J., Felber, E., and Gruner, J. (1973). *J. Antibiot.* **26**, 135.
Krasnobajew, V., and Böniger, R. (1975). *Chimia* **29**, 123.
Kutzbach, C., and Rauenbusch, E. (1974). *Hoppe-Segler's Z. Physiol. Chem.* **354**, 45.
Larsson, P. O., and Mosbach, K. (1974). *FEBS Lett.* **46**, 119.
Larsson, P. O., Ohlson, S., and Mosbach, K. (1976). *Nature* **263**, 796.
Lasch, J., Bessmertnaya, L., Kozlov, L. V., and Antonov, V. K. (1976). *Eur. J. Biochem.* **63**, 591.
Lecoq, D., Hervagault, J. F., Broun, G., Ghislaine, J., Kernevez, J. P., and Thomas, D. (1975). *J. Biol. Chem.* **250**, 5496.
Lee, D. D., Yee, Y. Y., and Tsao, G. T. (1975). *Staerke* **27**, 384.
Lewis, W., and Middleman, S. (1974). *AIChE J.* **20**, 1012.
Lilly, M. D., and Sharp, A. K. (1968). *Chem Eng. (London).* CE12.
Lilly, M. D., Hornby, W. E., and Crook, E. M. (1966). *Biochem. J.* **100**, 718.
Lilly, M. D., O'Neill, S. P., and Dunnill, P. (1973). *Biochimie* **55**, 985.
Løwenstein, H. (1974). *Acta Chem. Scand. Ser. B.* **28**, 1098.

Lowry, O. H., and Passonneau, J. V. (1972). "A Flexible System of Enzymatic Analysis." Academic Press, New York.
Maaløe, O., and Kjeldgaard, N. O. (1966). "Control of Macromolecular Synthesis, A Study of DNA, RNA and Protein Synthesis in Bacteria." p. 225. Benjamin, New York.
Maeda, H., Suzuki, H., Yamauchi, A., and Sakimae, A. (1975). *Biotechnol. Bioeng.* **17**, 119.
Marani, A., Marconi, W., Bartoli, F., and Morisi, F. (1975). *Chim. Ind. (Milan)* **57**, 455.
Marconi, W., Cercere, F., Morisi, F., Penna, G. D., and Rappuoli, B. (1973). *J. Antibiot.* **26**, 228.
Marconi, W., Gulinelli, S., and Morisi, F. (1974a). *In* "Insolubilized Enzymes" (M. Salmona, C. Saronio, and S. Garattini, eds.), p. 51. Raven Press, New York.
Marconi, W., Bartoli, F., Cercere, F., and Morisi, F. (1974b). *Agric. Biol. Chem.* **38**, 1343.
Marconi, W., Guinelli, S., and Morisi, F. (1974c). *Biotechnol. Bioeng.* **16**, 501.
Marconi, W., Morisi, F., and Mosti, R. (1975a). *Agric. Biol. Chem.* **39**, 1323.
Marconi, W., Bartoli, F., Cercere, F., Galli, G., and Morisi, F. (1975b). *Agr. Biol. Chem.* **39**, 277.
Marconi, W., Prosperi, G., Giovenco, S., and Morisi, F. (1976). *J. Mol. Catal.* **1,** 111.
Marshall, D. L. (1973). *Biotechnol. Bioeng.* **15**, 447.
Means, G. E., and Feeney, R. E. (1971). "Chemical Modification of Proteins." Holden-Day, San Francisco, California.
Melrose, G. J. H. (1971). *Rev. Pure Appl. Chem.* **21**, 83.
Messing, R. A. (1974). *Process Biochem.* **9**, [9] 26.
Messing, R. A. (1975). "Immobilized Enzymes for Industrial Reactors." Academic Press, New York.
Messing, R. A., and Filbert, A. M. (1975). *J. Agric. Food Chem.* **23**, 920.
Mogensen, A. O., and Vieth, W. R. (1973). *Biotechnol. Bioeng.* **15**, 467.
Morehead, F. F. (1971). In "High Polymers" (N. E. Bikales and L. Segal, eds.), Vol. V, Part 4, p. 213, Wiley (Interscience), New York.
Mori, T., Sato, T., Tosa, T., and Chibata, I. (1972). *Enzymologia* **43**, 213.
Mori, T., Tosa, T., and Chibata, I. (1974). *Cancer Res.* **34**, 3066.
Mosbach, K. (1976). *FEBS Lett.* **62**, E80.
Mosbach, K., Mattiasson, B., Gestrelius, S., and Srere, P. A. (1974). *In* "Insolubilized Enzymes" (M. Salmona, C. Saronio, and S. Garattini, eds.), p. 123. Raven, New York.
Mosbach, K., Danielsson, B., Borgerud. A., and Scott, M. (1975). *Biochim. Biophys. Acta* **403**, 256.
Nelboeck, M., and Jaworek, W. (1975). *Chimia* **29**, 109.
Ngo, T. T., and Laidler, K. J. (1975a). *Biochim. Biophys. Acta* **377**, 303.
Ngo, T. T., and Laidler, K. J. (1975b). *Biochim. Biophys. Acta* **377**, 317.
Pardee, A. B. (1969). *Ferment. Pap. Int. Ferment. Symp., 3rd, 1968* p. 3.
Peters, J., and Fromherz, P. (1975). *Biochim. Biophys. Acta* **394**, 111.
Peterson, E. A. (1970). "Cellulosic Ion Exchangers." North-Holland Publ., Amsterdam.
Pitcher, W. H., Jr. (1975). *In* "Immobilized Enzymes for industrial Reactors" (R. A. Messing, ed.), p. 151. Academic Press, New York.
Pitcher, W. H., Jr. and Weetall, H. H. (1975). *Enzyme Technol. Dig.* **3**, 127.
Quiocho, F. A. (1974). *In* "Insolubilized Enzymes" (M. Salmona, C. Saronio, and S. Garattini, eds.), p. 113. Raven, New York.
Regan, D. L., Dunnill, P., and Lilly, M. D. (1974). *Biotechnol. Bioeng.* **16**, 333.
Rigby, P. W. J., Burley, B. D., Jr., and Hartley, B. S. (1974). *Nature* (*London*) **251**, 206.
Rony, P. R. (1971). *Biotechnol. Bioeng.* **13**, 431.
Rony, P. R. (1972). *J. Am. Chem. Soc.* **94**, 8247.
Royer, G. P. (1974). *Chem. Technol.* **4**, 694.

Rusling, J. F., Luttrell, G. H., Cullen, L. F., and Apariello, G. J. (1976). *Anal. Chem.* **48**, 1211.

Sakai, T., Wehida, T., and Chibata, I. (1973). *Agric. Biol. Chem.* **37**, 1041.

Sato, T., Mori, T., Tosa, T., Chibata, I., Furui, M., Yamashita, K., and Sumi, A. (1975). *Biotechnol. Bioeng.* **17**, 1797.

Sato, T., Tosa, T., and Chibata, I. (1976). *Eur. J. Appl. Microbiol.* **2**, 153.

Schmidt, H. L., Krisam, G., and Grenner, G. (1976). *Biochim. Biophys. Acta* **429**, 283.

Self, D. A., Kay, G., Lilly, M. D., and Dunnill, P. (1969). *Biotechnol. Bioeng.* **11**, 337.

Shaltiel, S. (1974). *In* "Methods in Enzymology" (W. B. Jakoby and M. Wilchek, eds.), Vol. 34, p. 126.

Shimizu, S., Marioka, H., Tani, Y., and Ogata, K. (1975). *J. Ferment. Technol.* **53**, 77.

Simpson, R. T., Riordan, J. F., and Vallee, B. L. (1963). *Biochemistry* **2**, 616.

Sliwka, W. (1975). *Angew. Chem.* **87**, 556.

Solomon, B., and Levin, Y. (1974). *Biotechnol. Bioeng.* **16**, 1161.

Spencer, B. (1974). "Industrial Aspects of Biochemistry," Vol. 30. Am. Elsevier, Amsterdam.

Suckling, C. J. (1974). *Chem. Soc. Rev.* **3**, 387.

Sundaram, P. V., Tweedale, A., and Laidler, K. J. (1970). *Can. J. Chem.* **48**, 1498.

Tanford, C., and Hauenstein, J. D. (1956). *J. Am. Chem. Soc.* **78**, 5287.

Thornton, D., Flynn, A., Johnson, D. B., and Ryan, P. D. (1975). *Biotechnol. Bioeng.* **17**, 1679.

Tosa, T., Mori, T., and Chibata, I. (1971). *J. Ferment. Technol.* **49**, 522.

Tosa, T., Sato, T., Mori, T., Matuo, Y., and Chibata, I. (1973). *Biotechnol. Bioeng.* **15**, 69.

Tran-Minh, C., and Broun, G. (1975). *Anal. Chem.* **47**, 1359.

Urabe, I., and Okada, H. (1972). *Ferment. Technol. Today, Proc. Int. Ferment. Symp., 4th, 1972* p. 367.

Van Leemputten, E., and Horrisberger, M. (1974). *Biotechnol. Bioeng.* **16**, 385.

Vieth, W. R., and Venkatasubramanian, K. (1974a). *Chem. Technol.* **4**, 309.

Vieth, W. R., and Venkatasubramanian, K. (1974b). *Chem. Technol.* **4**, 434.

Vinogradova, N. N., Kozlov, L. V., Antonov, V. K., and Zhagat, R. A. (1975). *Khim. Prir. Soedin.* **11**, 113.

Wadiak, D. T., and Carbonell, R. G. (1975). *Biotechnol. Bioeng.* **17**, 1157.

Watson, J. (1965). "Molecular Biology of the Gene," p. 85. Benjamin, New York.

Weetall, H. H. (1973). *Chem. Ztg.* **97**, 611.

Weetall, H. H. (1974). *Adv. Exp. Med. Biol.* **42**, 191.

Weetall, H. H. (1975). *In* "Immobilized Enzymes for Industrial Reactors" (R. A. Messing, ed.), p. 99. Academic Press, New York.

Weetall, H. H., and Filbert, A. M. (1974. *In* "Methods in Enzymology" (W. B. Jakoby and M. Wilchek, eds.), Vol. 34, p. 59. Academic Press, New York.

Weetall, H. H., and Havewala, N. B. (1972). *In* "Enzyme Engineering" (L. B. Wingard, Jr., ed.), p. 241. Wiley (Interscience), New York.

Weetall, H. H., and Jacobson, M. A. (1972). *Ferment, Technol. Today, Proc. Int. Ferment. Symp., 4th, 1972* p. 361.

White, M. L., and Dorion, G. H. (1961). *J. Polym. Sci.* **35**, 731.

Widmer, F., Dickson, J. E., and Kaplan, N. O. (1973). *Anal. Biochem.* **55**, 282.

Wilchek, M., and Miron, T. (1974). *In* "Methods in Enzymology" (W. B. Jakoby and M. Wilchek, eds.), Vol. 34, p. 72. Academic Press, New York.

Wold, F. (1971). "Macromolecules: Structure and Function." Prentice-Hall, Englewood Cliffs, New Jersey.

Wolnak and Associates. (1972). "Present and Future Technological and Commercial Status of Enzymes," PB-219 636. Nat. Tech. Inf. Serv., U.S. Dept. of Commerce, Washington, D. C.

Wykes, J. R., Dunnill, P., and Lilly, M. D. (1972). *Biochim. Biophys. Acta* **286**, 260.
Wykes, J. R., Dunnill, P., and Lilly, M. D. (1975). *Biotechnol. Bioeng.* **17**, 51.
Yamana, T., Tsuji, A., and Mizukami, Y. (1974). *Chem. Pharm. Bull.* **22**, 1186.
Yamasaki, N., Hayashi, K., and Funatsu, M. (1968). *Agric. Biol. Chem.* **32**, 64.
Zaborsky, O. R. (1973). "Immobilized Enzymes." C.R.C. Press, Cleveland, Ohio.
Zaborsky, O. R. (1974). *In* "Enzyme Engineering 2" (E. K. Pye and L. B. Wingard, Jr., eds.), p. 115. Plenum, New York.

A Book Review of an Early (1781) Survey of Progress in Chemistry*

Arthur F. Scott

Reed College
Portland, Oregon

As the starting date for the so-called Chemical Revolution one might properly pick the year 1781; for it was in the summer of that year that Henry Cavendish conducted his famous series of experiments showing that the explosion, by means of an electric spark, of a mixture of "inflammable air" (hydrogen) and common air or "dephlogisticated air" (oxygen) produced "very pure water". Two years later, in 1783, Lavoisier, carrying this experiment to its logical conclusion, succeeded in demonstrating that water was a compound composed of hydrogen and oxygen, thereby writing *finis* to the concept of water as one of the traditional four elements.

In the critical year 1781, Tiberius Cavallo, a British natural philosopher (the word "scientist" was not coined until about 1840) published a compendious, 835-page, quarto volume summarizing the known facts of chemistry with special emphasis on developments in the field of pneumatic chemistry. The title page of the book reads as follows:

A
TREATISE
on the Nature and Properties of
AIR
and Other Permanently Elastic Fluids
To Which Is Prefixed
An Introduction to Chymistry
By Tiberius Cavallo, F. R. S.
London
Printed for the Author.

*This review originally was conceived as an Editor's Note, to appear next to the Preface, but then flowered into the present Essay.

Translated into twentieth century chemical language, the formidable title of Cavallo's book would read:

A Treatise on the Nature and Properties
of Atmospheric Air
and of
Eight Gases, recently differentiated from Air
With an introductory chapter on the state
of Chemistry

For the purpose of historical orientation it may be noted that in the year 1781 the British capitulated at Yorktown.

Cavallo, the author of the book under review, was born in Italy but at an early age settled in England. At the time his *Treatise on Air* was published, Cavallo was 32 years old; his standing as a natural philosopher had been recognized two years earlier by his admission to the Royal Society as a fellow. Cavallo went on to publish a number of articles on scientific subjects, and he was also the author of several scientific books, as follows: 1785–"History of Aerostation" (ballooning); 1786–"Complete History on Electricity"; 1787–"A Treatise on Magnetism in Theory and Practice"; 1798–"Medicinal Properties of Facticious Air"; and, in 1803, a large four-volume work, "Elements of Natural and Experimental Philosophy." It may be appropriate to note here that the scientific world of Cavallo's time was really quite limited. According to estimates in Frank Greenaway's book "John Dalton and the Atom," the world population of scientists during the period in question probably ranged from 200 to 1600, and of this number perhaps only 15 to 40 were of top rank. Exchange of information among this small group of investigators was not a simple matter. While there were a number of scientific societies which published journals, publication there was generally far from prompt. To get around this difficulty some philosophers reported the results of their investigations in the form of letters to other natural philosophers; others, like Priestley, waited until they had accumulated enough material to make a book. There was no established procedure for disseminating the results of new investigations. An interesting example of information exchange involves Cavallo himself. From 1791 on, Alessandro Volta, the Italian natural philosopher, turned his attention to the study of galvanism, and made a practice of communicating the results of his investigations to fellow philosophers in Europe. In the fall of 1792 he addressed two letters to Cavallo, in which he expressed the opinion that the electricity observed in the frog emperiment was not animal in origin—the view then current—but originated in the contact of two dissimilar metals which were present in the experiment. These letters (in French) were read to the Royal Society in 1793 and published in the Society's Transactions the same year. In 1794 the Royal Society awarded Volta its Copley medal.

From remarks by Cavallo it would appear that he undertook the preparation of his *Treatise* in the belief that there were individuals outside the select group of natural philosophers who had some curiosity regarding developments in chemistry and who might want to engage in experimental studies of their own. Thus, Cavallo identifies:

> my reader, whom I must suppose entirely unacquainted with the science (Chymistry), and desirous to acquire a knowledge of it.

The spirit of the book and the attitude of the author is set forth clearly in the preface to the *Treatise*:

> The numerous discoveries lately made relating to Air, and other permanently elastic Fluids; the scattered manner in which those improvements had been handed to the Public; and the importance of the subject, suggested the idea of compiling the present Work, and at the same time afford some excuse for the imperfections of the performance. The divers properties of physical, but invisible substances, which can be easily altered, with difficulty preserved, and with still greater difficulty examined, have been often misunderstood, and sometimes wilfully exaggerated. To select truth amongst a multiplicity of assertions, is the difficult though indispensable duty of the Author. Each particular writer on the subject thinks himself entitled to a full discussion of his observations, perhaps of his errors; and the least accidental or unavoidable mistake, is often construed as the most illiberal misrepresentation of his sentiments. Conscious of those difficulties, the Author has spared no pains to get information, and to repeat several experiments. He has endeavoured to take notice of every notable discovery and useful observation relating to his subject; . . . These discoveries and observations have been arranged in a methodical order, giving little attention to the time in which they were made; since this book was not intended as a history of the subject. But it is more than probable, that, in spite of every effort, this Work still contains many imperfections; and perhaps some useful discovery may have been overlooked, or improperly related. The Author, therefore, would think himself exceedingly obliged to any person, especially those who are conversant with subject, who would make him acquainted with any new observation, or any imperfection of his Work; it being his earnest desire to render it as perfect as possible, in case of a second edition.

Cavallo's *Treatise* is a blend of three ingredients: (1) what we today would call descriptive chemistry; (2) detailed laboratory directions; and (3) current speculations regarding the known chemical phenomena. One shortcoming of the book, in the reviewer's opinion, is the lack of a consistent policy with regard to references to sources of information. The reviewer's interest in the *Treatise* arises from the fact that it provides a close-up view of the state of chemistry as an art and a science at one of the important turning points in the history of chemistry. The review of the book that follows is intended to convey some sense of the status of chemistry on the eve of the Chemical Revolution.

The *Treatise* is divided into four parts. Part I bears the title "Introduction to Chymistry"; Parts II and III, which represent three-fourths of the text, are given over to a compendious review of the properties of Air (atmospheric) and of the other eight known gases. Part IV is a miscellany touching on still further "particulars relating to the subject with the account of some experiments by the Author." Cavallo views the material in Part I as necessary for an understanding of the material to be presented in the rest of the *Treatise.*

PART I—INTRODUCTION TO CHYMISTRY

This first section (156 pages) of the Treatise is given over to chemical items not related to the chemistry of Air or the other eight gases. Cavallo opens this section with his definition of chemistry:

> The science, which examines philosophically the various properties of bodies, so as to ascertain, by experiments and speculations, the methods of effecting the divers compositions and decompositions of the said bodies, is called Chymistry. In order to form a right idea of the principles of the science, it is necessary first to describe those simple primary bodies or elements, by the various combination of which the second class of bodies is formed and which, indeed, are the limits of Chymistry. The old philosophers making use of suppositions rather than experiments entertained various fanciful and often contradictory ideas relating to the elements of bodies. . . . The philosophical writers, however, of the same time generally acknowledged four elements, viz., Air, Water, Earth and Fire which opinion has also been adopted by most of the late chymists; . . .

A reading of Cavallo's discussion of the classical four-elements reveals that they had already lost, by 1781, much of their sanctified status. Cavallo rejects Air and Earth as possible elements on the ground that they had never been obtained—and, therefore, had never been known—in the pure state. Water is the only one of the four elements which Cavallo is willing to accept, but with a reservation:

> It may be therefore concluded that as no decisive experiment has hitherto demonstrated, that water is convertible into any other substance, or any other substance is convertible into water; we may consider it as a prime elementary principle; by which we imply not an impossibility of altering the nature of water but only acknowledge our incapacity of effecting it at present.

It may be recalled here that, as mentioned in the opening paragraph of this article, in 1781, the year in which the *Treatise* was published, Cavendish was initiating a series of experiments which were to write *finis* to the notion that water was an element.

As for Fire, Cavallo gives us the following definition:

> the other element of the Chymists; whose nature has occasioned innumerable controversies. . . . In order to give an useful idea of its nature and properties, it is necessary to observe, that by the word *Fire*, in common language, we understand not a simple substance, but a complex idea of something red, shining, which excites our sensation of heat, and rarifies all known bodies; now it is the

> cause of those effects that philosophers mean, under the appellation of fire considered as an element.

Cavallo's definition of Fire was framed to accommodate also the concept of phlogiston which the German chemist Stahl had introduced into chemical speculations in a textbook published in 1723. Regarding phlogiston Cavallo writes:

> Inflammability is not the property of some bodies arising from their nature, or from a particular arrangement of their constituent particle, but is the property of a particular substance, called *Phlogiston*, which, when combined with various substances in due proportion, renders them inflammable.

In a further elaboration of this idea, Cavallo writes:

> To explain the effects of fire is the greatest difficulty in the subject, and which indeed seems as yet to elude all philosophical investigations. Fire, or combustion, produces heat, rarefaction, and light. A sufficient quantity of light as rays of the sun collected into a small space by a burning glass, produces heat to make combustible bodies burn; and independent of light, a sufficient quantity of heat produced by friction, and other means, occasions combustible bodies to burn; yet it is difficult to assert whether phlogiston, either in motion or in a quiescent state, light, and heat are the same substance, or different principles, which act upon each other alternately, and according to various circumstances one or more of them become manifest to our senses.

The foregoing excerpt is typical of the murky phlogistic speculations which Cavallo offers later in his book whenever he discusses combustion or calcination of metals. In fairness to Cavallo, however, it should be noted that most of his contemporaries were also confirmed "phlogistonists" in spite of Lavoisier's reports of experiments which undermined the phlogiston theory. In fact, Cavallo summarizes some of these antiphlogistic experiments in Part III of his *Treatise* without being persuaded of Lavoisier's conclusions. It was not until 1783, two years after the *Treatise* appeared, that Lavoisier published his *Reflections on Phlogiston*, in which he summarized the evidence against the phlogiston theory and marshalled the evidence in support of his new oxygen concept.

The more important chemical topics which Cavallo reviewed in his Introduction to Chemistry are the following: acids, alkalies, salts, earths, inflammable metals, metals, semi-metals, vegetable and animal substances, and fermentation. Each topic is treated in some detail, and altogether Cavallo's discussion of each topic makes interesting reading. The reviewer, however, will mention only a few items, those that caught his eye for one reason or another.

Under the heading "Of Metals" Cavallo considers 14 metallic substances. After accepting "malleability" as a characteristic property of metals, Cavallo continues:

> The metals are again subdivided into *perfect* and *imperfect*. The perfect metals are Gold, Platina, and Silver; which are called perfect because they cannot be

decompounded; gold, for instance, if pure, will not lose any of its essential properties by being kept in a very strong fire. But the imperfect metals, which are copper, iron, tin, and lead, if they are kept a sufficient time in the fire, or exposed to the action of proper menstrums, besides fire, will lose their metallic properties, and appear in the form of earths; which are vitrifiable, not shining, not malleable, etc., and are called metallic calxes; which may again be reduced to metals or, as it is commonly said, may be *revivified*, by combining them with a sufficient quantity of inflammable principle; hence it appears that imperfect metals are principally, if not wholly, compounds, of earthy matter and phlogiston.

In view of the foregoing picture of the relation between a metal and its calx, one can understand Cavallo's puzzlement in trying to account for the properties of the red oxide of mercury which Priestley had used in 1774 in his oxygen-discovery experiment. We quote first Cavallo's directions for preparing this particular oxide.

If about two or three ounces of quicksilver be put into a glass vessel, whose capacity might contain twice or three times that quantity of quicksilver, and having a long neck, terminating in a capillary aperture; and if this vessel be exposed very little (less than that required to make the glass vessel red hot) to an equable but strong fire for several months, the mercury will be gradually converted into a red powder, which floats upon the surface of the yet liquid mercury, or adheres to the sides of the vessel. In this manner the mercury may be all converted into that red powder, which is improperly called *precipitate per se*, or *red precipitated mercury without additions*. This precipitate per se may again be reduced to fluid quicksilver, by heat only, without the addition of phlogistic matter; for which reason many imagine that the quicksilver, being turned into precipitate per se, is not calcined, as many others have thought, but is only disguised by acquiring or losing that superabundant quantity of phlogiston unnecessary to its metallic existence, which this and other facts seem to shew, that can easily adhere to or be separated from mercury. At last, it must be confessed, that our present knowledge of this singular phenomenon is very inperfect, and requires the assistance of farther experiments.

In fairness to Cavallo we note that later in Part III of the *Treatise*, in his discussion of dephlogisticated air (oxygen), he quotes as follows from a report by Fontana:

It being known that when quicksilver is turned into *precipitate per se* it augments in weight, it seems evident that the mercury, on the act of becoming *precipitate per se*, absorbs from the atmosphere the pure dephlogisticated part of it, which augments its weight, and which comes out in the form of dephlogisticated air, . . . when the *precipitate per se* is turned into running quicksilver. Though we are much in the dark respecting the nature of *precipitate per se*, and have no idea of the manner with which mercury can enter into combination with air, we are nevertheless certain, that the communication with the atmosphere is necessary for the conversion of mercury into *precipitate per se*; for it cannot be effected in vessels hermetically sealed;

That Cavallo wanted to separate himself as far as possible from any contemporaries who might still have harbored alchemical notions is quite evident from the following excerpted footnote:

The transmutation of one metal into another, or the making of a metal out of something else besides a metallic earth, may, perhaps, be possible in nature; but it has been attempted by so many people, whose trials have proved unsuccessful, after the expense of fortunes, and inconceivable labour; that whoever at present attempts any such thing, is laughed at, being considered a madman.

The reviewer has been impressed by the exquisite clarity with which Cavallo describes certain chemical reactions and laboratory operations. Consider some examples:

The solution of copper in volatile alkali (i.e., NH_3) has the remarkable property of appearing blue when exposed to the air, and of becoming colorless when the communication between it and the common air is intercepted: thus, if a bottle containing this solution be left open, the colour of the liquid will be blue, but if the bottle be well corked, the colour will gradually disappear; and if the bottle be uncorked, the surface of the liquid first and gradually the whole of it, will acquire a beatiful blue colour; and in this manner the colour of this solution may be made to appear and disappear as often as it is required.

* * *

Thin essential oils, especially ether, have a great affinity with gold; so that if ether be poured upon a solution of gold in aqua regia, and the mixture be shook, the gold will leave the latter and pass into the former.

* * *

When about two ounces of much concentrated vitriolic acid (i.e., H_2SO_4) is mixed with an equal quantity of pure water, it occasions a sudden ebullition, and a noise similar to that produced by a red-hot iron plunged into water.

* * *

When common acids are properly mixed with ice or snow, they produce cold. Indeed so great a degree of it has been produced by this means, that even quicksilver has been frozen by it.

* * *

Mercury may be separated from dust, and also in great measure from water, by straining it through chamois leather, or letting it pass several times through a paper funnel, that is a piece of paper rolled in a conical shape, having a very small hole in the bottom. It may be separated from grease by washing it in an alkaline lixivium; and from an alkaline mixture by vinegar.

* * *

Ether, like other thin and volatile matters, acts powerfully on the nerves, for which reason it has been often used in medicine as a sedative and antispasmodic. It possesses the dissolving property in a more eminent degree than spirit of wine. The vitriolic ether, especially, dissolves all essential and expressed oils, animal empyreumatic oils, resins, and in great measure even ambergrise, elastic gum or *caoutchouc*, petroleum, and oil of amber. Ether is very useful to take off spots of grease from silks, as it does not injure their colors.

Cavallo concludes Part I of his *Treatise* with two sections on experiments. He begins the first of these sections as follows:

> Chymistry being a science, which by aqtual experiments examines the nature, properties, and components of various substances; and forms new compounds out of divers principle; it does necessarily require a set of proper instruments; and of various substances, the use of which occurs more frequently in chymical operations. It also requires that those instruments, etc. be kept in a convenient place, called a Laboratory, where the various required processes may be carried on, and where the effects of some operations may not occasion any disagreeable accident as would be the case, if they were conducted in a close room of a private apartment.

Cavallo then describes a suitable Laboratory and the more important types of furnaces that may be needed. He then proceeds to list "other instruments, besides furnaces"; for example, "Alembics, either of glass or of copper;... Funnels of glass and of tin ... a portable pair of bellows, exact scales for weighing, ... a good steel for striking a fire, and for trying the hardness of certain substances...."

The final section of Part I contains directions for nine experiments. Noting that since Part I, "An Introduction to Chymistry," was "intended as a necessary preliminary to the subject of air and other permanently elastic fluids," the processes and experiments to be described "are mostly subservient to that interesting branch of natural philosophy." But

> there are added a few others which, although not immediately relating to that subject, are yet mostly new and diverting; for which reason they were thought deserving of the attention of the reader.

The nine experiments are described in ample detail. Their titles are listed below in the belief that they provide a clue to the interests of natural philosophers in 1781. It is left to the reader to decide which experiments might relate to the study of permanently elastic fluids and which might be "mostly new and diverting."

- I To make Lime-water.
- II To make Liver of Sulphur.
- III To inflame Regulus of Antimony by means of Aqua-regia.
- IV To make Hemberg's *Pyrophorus*.
- V Dr. Higgin's Experiment of detonating cuprous Nitre by contact with Tin.
- VI To make Mr. Canton's calcareous Phosphorus.
- VII To make Vitriolic Ether.
- VIII To make Kunckel's Phosphorus.
- IX To make Fulminating Powder.

Cavallo cautions his reader who might undertake these experiments

> not to be surprised if he fails in performing any experiment, ... the circumstances concurring to the success of any experiment in chymistry, being so many, that they can hardly be minded by the most experienced practicioner, and may therefore more easily escape the notice of a beginner. Mistakes and perseverance will warn the experimental philosopher against many useless and dangerous practices, and will gradually enable him to examine the powers of nature, and to perform the experiments with dexterity and precision.

Parts II & III—Of the Nature of Air and Other Gases

For pedagogical reasons Cavallo divides his treatment of gases into two parts: Part II (210 pages) gives an introduction to the subject, with detailed directions for laboratory experiments and the preparation of necessary apparatus; and Part III (415 pages) is given over to a review of the subject at a more advanced or more sophisticated level. Since the pedagogical approach to the subject is not of importance to this review, we shall treat Parts II and III together, hoping thereby to develop a clearer picture of the science of chemistry as it existed in the transition period between the discoveries of the pneumatic chemists and the final overthrow of the phlogiston theory.

That the thinking of natural philosophers regarding "elastic fluids" was in a state of flux at the time Cavallo was writing his *Treatise* is evident from the fact that, at the outset, he had to deal with the question of naming these fluids. The older, traditional system of nomenclature gave the label "air" to any gaseous substance on the assumption that it was a modification of ordinary air. The newer system, which was really a revival of century old suggestion, substituted the word "gas" for the traditional "air," reserving the word Air for atmospheric air. Thus "inflammable air" and "inflammable gas" were practically interchangeable during the transition period. Cavallo considered the "gas" terminology to be "most rational." But, he writes:

> Were I the first, or the principal writer on this subject, it would in that case be incumbent on me to adopt such names as are unexceptionable: but when the subject is pretty well known in the philosophical world, to adopt any new or less-used names, would be certainly displeasing and perplexing to an ear otherwise accustomed.

The gases which form the primary subject of the *Treatise* are listed in

TABLE I

List of Gases Which Form the Subject of Cavallo's Treatise

Date of discovery	Cavallo's name for gas	Present name and chemical formula
—	Air	Atmospheric air
1754	Fixed air	Carbon dioxide (CO_2)
1766	Inflammable air	Hydrogen (H_2)
1772	Nitrous air	Nitric oxide (NO)
1772	Marine acid air	Hydrogen chloride (HCl)
1774	Alkaline air	Ammonia (NH_3)
1774	Dephlogisticated air	Oxygen (O_2)
1775	Vitriolic acid air	Sulfur dioxide (SO_2)
1775	Fluoracid air	Silicon fluoride (SiF_4)
1777	Nitrous acid air	Nitrogen dioxide (NO_2)

Table I in the order in which they were "discovered," that is to say when they were differentiated from ordinary air. Credit for the discovery of "fixed air" is given to Joseph Black, and for the discovery of "inflammable air" to Henry Cavendish. The other gases listed were all discovered by Joseph Priestley. It may be noted here that Cavallo gives detailed direction for the preparation of each of the gases listed in Table I.

Cavallo gives what appears to be adequate information for a novice to set up a laboratory and to perform the basic experiments which are described or outlined in the *Treatise.* To give some idea of the detailed directions which are given for the construction of necessary pieces of equipment and for certain important laboratory operations we offer the following notes:

Tubs:—According to Cavallo:

> The principal instruments that are necessary, to make experiments on air and other permanently elastic fluids, are, a tub filled with water, and another filled with mercury.

Cavallo then gives the design, dimensions, and construction (wood, preferably mahogany) of these tubs:

> When a person has the advantage of a pipe that conveys abundance of water into his experimental room, he may use as large a tub as he pleases.

For the mercury tub Cavallo recommends a wooden box

> at least 12 inches long, 4 inches wide, and 5 inches deep; which when filled with quicksilver till within about half an inch of the top or edge contains about 100 pounds weight of quicksilver.

Vessels for production, collection, and storage of gases:—Cavallo describes these in minute detail. Of especial interest historically is his reference to

> bottles of elastic gum, or *caoutchouc.* They answer both the purpose of retaining elastic fluids, like bladders or receivers, for which use they are preferable to bladders; . . . When these bottles are required to be used instead of bladders, they should be as large as may be got, and should be rather thin, so as to be easily compressed. . . .

Sources of heat:—Cavallo discusses not only means of producing different degrees of heat (candle flame, charcoal fire, "strong red or white heat"), but also various ways of protecting the glass vessels being heated. He notes that retorts of cast iron or gun barrels

> cannot be safely used for nice experiments, because heated iron emits some inflammable air or phlogiston.

Instruments necessary to show the properties of the gases:—Cavallo recognizes that certain pieces of apparatus are necessary "to shew the various properties of different sorts of elastic fluids." He therefore gives detailed instructions for the construction of the more general and important items, which he specifies as follows:

> The apparatus for impregnating water or other fluids with fixed air; the receiver necessary to transmit electric sparks through any particular sort of elastic fluid; the receiver to fire inflammable air by means of an electric spark,

> and to retain it after the explosion; the pistols which shew the force of inflammable air exploding together with common or dephlogisticated air; the eudiometer, or instrument necessary to observe the degree of purity of respirable air by the diminution arising from its combination with nitrous air; and the instrument necessary to examine the specific gravities of the various elastic fluids.

Cavallo describes each of the foregoing special pieces of apparatus clearly and in some detail, referring advantageously to the drawings shown in the "three copper plates" which are part of the *Treatise*. A glance at the many drawings reveals that either the natural philosophers themselves were skillful in designing and constructing simple apparatus, especially those pieces which involved simple glassblowing operations, or that they could count on skillful assistants to help them.

Though he made frequent reference to certain pieces of laboratory equipment, viz., thermometer, chemical balance, air (vacuum) pump, the electrical machine, barometer, and hygrometer, Cavallo made no attempt to describe their construction, partly, it is implied, because some of them were available in instrument shops but mainly because they were fully described in other treatises.

To conclude his account of laboratory equipment and operations, Cavallo devotes an entire section to "Practical rules concerning the use of the apparatus, and the performing of experiments." His point is made in the following excerpt from the first paragraph of the "Practical rules":

> Experiments relating to other works, as optics, electricity, etc. although liable to be mistaken, yet can seldom produce any dangerous effect; but the errors which happen in performing experiments on the subject of this work, may be productive of the most serious consequences.

From this point on in our review of the *Treatise* we limit our discussion of the material to just two gases: Air and Hydrogen. The former had been the most extensively investigated; the latter, though it had been known for 15 years, was still the object of very considerable curiosity. We turn first to the experiments described by Cavallo to provide his readers with hands-on experience.

Cavallo's Experiments on Air

Following some general introductory comments on the properties of air, especially those bearing on combustion and animal life, five experiments are described in some detail. These experiments are extremely simple and, according to Cavallo, do not require any particular apparatus. The nature of those experiments is worth noting because of the light they throw on the state of chemistry and the experimental techniques in 1781.

Expt. 1. Burning of a lighted taper in the air of a bottle, inverted with its mouth in a basin of water.

Expt. 2. Burning of sulfur on cotton in the air of a tumbler in inverted position in a basin of water. Ignition achieved by means of a burning lens "which needs not be larger than three inches in diameter." The reader is cautioned by the remark that "some combustible substances, as gunpowder, . . . discharge a considerable quantity of elastic fluid in the act of burning, for which reason they cannot be used for this experiment."

Expt. 3. Uses same tumbler as in Expt. 2; this time a mouse or a bird is placed under it. After dead mouse or bird is withdrawn a second live animal is introduced.

Expt. 4. Expt. 3 is repeated with the intent of noting contraction in volume of air due to respiration.

Expt. 5. Another tumbler experiment, "By this experiment is to be shewn, that air, contaminated by a burning candle, and wherein a lighted candle is immediately extinguished, is not so noxious to animal life, as the air contaminated by animals breathing and dying in it.

Cavallo postpones until his experiment on nitrous air (nitric oxide) the description of a very important chemical property of the atmosphere and an experimental demonstration thereof. The experiment is designed to show that the final volume of a mixture of nitrous air and common air is less than the sum of the volumes of the gases taken. If the experiment is repeated using fixed air (carbon dioxide) or inflammable air in place of common air, no diminution in volume takes place. According to Cavallo

> diminution will take place only when respirable air is mixed with nitrous gas, and it is greater when the air is purer, and contrarywise: hence this method is used to try the purity of the air.

Credit for this means of measuring the "goodness" of air belongs to Priestley. Although the reaction was not understood chemically at the time, it formed the basis of an instrument which became known as "eudiometer," which was believed to observe the purity of air. In the section dealing with Laboratory Equipment, Cavallo describes, with the aid of drawings, the design, construction, and operation of *seven* different eudiometers, one of them by Cavallo himself. As will appear later, the eudiometer played a very important role in the studies of the atmosphere by natural philosophers prior to 1781. In passing it may be noted that Cavendish, in the fall of 1781, undertook a study of the eudiometer which was published in 1783 under the title "An Account of a New Eudiometer." At about the same time Volta invented a new type of eudiometer which involved the explosion, by means of an electric spark, of a mixture of a sample of the air to be tested with "inflammable air." From the resultant contraction in volume of the gas mixture, the "goodness" of the air could be estimated. Another parenthetical note is of interest. Sir Edward Thorpe, the historian, has pointed out that the word eudiometer "survives in our literature as the only remnant of the terminology of phlogistonism."

Experiments Relating to Hydrogen

In this series of experiments Cavallo employs the method of production of the gas (iron filings or zinc with sulfuric acid) and the method of collecting the gas which were prevalent 25 years ago in high school and college laboratory manuals in this country. The behavior of a test tube containing different proportions of inflammable air and common air, when brought near a flame, is carefully described. Cavallo describes two special experiments which must have been quite novel at the time. One is based on the fact, established by Cavendish in his 1766 hydrogen-discovery experiment, that the specific gravity of hydrogen is about one-tenth that of air; the other experiment makes use of the explosive property of Inflammable Air – Air mixtures to contrive a pistol.

The first of these two experiments can best be described in Cavallo's own words:

> When a slender pipe is tied to the neck of a bladder, and the bladder is filled with inflammable air, after the manner described in the preceding experiment (viz. when the bladder was required to be filled with fixed air) two very pleasing experiments may be performed with it. First the inflammable gas may be inflamed by applying the flame of a candle to the extremity of the pipe; and squeezing at the same time the bladder, a stream of fire will be formed in the air, which will last as long as the bladder contains any inflammable air; for this air coming out of the pipe with violence, will continue inflamed a considerable way into the air. Secondly, the extremity of the pipe may be dipped into a solution of soap, then removing it from the solution, and squeezing the bladder very gently, a ball of soap water may be formed, including inflammable air; which ball, on account of the inflammable air being much lighter than common air, as soon as it is detached from the pipe, will ascend upwards, and will break by dashing against the ceiling, contrary to those commonly made by children, which in still air go downwards.—Whilst the ball is ascending, if the flame of the candle is approached to it, the film of soap water will be instantly broken and the inflammable air will take fire; thus a flame may be shewn to be seemingly produced from a soap-ball.

Subsequent to the experiment just quoted, Cavallo describes the fate of an animal introduced into an atmosphere of hydrogen. He then expresses a principle of ethics:

> but except it is for the sake of ascertaining some new fact, diverting experiments in general should be rather made with inanimate things, than with the lives of animals, and at the expense of humanity.

In his section on special laboratory equipment, Cavallo describes, with the aid of diagrams, two experiments involving the explosion of a mixture of hydrogen and air by an electric spark. In one experiment the explosion takes place in a glass vessel under conditions such that the gases are "retained after the explosion." In this experiment the containing fluid is water; if it had been mercury, Cavallo might have observed the dew which led to

Cavendish's great discovery that the product of the explosion is water. Cavallo's second explosion experiment is designed to show the explosive force produced when a mixture of hydrogen with air or oxygen is fired by an electric spark. The device was in the form of a pistol. We quote from the *Treatise*:

> Since these instruments were introduced, their form has been very much diversified; some having been made of brass, and so large as to drive a leaden bullet; others have been made of tin, etc.; but that shown by this figure is of glass, and so simple that it may be easily constructed, even by persons of no great genius.

Cavallo describes also a more elaborate pistol in which "a small leaden bullet, wrapped up in soft leather" is used and the necessary inflammable air is supplied from a bladder!

Cavallo continued his soap-bubble experiments and in June 1782 he read a report before the Royal Society: his paper was entitled: "An Account of Experiments relating to the property of common and inflammable air pervading the pores of paper." In this paper Cavallo describes his attempts to construct a vessel which, when filled with inflammable air, would ascend into the atmosphere. In his paper Cavallo states that he discontinued his experiments as he "tired of the expense and loss of time."

It seems appropriate to digress briefly at this point to consider the place of Cavallo's soap-bubble experiments in the history of the balloon invention. The historical record shows that the first balloon ascent was that of a hot-air balloon on June 5, 1783 at Annanay, France, under the direction of the Montgolfier brothers; the second ascent was a hydrogen-filled balloon, just 12 weeks later in Paris. This second ascent was planned and conducted by a group of French natural philosophers headed by J. A. C. Charles whose name is associated with the famous Charles gas law. Historians, in trying to trace the origin of the balloon concept have established that Joseph Black, the leading chemist in Europe in Cavallo's day, made mention in his lectures, in the late 1760s, of the possibility that a bladder filled with hydrogen would ascend in the air due to its buoyancy. But Black admits that he never actually performed this particular experiment. It would appear, therefore, that Cavallo, with his soap-bubble experiments, was the first to have tested "a sort of inflammable balloon," to use Cavallo's own phrase. But Cavallo never pressed any claim respecting the invention of the balloon. The story of the Charles' experiment with the first hydrogen balloon provides another glimpse of the state of chemistry in the early 1780's. One can estimate that to generate sufficient hydrogen to fill a balloon bag eight feet in diameter (the size of the first hydrogen-filled balloon) would require approximately the following quantities of materials: 78 lbs. iron, 78 lbs. conc. sulfuric acid, and 370 lbs. water. That this was probably the first time hydrogen had been produced on such a large scale is shown by the log of operations:

Aug. 23	Started at 8 am; changed apparatus at 2 pm. The new apparatus was a cask. Many difficulties due to temperature and effervescence. At 9 pm balloon bag only one-third filled. Operation shutdown.
Aug. 24	Daybreak—discovered that stopcock had been left open and that air had diffused into the bag. Started over again. By 7 pm the balloon bag manifested definite levity. Operation shutdown for the night.
Aug. 25	Daybreak—produced more hydrogen. At 6 am a test showed that balloon could lift 21 lbs. No more hydrogen. At 9 pm it could lift only 18 lbs.
Aug. 26	Daybreak—further loss of levity. More hydrogen produced. At 8 am the levity of balloon was such that it was allowed to ascend repeatedly to 100 feet. Attracted crowd of spectators.
Aug. 27	Before daybreak the balloon, on a cart, was transported about 2 miles to the Champ de Mars where the exhibition was to take place. The balloon was escorted by troops; the procession was lighted by lanterns and torches. At Champ de Mars more hydrogen was introduced into the balloon. At 5 pm, the discharge of a cannon gave signal to release the balloon. The balloon rose to a height estimated to be 3,123 feet; it remained aloft only three-quarters of an hour, having travelled 15 miles.

Cavallo's Survey of Pneumatic Chemistry

The greater part of Cavallo's *Treatise* is given over to a summary of developments in the 25 year old field of pneumatic chemistry. The reviewer has read Cavallo's account with mixed feelings: with great pleasure and admiration when it appears that the natural philosophers had readily established facts regarding some new (or old) phenomenon, and yet, with a sense of discouragement to see how easy it was for the natural philosophers to choose the wrong interpretation or explanation of some established fact. In most instances they were led astray by the concept of phlogiston. But not always, as in the case of humidity, which will be mentioned below.

It is out of the question for the reviewer to undertake anything like a complete summary of the speculations on Air and Inflammable Air which appear in Cavallo's *Treatise*. It can be stated, however, that he quotes extensively from the works of the outstanding philosophers of the time: Black, Priestley, Lavoisier, Cavendish, Scheele, Ingenhousz, and Fontana.

Notes on Cavallo's Summary of Information on Air

We shall first take notice of some of the interesting material that appears in Cavallo's review of Air. There is no question that the so-called Boyle's

Law regarding the relation between the volume and pressure of a sample of air was known and utilized in the 1780s. On the other hand, while it was known that warming a sample of air caused expansion or rarefaction, as it was commonly called, and that cooling had the contrary effect, the relation between the volume of a gas sample and temperature was unknown in Cavallo's time.* Notwithstanding this limitation, Cavallo is able to point out its general meteorological significance:

> This admirable property of the air, viz. of being easily expanded by heat, and contracted by cold, is the principal if not the sole cause of winds; . . . If the air in one place becomes colder, by the setting of the sun, or other causes, it contracts into a smaller space, and the hotter air from the circumjacent places runs toward that part.

Phenomena involving the vapor pressure of water were not understood because of the view, expressed by Cavallo, that: "It is certain that water is chymically combined with air." Consider Cavallo's dictum regarding hygrometers:

> It is a vulgar error to suppose that hygrometers show the quantity of water combined with air; for they indicate only the quantity of water which cannot be combined with air; or (in chymical phraseology) that quantity of water which is precipitated from the air.

Cavallo looks upon the following experiment as proof of his concept:

> Introduce an hygrometer of any construction, provided it be pretty sensible, into a large glass bottle, containing only common air, which bottle let be stopped, so that outward air may not communicate with the air within it. Then let the bottle be brought into a cold place, or let any freezing mixture be put around it, and you will perceive, that on the accession of cold, the index of the hygrometer immediately moves toward *moist*, because cold air cannot keep as much water dissolved as when it is hot; but if then the apparatus, being removed from the cold, be brought in a warmer place, the index of the hygrometer will move towards *dry*.

Cavallo, since he was a confirmed phlogistonist, could make the following dogmatic statement:

> The invisible substance, which combines with air more easily, and by its combination alters remarkably the nature of the elastic fluid, is the inflammable principle or phlogiston. In all those processes called phlogistic, in which the inflammable principle escapes from bodies as in respiration, combustion, putrefaction, dissolution, etc . . . the air does readily combine with it and thence becomes phologisticated. . . . But this combination is limited; for according as more and more phlogiston is combined with the air, its capacity of absorbing more phlogiston decreases, and at last ceases; in which state the air is said to be *completely*

*In 1787 J. A. C. Charles by his experiments found that gases expand equally between 0° and 80°, but he did not measure the expansion. Charles did not publish his results but communicated them verbally to Gay-Lussac, a young engineering student who was engaged in determining the amount of the expansion. In reporting his experiments in 1802, Gay-Lussac gave credit to Charles for the first statement of the Law that bears the latter's name.

phlogisticated . . . and then it becomes quite unfit for respiration, combustion, or, in short, for any phlogistic process.

The great breakthrough in the early years of pneumatic chemistry was Priestley's discovery of the reaction of nitrous air (nitric oxide) with dephlogisticated air (oxygen), which could be used to estimate the degree of goodness of the air. As has already been pointed out, Priestley's reaction led to the invention of the *eudiometer*, a device for estimating the goodness of air (i.e., in today's terms a measure of oxygen in the air).

Cavallo reviews the data available regarding the three phlogistic processes—respiration, combustion, and calcination—in considerable detail, with special emphasis on information provided by eudiometric tests. While the facts presented are substantially correct, their interpretations are badly warped to bring them into accord with the phlogiston hypothesis.

The eudiometer was the key to many new investigations of the water–air relationship. Cavallo covers this topic at length, but only one feature of the investigations will be touched on here. Here is one of the findings reported by Cavallo:

The air which is imbibed by distilled and boiled water is not the air of the atmosphere in its usual state: but it seems that water absorbs the purest part of it and leaves the worst; for if a greater quantitiy of common air than can be absorbed by a given quantity of boiled water, is exposed to it, part of that air will be found to be absorbed by the water, and the remainder will appear to be considerably phlogisticated.—That this phlogiston cannot wholly come from the water, seems evident from this, viz. that if, when the boiled water has absorbed as much air as it can, more new common air be added to it, this will not be at all phlogisticated, or at least in an exceedingly small degree, and after a very long time. But Mr. Scheele's ingenious experiments put this matter out of doubt; for he not only observed that the remaining quantity of air, after that boiled water had absorbed part of it, was so far injured as not to admit a candle to burn in it; but, boiling the water again, and thus extracting the air it had imbibed, found that this air was much purer than common air, so that a candle burned brighter in it than in common air.

The foregoing experiments must have been the first demonstration that oxygen is more soluble than nitrogen in water.*

We turn next to a question which was of interest to natural philosophers of the 1780 period and which continues to be of interest to scientists of today. Cavallo presents the question in this way:

The first consideration that naturally occurs is, that the whole mass of atmospherical fluid, which surrounds the planet we inhabit, being continually contaminated by the respiration of many myriad of animals, by the burning of a great many artificial and natural fires, by the putrefaction and dissolution of

*It is now known that oxygen is approximately twice as soluble as nitrogen in water. In 1869 J. T. A. Mallet took out a patent for the commercial preparation of oxygen based on the difference in solubilities. The final product of air was 97.3% oxygen.

> an infinite number of substances, and by many other phlogistic processes, must gradually become more and more infected, till it be rendered absolutely unfit for the purposes for which it is destined, except nature has provided some means to purify it, in proportion as it is contaminated, so as to keep it in a proper state of salubrity, without growing continually worse.

As part of his discussion of this subject Cavallo presents the findings of six natural philosophers who investigated, by means of the eudiometer, the purity of air in different parts of England, in different countries on the Continent, and also at sea. The last investigator in Cavallo's series, Fontana whom Cavallo characterizes as "that judicious and accurate philosopher," concludes his observations as follows:

> Nature is not so partial as we commonly believe. She has not only given an air almost equally good everywhere, at every time, but has allowed us a certain latitude, or a power of living and being in health in qualities of air which differ to a certain degree. By this I do not mean to deny the existence of certain kinds of noxious air in some particular places; but only say, that, in general, the air is good everywhere, and that the small differences are not to be feared so much as some people would make us believe. Nor do I mean to speak here of those vapours and other bodies which are accidentally joined to the common air in particular places, but do not change its nature and intrinsical property. This state of air cannot be known by the taste of nitrous air, and those vapours are to be considered in some manner as we should consider so many particles of arsenic swimming in the atmosphere. In this case, it is the arsenic, and not the degenerated air, that would kill animals who ventured to breath it.

Cavallo introduces his next important topic respecting the properties of air as follows:

> The variation of the air's purity, and the concomitant phenomena, are highly interesting to mankind in general; by the discoveries made principally in this age, several methods are deduced by which this subject may be examined, and many useful observations have been actually made concerning it. But the essential part seems to be still in the dark; it is therefore requisite that philosophical people, in various parts of the world, would make as many and as various experiments, concerning the purity of the air at different times, . . . in order to investigate the laws of those changes; which subject is perhaps the most interesting part of the study of elastic fluids.

Cavallo then proceeds "to examine the natural and artificial means of purifying contaminated air, and also the effects of plants upon that element." The first "means" mentioned by Cavallo is an amusing one:

> Mr. Boyle relates of one *Drebell*, who pretended to have found a liquor, the effluvia of which could render contaminated air again good for respiration.

Cavallo quotes verbatim Boyle's account of the Drebell invention, and concludes as follows:

> The nature of this liquor was never publicly known, and probably it did never in reality exist; since it is now ascertained, that the air's unfitness for respiration, etc. is owing, not to its having lost a vital principle, but to its having imbibed a substance, to wit, phlogiston, which cannot be easily separated again from the air; especially by the fumes of a liquor. Perhaps the above-mentioned Drebell

made the people believe that he had a liquor capable to restore the air, though, in reality, they lived in that confined air only because its quantity was sufficient to support their lives for the short time . . .

Cavallo next relates a number of unsuccessful attempts by Priestley, Hales, Cavallo, and others to find a means to mend unwholesome air. This recital leads to the following passage:

Dr. Priestley's unsuccessful experiments concerning this subject, and his unwearied labour, were at last rewarded, by the discovery of a natural method of meliorating contaminated air. He found that air, which had been rendered perfectly noxious by animals breathing and dying in it, was so far restored by the vegetation of plants, that, after some days, an animal would live in it nearly as well and as long as in an equal quantity of common air. This, like other great discoveries, was imperfect at first; . . . And the Doctor himself, with his usual candour, seems, in some part of his publications, to doubt the general truth of this law: By further observations, and especially by the late experiments of Dr. Ingenhousz, this subject has been much elucidated.

Following this introduction, Cavallo outlines Priestley's key experiments and then takes up those of Dr. Ingenhousz. A condensed abstract of Cavallo's presentation (itself an abstract) of Ingenhousz' paper is of interest. We shall use Cavallo's numbered paragraph, omitting completely the detailed back-up statements.

The numerous experiments of Dr. Ingenhousz, published in 1779, not only throw great light upon this subject of the power of plants upon air, etc. but reconcile several contradictions seemingly established by thc preceding experiments of other persons.–The following is a concise account of the same . . .

I The Doctor has observed that plants have a power to correct bad air, and to improve common air, in a few hours, when exposed to the light of the sun; but in the night-time, or in general when they are not influenced by the rays of the sun, they contaminate common air.

II Plants not only mend bad air, but yield, out of their substance pure, respirable, or dephlogisticated air, when they are exposed to the sun; but in the night-time, or when they are screened from that luminary, they yield very impure air. . .

III It is not every part of the plant that improves common air, and yields dephlogisticated air, but only the leaves and the green stalks. The full-grown leaves yield the purest dephlogisticated air, and in great quantity, especially from their under surface. . . .

IV The office of plants to yield dephlogisticated air, begins a few hours after the sun has made its appearance above the horizon, or rather after it has passed the meridian, and ceases with the close of the day, excepting some plants which continue it a short time after sun-set. . .

V Lastly, it must be observed that plants do not seem to produce dephlogisticated air by a kind of generation, or by converting some other principle into that elastic fluid, but only by filtrating the common air (for plants absorb common air) and separating phlogiston from it; which phlogiston, by the joint experiments of Dr. Ingenhousz and others, has been evidently shewn to be absorbed by plants and incorporated in their nature; in which operation they do just the contrary of what is performed by animals, viz. plants, in their vegetation, absorb phlogiston from the air; whereas animals, by their respiration, separate the phlogiston from their bodies and give it to the air...

* * *

Yet it must be confessed, that Dr. Ingenhousz, pursuing the discoveries of Dr. Priestley, has clearly shewn that the vegetation of plants is one of the great means employed by nature to purify the atmosphere, so as to counteract, in great measure, the damage done by animal respiration, by combustion, etc. It may only be said, that vegetation does not appear to be sufficient to remedy intirely the damage; and that probably nature employs other means, hitherto undiscovered by human industry, which, together with vegetation, and agitation of air in water, do all contribute to that great end.

Notes on Cavallo's Summary of Information on Inflammable Air

When Cavallo wrote his *Treatise* natural philosophers were acquainted with several "airs" which they referred to as inflammable airs, though they were not able to distinguish them chemically. Thus

> The permanently elastic fluid called *inflammable air*, or *inflammable gas*, is, like fixed air, both the production of natural and artificial processes. It being abundantly produced by putrid animal and vegetable substances, and, in general, by those inflammable substances, the phlogiston of which is easily separable from them; it follows, that in all those places in which like processes are continually going on, there the inflammable gas is produced; hence, in caverns and mines, this elastic fluid is often found...

Ten pages of the *Treatise* are required to identify the various inflammable airs in a general way.

Regarding the special inflammable air, identified by Cavendish in 1766, which we know as hydrogen gas, Cavallo writes:

> By means of acids, inflammable air is obtained in greater abundance and more readily. Iron, zinc, or tin yield plenty of inflammable air when acted on by diluted vitriolic or marine acids.

As a footnote to this statement, Cavallo quotes Priestley's 1777 statement:

> It is something remarkable that all the acids that produce any air by the solution of metals, give inflammable air, except spirit of nitric (nitric acid)...

In apparent acceptance of this view of things Cavallo refers to the Cavendish gas as "The inflammable air extracted from metals."

It may be recalled here that Cavendish in his 1766 paper used two tests to characterize the inflammable air generated by the action of two different acids on zinc, iron, or tin, viz., the specific gravity; and the explosibility of mixtures of the inflammable air and ordinary air. By means of these two tests Cavendish showed that the different samples of inflammable air generated by the action of the two acids on the three metals were essentially identical. Cavallo refers to these two tests as follows:

> The specific gravity of inflammable air, according to Mr. Cavendish, is to the specific gravity of common air, as 1 to 12 nearly. . . . According to Mr. Fontana, when the barometer is at mean height, and the thermometer about temperate . . . its specific gravity is to that of common air as 1 to 15 nearly.

Regarding the explosibility test, Cavallo gives more detailed information:

> Mr. Cavendish, in order to examine the effects arising from the firing of inflammable air, when mixed with different proportions of common air, made several experiments, the results of which I shall, for shortness sake, exhibit in the following table:*

Parts of inflammable and of common air in the mixture	Effects
1:9	Was fired with difficulty, and made very little noise.
2:8	Fired easily—the report moderately loud.
3:7	Loud explosion.
4:6	Explosion little louder than the former.
5:5	Exploded like the last.
6:4	An explosion not loud.
7:3	Gentle explosion.
8:2	Burned without noise.

A condensed summary of the more interesting material presented by Cavallo in his review of the properties of inflammable air appears below:

• A mixture of respirable and inflammable air, when fired in a closed vessel "exerts a great force, though not to be compared to the force of gunpowder." Cavallo explains:

> A quarter of an ounce of gunpowder, or even much less, in a proper pistol drives

*Cavallo does not make it clear that Cavendish's gas mixtures were contained in a quart bottle and were exploded by means of a flame. Nor does he mention Cavendish's safety precautions. Cavendish wrote: "In order to avoid being hurt, in case the bottle burst by the explosion, I have commonly, in making these sort of experiments, made use of an apparatus contrived in such a manner, that, by pulling a string, I drew the flame of a lamp over the mouth of the bottle, and at the same time pulled off the cap, while I stood out of the reach of danger. I believe, however, that this precaution is not very necessary, as I have never known a bottle to burst in any of the trials I have made."

a bullet with a greater violence than ten cubic inches of a mixture of inflammable air and dephlogisticated air in the best proportions.

• Cavallo gives an account of an explosion experiment. Knowing what we know today we can see that the gases in the explosive mixture were roughly in the proportion $H_2:O_2:N_2 = 1:0.4:1.6$. The result of the experiment would therefore appear to be quite good, considering the uncertainties involved.

If a mixture of one part inflammable and two parts respirable air, is fired in a close vessel . . . the elastic fluid remaining will be found to be phlogisticated air, which seems to contain no fixed air.

• Regarding the effect of inflammable air on plants, Cavallo writes:

Plants in general grow pretty well in inflammable air, and their growth diminishes little or nothing the inflammability of the gas. The willow plant absorbs inflammable air . . . and the remainder seems to be as inflammable as before.

• Cavallo's discussion of the effect of inflammable air on animal life reveals a disagreement among Priestley, Scheele, and Fontana. Since there is no point in tracing here the development of this debate, we restrict ourselves to Cavallo's description of one dramatic experiment:

Mr. Fontana having applied the neck of a bladder, containing 350 cubic inches of inflammable air to his mouth, after a very violent expiration, began to breathe it in my presence. The first inspiration of the inflammable air occasioned a great oppression upon his lungs; the second, affected him so much that he looked very pale; the third and last inspiration of the inflammable air was scarce accomplished, when, for want of strength he fell upon his knees, and thus testified the noxious quality of the gas.

Experiments similar to those were by the same philosopher repeated upon birds and quadrupeds, but they all tended to confirm the same general observations. He even suspecting, that possibly the inflammable air might act upon some other part, besides the lungs, of those animals which were confined in it with their whole bodies, stopped the nose of some animals with soft wax, and thus introduced them into receivers containing inflammable air; to the mouths of other animals he applied bladders containing inflammable air, and also by means of glass tubes, which reached a good way down the windpipe, introduced the inflammable air into their lungs, without exposing any other part of their bodies to it; but the unfortunate animals constantly perished after a few inspirations.*

*Only recently have the conditions necessary for the maintenance of live mice in a hydrogen atmosphere been realized experimentally. Dole, Wilson, and Fife have recently reported [*Science* **190**, 152 (1975)], in a paper entitled "Hyperbaric Hydrogen Therapy—A Possible Treatment of Cancer," a series of experiments in which mice were exposed for periods of up to 10 days, at a total pressure of 8.28 atm, to a gas consisting of 2.5% O_2 and 97.5% H_2. In this system the oxygen percentage was outside the explosion limits, and its partial pressure was equal to that of the normal atmosphere. Carbon dioxide was removed by circulating the gas mixture through a soda lime scrubber device.

• Cavallo concludes his summary of information on inflammable air with a review of the various ideas regarding the nature of this gas. His opening paragraph tells the story:

> In regard to the intimate nature of this inflammable elastic fluid, various have been the opinions of those who have endeavored to investigate it. Some consider it to be the pure inflammable principle, i.e. phlogiston; others take it to be a sulphur, viz., a combination of phlogiston with an acid; and others suppose, that it is a compound of several ingredients.—That inflammable air contains phlogiston, is evident; and that some principle or principles besides phlogiston are contained in it, seems to be highly probable; but it must be confessed that we are as yet ignorant of the number, quantity, and reciprocal bonds of those principles.

Part IV—Containing the Uses, Theory, and Various other Particulars Relating to, and some Experiments of the Author on the Subject of Fluids Permanently Elastic.

This Part is the shortest (54 pages) of the four Parts. It is subdivided into eight chapters, with a small section entitled "Additions and Alterations." For the purpose of this review we will cite the chapter titles, and for each chapter a sentence or two to indicate the nature of the content of the chapter.

"Chap. I. Of the elastic Fluids of Mineral Waters."

Cavallo describes the content of this chapter: "I shall only mention, in a succinct and summary way, the more interesting particulars relating to the elastic fluids contained in many mineral waters, and to which some of those waters owe their greatest medicinal powers." Cavallo identifies the gases extracted from mineral waters as "three sorts, namely *fixed air*, *inflammable air*, and *phlogisticated air*.

"Chap. II. Of fulminating substances."

Apparently many of the leading chemists of the day were interested in the chemistry of gunpowder and fulminating substances. Regarding the results of experiments on gunpowder Cavallo writes: "Notwithstanding the great light lately thrown upon this subject, it must be confessed, that the nature of gun-powder has not been yet thoroughly investigated." Cavallo puts down Ingenhousz' explanation of the nature of "fulminating powder" with this comment: "My reader will perceive that these lame explanations are not convincing."

"Chap. III. Advantages derived from the subject of permanently elastic fluids."

A few sentences excerpted from this short chapter reveal Cavallo's enthusiasm for the subject:

> The improvements that the subject of air and other permanently elastic fluids has received within these last ten years, are perhaps of more advantage to mankind, and reflect more honour upon the philosophers that made them, than those which any other branch of natural philosophy ever received in the like number of years.
>
> Besides the immediate use to which those capital discoveries may be applied, they throw a great light upon many branches of knowledge ... a great many perplexing processes in chemistry have been easily explained since those discoveries have been made.

Cavallo also discusses the importance of taking advantage of bladders filled with common air or with dephlogisticated air by persons "who are obliged to work in mines, or in general in those places wherein noxious elastic fluids are often met with."

"Chap. IV. Conjectures relating to the theory of Fluids permanently elastic, and the constitution of the Atmosphere."

Cavallo begins this chapter with a word regarding the temptation for speculation: "The changes which bodies are observed to undergo, in performing experiments on elastic fluids, when duly considered, afford matter of admiration, and of the profoundest contemplation, for a philosophical mind." Reading Cavallo's succinct review of these spculations persuades the reviewer that in Cavallo's time the art of conjecture was not keeping up with the art of experimentation. The major block to making speculative progress appears to have been the concept of phlogiston or the inflammable principle. According to Cavallo: "The inflammable principle, though in various quantities, seems to enter into the composition of every elastic fluid."

Chaps. V, VI, VII and VIII

These four chapters describe the following experiments by Cavallo: "On Gun-powder"; "Observations relating to Inflammable and Fixed Air"; "Experiments and Observations on Plants"; and "Promiscuous Experiments." The experimental material in these chapters is good, but inasmuch as it adds little to what has been presented previously, it is not dealt with further here.

* * *

In concluding this essay-review of Cavallo's *Treatise* the reviewer can only hope that he has succeeded in imparting something of its flavor. It was published when all the leaders of the pneumatic chemistry period (Black,

Cavendish, Priestley, Scheele, and Lavoisier) were still alive and active as natural philosophers. From his own reading of the book, the reviewer has gained three distinct impressions. First of all, it seems abundantly clear that the small group of chymists of the period possessed an enthusiasm for their field of study which must have been very similar to that which characterizes chemists of today. Secondly one can have only admiration for the number of variety of solid chemical facts which the early chemists had accumulated in the course of their unsystematic investigations. And finally, the reviewer must confess to a mixed feeling of amazement and amusement at the ease with which a confirmed phlogistonist, which Cavallo was, could manipulate the phlogiston concept to account for almost any given set of facts. In the end, of course, it was the results of experiments within the field of pneumatic chemistry that led to the undoing of the phlogiston theory.

Since Cavallo's *Treatise* aimed primarily to describe the front-line activities of pneumatic chemists, it failed to record the reactions of those natural philosophers who were not pneumatic chemists to the great discoveries in that field. Some idea of the pride and excitement of these other natural philosophers can be found in the following statement, made by Tobern O. Bergman circa 1779:

> During the last ten years chemistry has not only soared into the region of invisible substances, but it has dared to explore the nature of these substances, and to search into their constituent principles.

Bergman was Professor of Chemistry at Upsala; he was the preeminent analytical chemist of the day and had not played a role in the discovery of gases.

Subject Index

G

H

I

A
B 7
C 8
D 9
E 0
F 1
G 2
H 3
I 4
J 5